Instrumentation Applications

for the Pulp and Paper Industry

Also published by Miller Freeman Publications for the pulp and paper industry:

An Introduction to Paper Industry Instrumentation
by John R. Lavigne

Pulp Technology and Treatment for Paper
by James d'A. Clark

Synthetic Polymers and the Paper Industry
by Vladimir M. Wolpert

Manual of Supercalender Operations
by Francis E. Schiller

Metrication for the Pulp and Paper Industry
by Kenneth E. Lowe

Practical Computer Applications for the Pulp and Paper Industry
edited by Kenneth E. Lowe

Fiber Conservation and Utilization
Proceedings of the 1974 Pulp & Paper Seminar, Chicago, Illinois

Transport and Handling in the Pulp and Paper Industry, Volume 1
Proceedings of the first Pulp & Paper International Symposium, Rotterdam, Netherlands, 1974

Transport and Handling in the Pulp and Paper Industry, Volume 2
Proceedings of the second Pulp & Paper International Symposium, Rotterdam, Netherlands, 1976

Transport and Handling in the Pulp and Paper Industry, Volume 3
Proceedings of the third Pulp & Paper International Symposium, Vancouver, British Columbia, Canada, 1978

International Glossary of Technical Terms for the Pulp and Paper Industry
edited by Paul D. Van Derveer and Leonard E. Haas

Instrumentation Applications
for the Pulp and Paper Industry

John R. Lavigne
CONSULTANT
PULP AND PAPER INDUSTRY DIVISION
THE FOXBORO COMPANY
FOXBORO, MASSACHUSETTS

To my grandchildren, their children,
and their children's children

Library of Congress Catalog Card Number: 77-93837
International Standard Book Number: 0-87930-074-4

Printed in the United States of America.

Contents

Preface

This book on instrumentation applications can be considered as a sequel to my book, *An Introduction to Paper Industry Instrumentation*, which is a comprehensive discussion of the fundamentals of instruments and controls used in the pulp and paper processes.

The introductory book exposes the uninitiated technician or process engineer to the four major areas in an automatic control system—primary measurements, signal transmission, the automatic controller, and final control elements—and gives pertinent information on computers and their use in the pulp and paper industry. It is also useful as a refresher or a reference source for those with limited and advanced knowledge in the instrumentation field.

Logically, the next step after acquiring this basic knowledge of the instrumentation used in the paper industry is to find out how to apply it to the process for the purpose of measuring and controlling the variables to achieve the final objectives of increased operating efficiencies: reduction in production costs and improvement of product quality. The intention of this book is to provide assistance to the pulp and paper industry's instrument technician, instrument engineer, and process control engineer. For others associated with the pulp and paper industry or those who become interested and involved in the area of instrumentation applications to the process this book, like the first one, can serve as a source of reference.

To accomplish this purpose, the pulp and paper industry was surveyed and typical instrumentation installations from the various areas of the pulp and paper manufacturing process were selected as representative of the many types of applications that exist in the industry today.

Due to the many and varied ways of applying instrumentation to achieve specified goals, it would be an impossibility to cover all aspects and configurations that have been or are being implemented today. Therefore, there was no attempt to make this book all-inclusive and to discuss all known applications in the pulp and paper industry. Rather, the more widely used applications were selected in order to provide the reader with the fundamentals involved in applying instruments and controls to the unit processes normally encountered in the manufacture of pulp and paper. By doing this, the text provides a source of ideas from which more advanced instrumentation and control concepts can be conceived and developed to suit particular needs and requirements.

With this use of the book in mind, a series of simplified process instrument diagrams was compiled and arranged chronologically with respect to material flows. The instrument symbology developed by the Instrument Society of America (ISA) was used insofar as possible. In some cases, symbols developed by the Technical Association of the Pulp and Paper Industry (TAPPI) and the Canadian Pulp and Paper Association (CPPA), with some improvisations, were used. The actual implementation of instrumentation can be done either pneumatically or electrically. Therefore, the instrument lines used are all solid with no designation distinguishing them as pneumatic or electric as shown on symbology standards.

In general, instrumentation systems assume transmission of signals to centralized control stations with measurement transmitters shown in lieu of direct connections, as applicable. The diagrams in this book do not represent design or installation recommendations. They are to be taken only as simplified reference-type diagrammatic indications of typical process instrumentation and, in most cases, they neglect minor or optional secondary instrumentation such as machinery running lights, automatic interlock devices, and incidental measurements.

Without the cooperation of my many acquaintances in the pulp and paper industry and fellow Foxboro personnel, this book would have been impossible to prepare. To all those people and especially to Norma Blount for typing the manuscript, David H. Fuller for the technical editing of the text, Diane T. Laliberte for the illustrations, and John B. Prendergast for his aid in coordinating the many activities involved in this endeavor, I would like to express my sincerest appreciation.

John R. Lavigne

Introduction 1

The pulp and paper manufacturing process employs many different and varied operations which, historically, have been primarily batch or semibatch. For many years, the trend has been toward continuous operations until today this method is by far the most dominant one used. A wide variety of instrument applications are involved in these operations. Although all possible instrument applications for all operations cannot be covered, it would be helpful to have a general overall concept of the process. Then, where representative installations are discussed, they can be associated with the area in which they are applied.

Paper is basically composed of a mat of fibers. Since its invention, many fibers have been used for its manufacture. These have included the bast fibers of flax and mulberry, the stalks of bamboo and other grasses, various leaf fibers, cottonseed hair, wool, asbestos, and the woody fibers of trees. Cotton, linen rags, and straw have also been used. Many research projects have been conducted on the possible use of glass fibers, rayon, nylon, and other synthetics to impart unique properties to the sheet of paper. Today, wood is the primary raw material from which most paper pulp is made.

The actual process begins in the woodlands where trees designated as pulpwood are cut into prescribed lengths. The logs are either hauled directly from the forests to the mill woodyards or, during certain times of the year, the bark is removed by peeling it before shipment to the mill. Also, during certain times of the year, the logs can be chemically debarked. A strip of bark is peeled off around the base of the tree and the exposed area is painted with a special chemical solution. The tree begins to die and, by the following year when it is cut, the bark is loose enough that it can be removed during the harvesting operation. Logs are also reduced to chip form before being hauled to the mill woodyard for storage.

WOOD PREPARATION

In the mill woodyards, the unbarked logs are fed into a giant, revolving drum barker in which their bark is stripped as they tumble against each other and the steel channeled wall of the drum. Bark is also removed by moving the logs past streams of high-pressure water in a hydraulic-type barker, or by sets of mechanical knives. Depending on the type of pulping process they are going to be used in, the debarked logs from the barker and/or the woodlands are either sent directly to the process or to chippers where, by dropping against a revolving disc with heavy, sharp knives set at an angle, they are reduced to small chips approximately one-half to three-quarters of an inch in size. These wood chips and the unscreened wood chips from the woodlands are conveyed to vibrating screens. Oversized chips are removed, or removed and sent to a rechipper and returned for another pass through the screens. Usually, the undesirable fine chips and sawdust that are removed at the screens are burned in the power boiler as fuel.

The screened chips are then either transported by air conveyors to outside chip storage piles or belt-conveyed to huge tanks or silos for storage according to the type or species. Then they are transported from there to smaller storage bins located in the pulp mill, usually directly over the digesters in which the wood is to be processed.

Up to this point in the process, very little process-type instrumentation, as we know it, is used. Any automation involved in these operations is strictly mechanical.

PULP MANUFACTURE

Before wood can be made into paper, paperboard, plastics, and other products, it must be reduced to its basic components to form pulp.

Wood is made up primarily of cellulose fibers bound together with lignin, a glue-like binder, plus sugars, gums, resins, and mineral salts in lesser quantities. The task of pulp manufacturing is to separate the wood into fibers and other components, remove the undesirable components, and provide a means of treating the fiber to produce a suitable pulp for the paper mill.

The conversion of wood into wood fibers is usually carried out by one of three general methods, namely: mechanical, chemical, and semichemical, depending on the type of wood used and the requirements of the end product.

Mechanical Pulping

The simplest pulping method is generally referred to as mechanical pulping. It differs from other pulping methods in that the reduction of wood to fibers is essentially a physical operation in which the fibers are actually pulled away from each other by the application of some type of mechanical force. The three most prevalent forms of mechanical pulping are: groundwood or stone groundwood, refiner groundwood or refiner mechanical, and thermomechanical.

Groundwood Mechanical Pulping. In mills using the groundwood mechanical pulping method, barked logs from the barking drums or from the woodlands are transported to the grinders. The grinders consist of huge, rough-faced grinding stones driven by an electric motor or hydraulic turbine. The stone is enclosed in a housing on which pockets are mounted. The logs are placed in these pockets and their sides are pressed against the revolving groundstone by the use of hydraulic pistons or gravity. In this way, the fibers are literally torn from the log.

The resulting pulp is screened to remove slivers, knots, and oversized pieces which are sent through a refiner to be broken up and returned to the screens. After washing, the cleaned suspension is sent to the paper mill to be used for papermaking, as shown in the simplified flow diagram of a typical groundwood mill (Figure 1-1).

Since lignin is not removed, paper made with pulp produced by this method does not generally maintain its brightness and strength as long as papers made from chemical pulps. However, groundwood paper has certain qualities desired for various printing processes. Paper on which newspapers are printed consists largely of groundwood pulp. So do the lower grades of tablet paper and other papers which do not require unusual strength or are intended to be used only for a short time. By the development of highly successful bleaching methods, groundwood pulp has now become one of the important raw materials for making papers such as those used for color printing in magazines, as well as other quality papers.

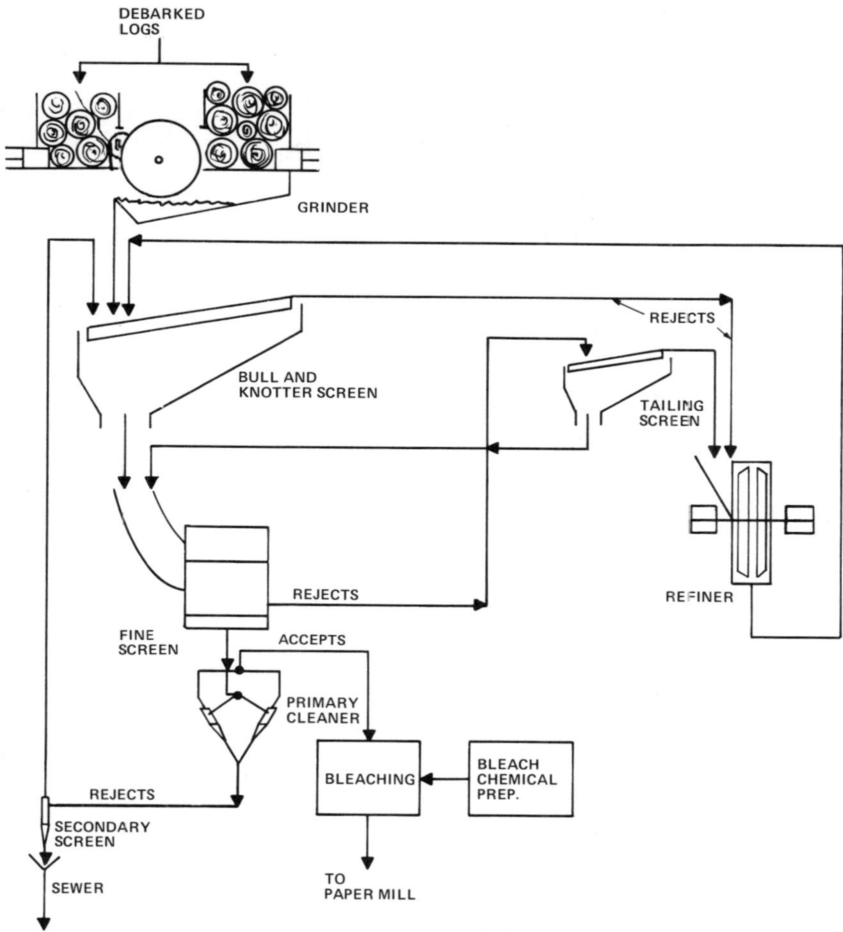

Figure 1-1. Simplified flow diagram of a typical groundwood mill.

Refiner Mechanical Pulping. This method involves the physical separation of fibers after the logs are reduced to chip form. A number of techniques have been developed to accomplish this by using refiners of various designs. The fundamental operation of refiners is based on the passage of wood chips between two disc-shaped rough surfaces moving in opposite directions and in close proximity to each other. The number of refiners used varies, depending on the operating practices of the mill. A common configuration consisting of the refiner stages is illustrated in Figure 1-2.

Thermomechanical Pulping. This process differs from the refiner mechanical pulping process in that with the thermomechanical pulping system wood chips are preheated and then subjected to mechanical forces in a disc refiner under pressure. The chips are subjected to repeated compressions and stress relaxations in the gap between the opposing bars and grooves. A second-stage mechanical treatment is performed on the pulp from the pressurized refiner in a nonpressurized, atmospheric-type refiner located downstream in the process, as shown in Figure 1-3.

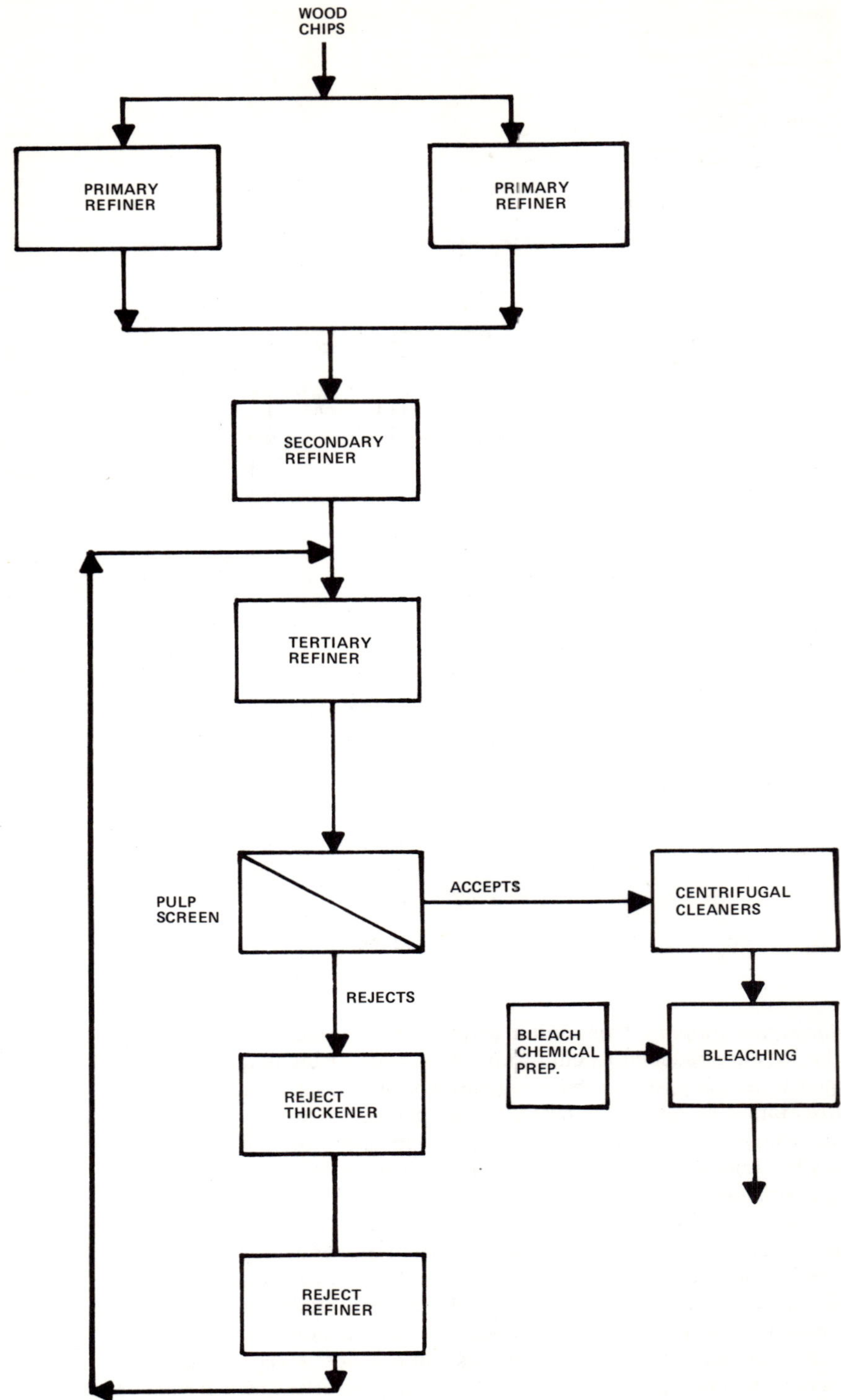

Figure 1-2. Block diagram of a typical refiner mechanical pulping process.

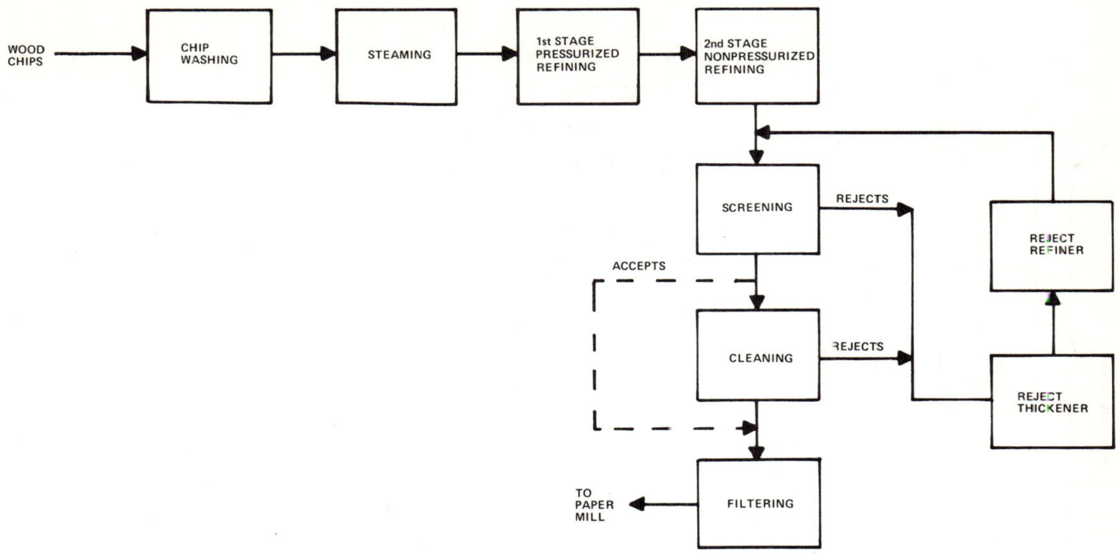

Figure 1-3. Block diagram of the thermomechanical pulping process.

Chemical Pulping

In these types of processes, the wood chips are cooked in a chemical solution under pressure until the fibers are separated from their lignin binder and fall apart without mechanical action. There are several chemical pulping processes using one of several solutions, depending on the type of wood used and the kind of pulp desired. The most common are the sulfate and sulfite processes.

Sulfate Pulping Process. The most important chemical process is the sulfate or kraft process. Practically any species of wood can be used. The cooking chemical is a solution of sodium hydroxide and sodium sulfide. As shown in Figure 1-4, in sulfate pulping the wood chips are boiled under pressure in a strong solution of sodium hydroxide and sodium sulfide. After cooking, the liquor which is separated from the pulp suspension at the washing stage is concentrated in multiple-effect steam-heated evaporators. Sodium sulfate is added and the mixture is burned in the furnace. The molten smelt issuing from the bottom of the furnace is dissolved in water to form green liquor which is causticized by adding slaked lime in the causticizing plant. The lime mud formed is removed and sent to the lime recovery area for recalcining into lime which is slaked and reused in the causticizing area. The sodium sulfide-caustic soda solution, which is now called white liquor, is sent to the digester area for reuse in cooking new wood chips.

The washed brown stock is screened, bleached, and either pumped directly to an integral on-site paper mill or processed for shipment to other paper mills or pulp processing plants.

Sulfite Pulping Process. Another common pulping process is the sulfite process which uses mainly softwood species such as spruce, hemlock, and fir. The chip cooking liquor consists of sulfurous acid and a salt of this acid produced by burning sulfur to sulfur dioxide. The desired cooking liquor is produced by the absorption of the sulfur dioxide gas with the hydroxide of the base chemical. Formerly, calcium was used but attempts at chemical recovery were a

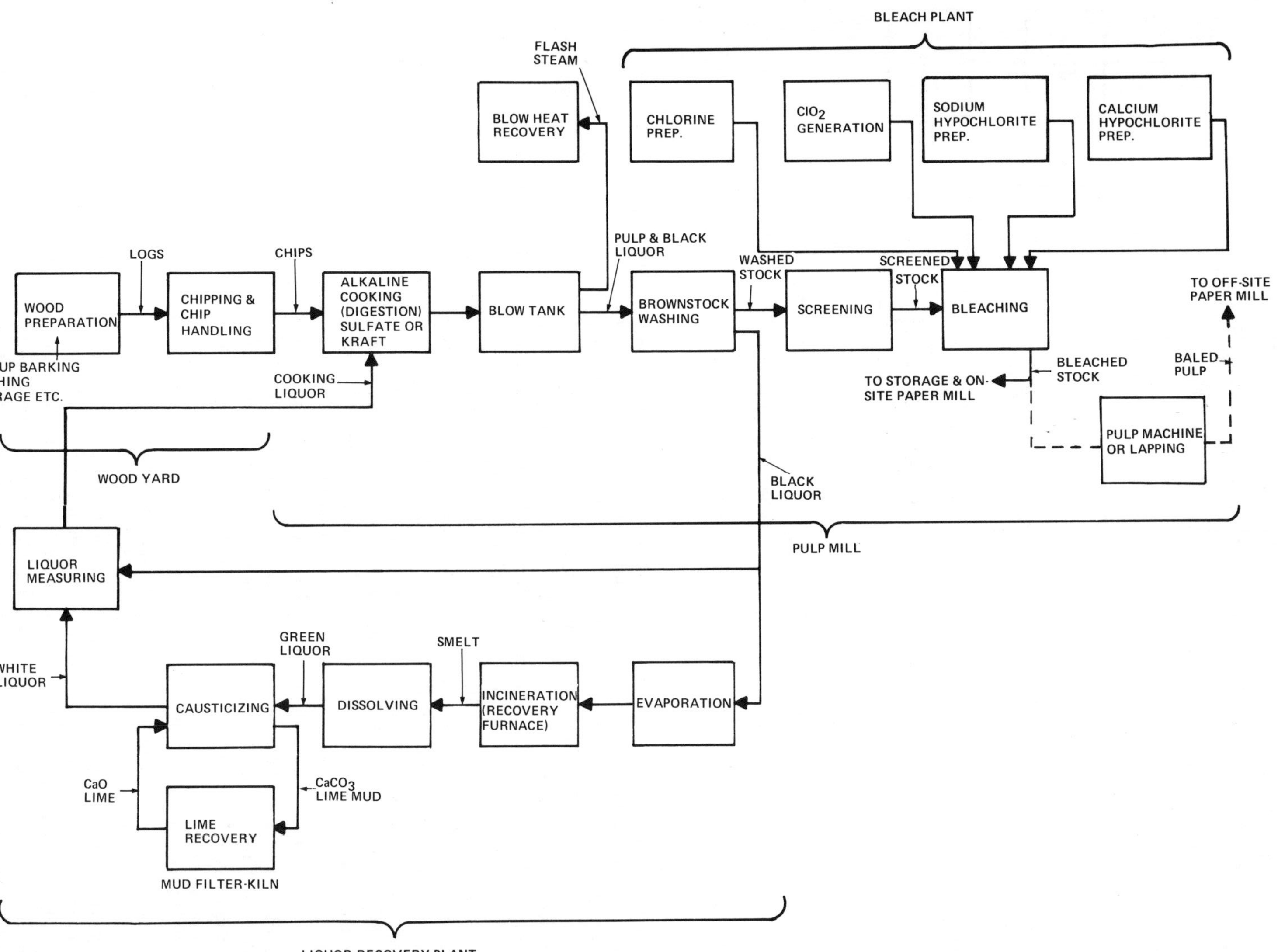

Figure 1-4. Block diagram of the sulfate pulping process.

problem and never became feasible. Newer mills are using recoverable bases such as magnesium, sodium, and ammonium. Figure 1-5 illustrates the cyclic nature of chemical recovery in a typical magnesium-base sulfite pulping process. It is similar to the sulfate process in configuration.

Pulp and spent sulfite liquor are discharged into a blow tank from which the digester's contents, diluted with additional weak liquor, are pumped to stock washers. The first-stage washer filtrate enters the weak red liquor storage tank.

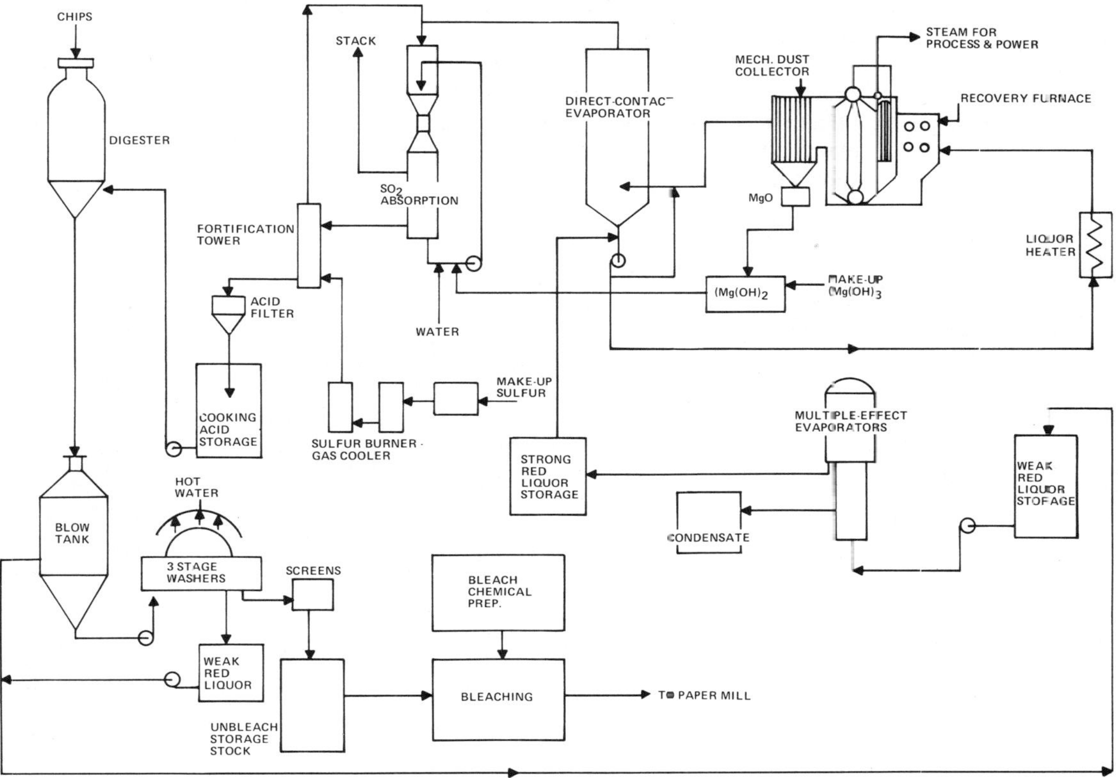

Figure 1-5. Flow diagram of a typical magnesium-base sulfite pulping process.

The liquor is concentrated in multiple-effect evaporators and transferred to the strong liquor tank. The strong liquor is then fired in a recovery furnace. The combustion products of the sulfur and the magnesium in the liquor are discharged from the furnace in the gas stream as sulfur dioxide and solid particles of magnesium hydroxide. The sulfur dioxide is recovered by reaction with the magnesium hydroxide to produce a magnesium bisulfite acid in an absorption system. This acid is passed through a fortification, or bisulfiting, system and fortified with makeup sulfur dioxide. The finished cooking acid is filtered and placed in storage for reuse in the digester.

The unbleached pulp from the washers is screened, bleached, and sent to the paper mill.

Sulfite pulp is used in the manufacture of bond, writing, high-grade book, and other fine papers.

Semichemical Pulping

Another way of making wood pulp combines mechanical and chemical methods and is called the semichemical process. In semichemical pulping, the wood is given only a mild chemical treatment, softening or removing just enough lignin to loosen the fibers, but not enough to separate them. Separation is then effected by mechanical means. The process was developed particularly for the pulping of hardwoods and has many variations. Three of the important methods are the neutral sulfite, cold soda, and chemigroundwood processes. Semichemical pulp produces stiff, resilient products and is used in making corrugated paperboard, egg cartons, and similar items.

Neutral Sulfite Semichemical Pulping. Commonly referred to as the NSSC process, this method uses a sodium sulfite liquor buffered with sodium bicarbonate to cook the wood chips. Although the design configuration variations are many, a typical process for the production of corrugating paperboard pulps is shown in Figure 1-6.

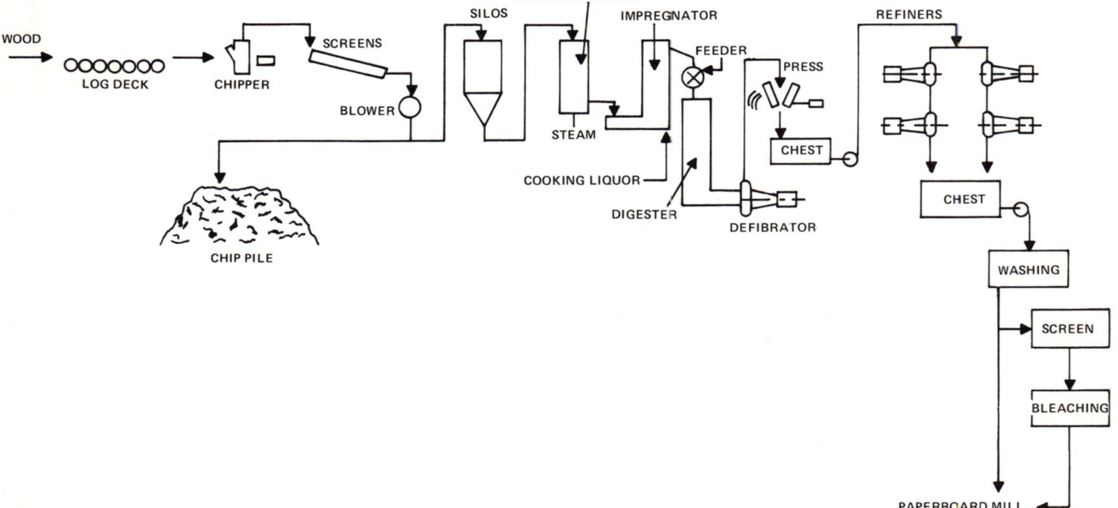

Figure 1-6. Simplified flow diagram of a typical NSSC pulping process.

The process begins with the heating of screened wood chips in a steaming vessel from which they enter a screw conveyor where they are impregnated with cooking liquor. After cooking in a digester, the chips are broken up by an impeller or defibrator under full pressure. The coarse pulp produced is compressed by a disk or screw press, diluted, and sent to the refiners. After washing on filters or screw presses, the refined stock is sent to the paperboard mill. In the case of bleached grades, the pulp is screened and bleached, and then is sent to the paperboard mill.

Cold Soda Pulping. In the cold soda process—also referred to as the chemical-mechanical process because there is no major change in the lignin—chips are treated with cold sodium hydroxide using hydrostatic, mechanical, or atmospheric pressure. Like the NSSC process, the cold soda process entails many variations in design. However, the major steps are shown in Figure 1-7.

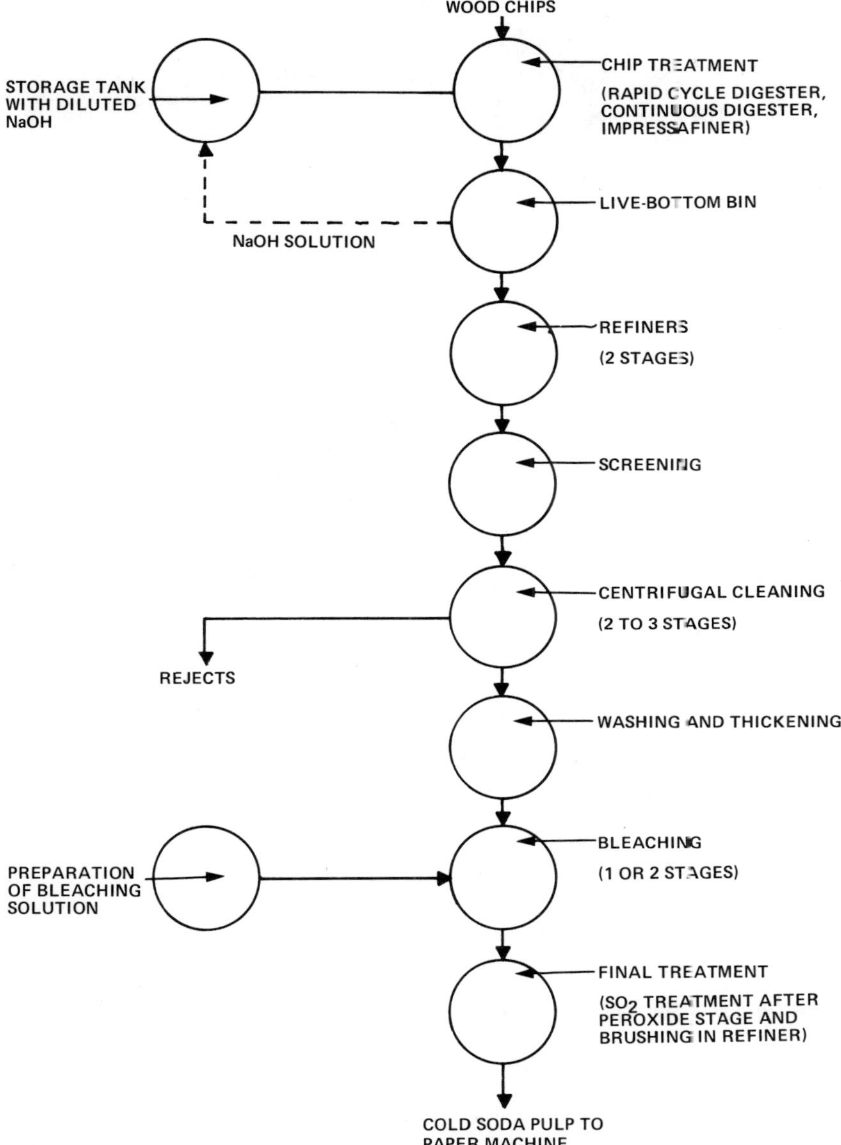

Figure 1-7. Major steps in the cold soda pulping process.

After treatment, the liquor is usually collected, fortified by fresh sodium hydroxide solution, and used again. The pulp is refined in a two- or three-stage disk refining operation, screened, centrifugally cleaned, washed, and thickened When the bleached pulp is used for printing papers or board, one- or two-stage high-density bleaching is applied. After the final treatment, the pulp is ready to be used on the paper machine. In some cases, unbleached cold soda is used, eliminating the bleaching operation.

Chemigroundwood Pulping. Another process, which is also referred to as chemical-mechanical for the same reason as cold soda pulping is called the

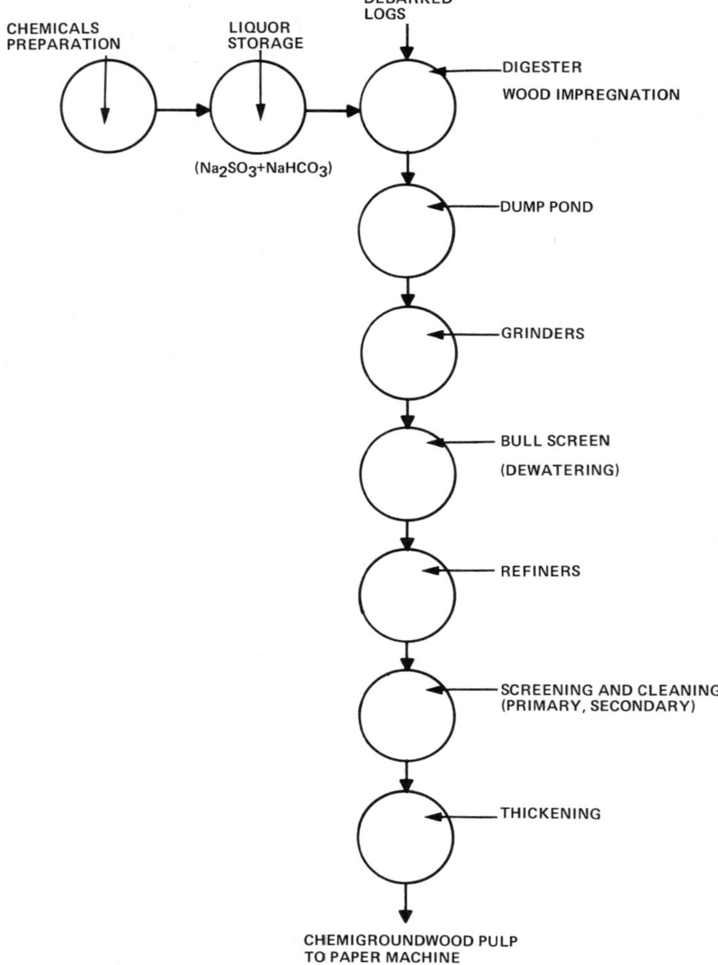

Figure 1-8. Major steps in the chemigroundwood process.

chemigroundwood process. In this process (Figure 1-8) four-foot logs are cooked in digesters in a mixture of sodium sulfite and sodium bicarbonate under pressure. The digester is evacuated for a short period. Following evacuation, the digester is filled with heated liquor. After treatment under pressure for a period, the logs are dumped in a pond and ground on conventional wood grinders. The stock is then directed through a bull screen, refiners, primary and secondary screens, thickeners, and then to the paper mill.

PAPER MANUFACTURE

Pulp, produced by any of the foregoing or miscellaneous processes, is made into paper in a mill that is at the same location as the pulp mill (referred to as an integrated mill), or it is dried and shipped to a paper mill located remotely from the pulp mill. Paper mill layouts vary somewhat, based on the pulps used and grades of paper produced. Figure 1-9 shows the basic operations carried on in a

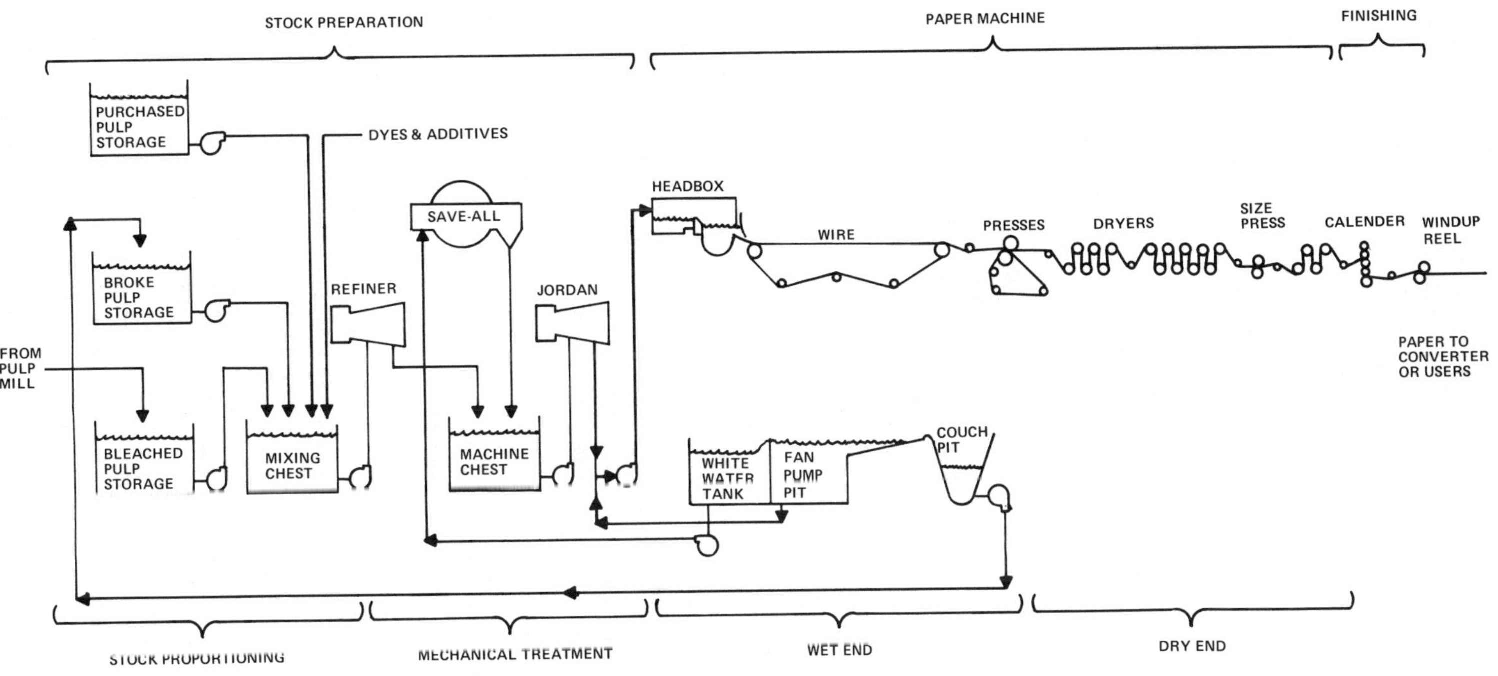

Figure 1-9. Simplified flow diagram of a typical paper mill.

typical paper mill. They are generally divided into stock preparation, paper machine, and finishing operations. The stock preparations can be further divided into stock proportioning or blending and mechanical treatment. Paper machine operations can also be subdivided into wet end and dry end.

Stock Preparation

Most pulp cannot be used for papermaking as it comes from the pulp mill. Therefore, other types of pulp having different characteristics must be purchased from other pulp mills to give the final qualities needed for the paper to be made. Dyes and additives are also added to achieve the desired color and physical properties of the sheet. These operations are usually referred to as stock proportioning or stock blending.

To impart mechanical strength to the final sheet, the pulp is refined in a variety of machines, typical of which are the refiners, jordans, and beaters. Basically, the operation consists of passing the pulp repeatedly between sharp moving bars that cut and abrade the fibers. The sets of bars can be adjusted to turn out various lengths of fibers with rougher or smoother edges. This improves fiber-to-fiber bonding, making it more uniform, more dense, less porous, or more transparent, depending on the kind of paper to be made.

Before going to the paper machine, the resulting pulp is screened and, in some cases, cleaned by passing through centrifugal-type cleaners to remove heavy particles of dirt that have not as yet been removed.

Paper Machine

Wet End. The most common major component of the wet end portion of the paper machine on which the paper is formed is a fourdrinier. It consists mainly of a continuous fine screen, called a wire, on which the pulp suspension is spread. The wet end section varies in widths in accordance with the size of the machine and moves at different speeds as determined by the type of paper being produced. Most of the water drains at the top end of the wire to form a mat of fibers. The wire then passes over a series of vacuum suction boxes which suck more water from the wet mat through the wire. The wet paper leaves the fourdrinier machine at a consistency of about 20 percent (20 percent fiber and additives, 80 percent water).

Paperboard, which consists of several layers of paper, is usually made on a cylinder-type machine instead of a running flat screen. Pulp is pumped to several vats. In each vat, a cylinder covered with a fine wire screen turns at the same speed as the rest of the machine. Water passes from the vat through the screen and is removed from the center of the cylinder, leaving a mat of paper on the wire screen. A continuous felt blanket in contact with all cylinders picks up layers of paper, forming a laminated, layered board. A good grade of pulp is frequently used on the end cylinders to give the outside of the board a good appearance. The second sheet is often made of a slightly poorer grade, while the inner layers are normally reclaimed paper of poor grade.

Dry End. After leaving the wet end section of the paper machine, the wet paper is sent to the presses where it is supported by endless woolen or synthetic loops called felts. The paper on top of the felts is then passed between heavy press rolls to press out as much water as possible. The paper leaves the press section at approximately 35 percent consistency. The rest of the water is then evaporated on steam-heated rolls located in the dryer section. Endless felts again carry the paper through and press it against steam-heated rolls on opposite sides.

Finishing

The dried paper then passes on to the finishing stage of the process. Most papers go through one or more additional processes, one of which is calendering. This process consists of ironing the paper between heavy, polished steel rollers, giving it a much smoother surface. Some paper is wound in large rolls as it comes from the calenders. These are later rewound and cut into smaller rolls or sheets as required by the user. Some papers are produced specially for further processing by converter plants which make envelopes, milk containers, grocery bags, paper cups, and many other consumer products.

Paperboard is used for corrugated cartons, folding boxes for frozen foods, and many other items. Large quantities of pulp for some types of paperboard are made up of repulped waste paper from which ink and other impurities have been removed.

INSTRUMENTATION SYMBOLS

Now that we have a general appreciation of the process in which instrumentation and control applications will be discussed, it will be helpful to review some aspects of the instrument diagrams which will be presented along with the text. These diagrams use a system of simplified letter and pictorial symbols in place of actual instruments. A summary of pictorial symbols used in the flow diagrams is shown in Figure 1-10.

A combination of letters is used to identify each function. This system is shown in Table 1-A.

A number is usually assigned to each instrument or loop, depending on user practice. Loop numbers are used here. Combining the symbols and numbers defines a particular circuit or device as illustrated by the basic loops shown in Figure 1-11. Although this general concept of depicting instrumentation is generally used by process control and design engineers and accepted in the industry,

Figure 1-10. Typical pictorial instrumentation symbols.

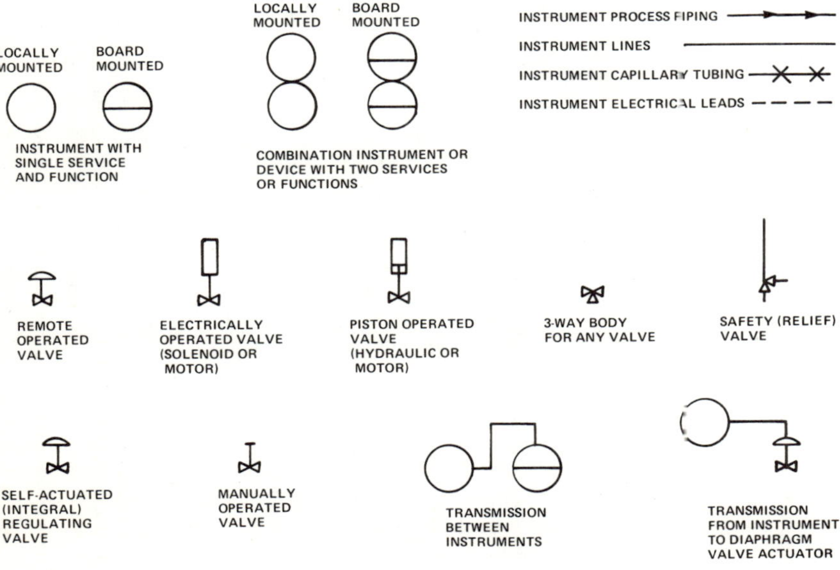

Table 1-A. Typical Letter Instrumentation Symbols

Process variable	Measuring					Transmitting			Controlling							Alarm[5]			
	First letter	Indicating	Recording	Totalizing or integrating[4]	Glass device for observation only	Blind	Indicating	Recording	Blind	Indicating	Recording	Self-actuated valve	Safety valve	Control valve	Powered operator	Blind	Indicating	Recording	Primary elements
		–I	–R	–Q	–G	–T	–IT	–RT	–C	–IC	–RC	–CV	–SV	–V	–P	–A	–IA	–RA	–E
Analysis[1]	A–	AI	AR	AQ		AT	AIT	ART	AC	AIC	ARC			AV	AP	AA	AIA	ARA	AE
Burner flame	B–	BI				BT	BIT		BC	BIC				BV	BP	BA	BIA		BE
Consistency	C–	CI	CR			CT	CIT	CRT	CC	CIC	CRC			CV	CP	CA	CIA	CRA	CE
Conductivity	Cd–	CdI				CdT	CdIT	CdRT	CdC	CdIC	CdRC			CdV	CdP	CdA	CdIA	CdRA	CdE
Density or specific gravity	D–	DI	DR			DT	DIT	DRT	DC	DIC	DRC			DV	DP	DA	DIA	DRA	DE
Electric[2]	E–	EI	ER	EQ		ET	EIT	ERT	EC	EIC	ERC	ECV		EV	EP	EA	EIA	ERA	EE
Flow[3]	F–	FI	FR	FQ	FG	FT	FIT	FRT	FC	FIC	FRC	FCV	FSV	FV	FP	FA	FIA	FRA	FE
Hand	H–	HI	HR						HC					HV	HP				
Time or time sched.	K–	KI	KR	KQ		KT	KIT	KRT	KC	KIC	KRC					KA	KIA	KRA	KE
Level	L–	LI	LR		LG	LT	LIT	LRT	LC	LIC	LRC	LCV		LV	LP	LA	LIA	LRA	LE
Moisture or humidity	M–	MI	MR			MT	MIT	MRT	MC	MIC	MRC			MV	MP	MA	MIA	MRA	ME
Pressure or vacuum[3]	P–	PI	PR			PT	PIT	PRT	PC	PIC	PRC	PCV	PSV	PV	PP	PA	PIA	PRA	PE
Freeness	Q–	QI	QR			QT	QIT	QRT	QC	QIC	QRC			QV	QP	QA	QIA	QRA	QE
Resistance	R–	RI	RR			RT	RIT	RRT	RC	RIC	RRC					RA	RIA	RRA	RE
Speed or frequency	S–	SI	SR	SQ		ST	SIT	SRT	SC	SIC	SRC	SCV	SSV	SV	SP	SA	SIA	SRA	SE
Temperature[3]	T–	TI	TR			TT	TIT	TRT	TC	TIC	TRC	TCV	TSV	TV	TP	TA	TIA	TRA	TE
Viscosity	V–	VI	VR		VG	VT	VIT	VRT	VC	VIC	VRC			VV	VP	VA	VIA	VRA	VE
Weight, load force, basis weight	W–	WI	WR	WQ		WT	WIT	WRT	WC	WIC	WRC	WCV		WV	WP	WA	WIA	WRA	WE

Notes:

1. Readily recognized self-defining symbols such as CO_2, O_2, pH, and ORP may be used in place of A.
2. Letter subscripts c (current), v (voltage), or w (power) may be used after E to designate type of measurement.
3. Lower case letters r, d, and t may be inserted to distinguish ratio, difference, and time respectively.
4. Q may also be used with letters I, R, and C where indicating, recording, or control functions also exist.
5. Lower case letter subscripts l and h may be used to denote low or high alarm functions.

If required note (1) could be expanded to cover any first letter, not in conflict with those listed, as long as a description for such a designation is part of the flow diagram.

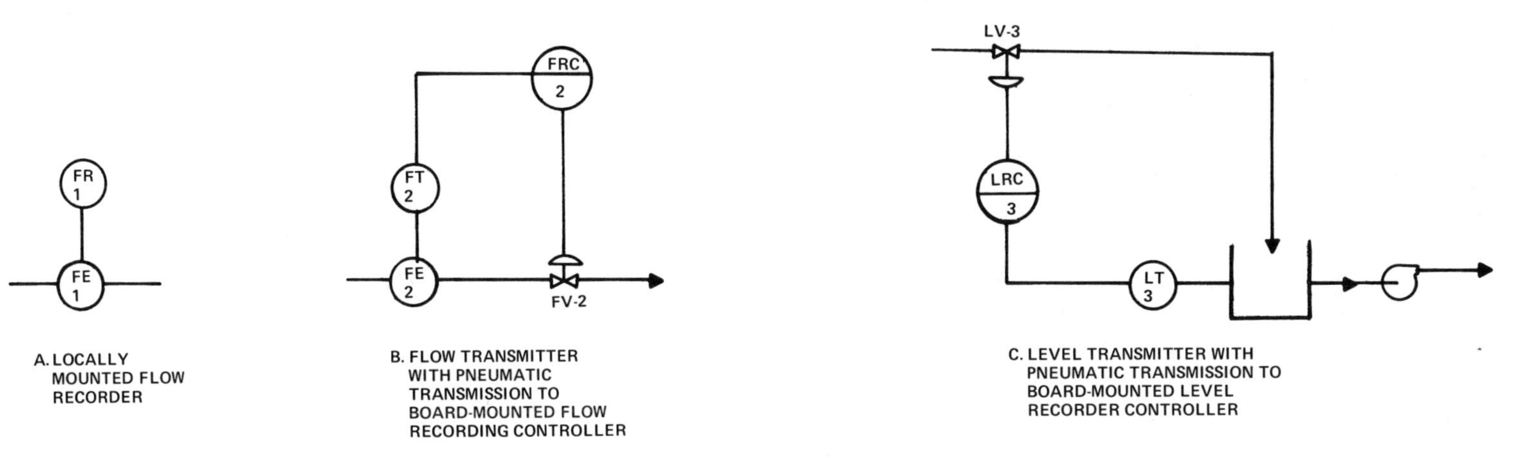

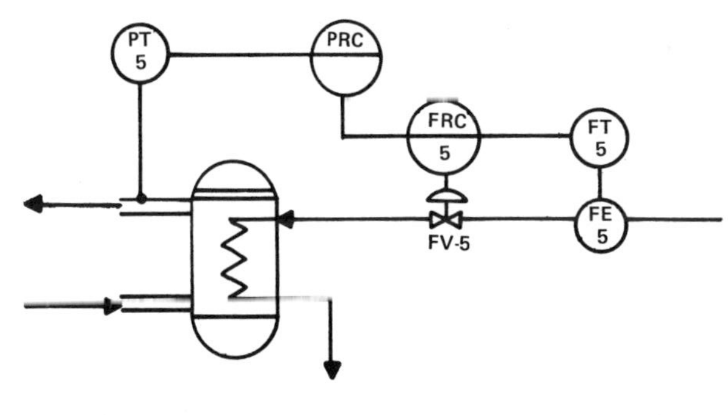

Figure 1-11. Sample basic instrument diagrams.

a variety of modifications are made by users to satisfy special circumstances. For example, when a digital computer is involved some symbols must be changed or supplemented to accommodate this function.

All standards developed for instrument diagrams differentiate between electrical and pneumatic signal and impulse lines. A general instrumentation can be applied to the pulp and paper process with the implementation of pneumatic and/or electrical devices; the diagrams in this text have not differentiated between the different types of instrument lines that could be used. They are all shown as a single solid line. Where the signal is exclusively electric, a broken line is used.

Modern installation practices invariably utilize long-distance measurement and control techniques with display and operating instruments located on panelboards centralized in elaborate, efficient control centers. Therefore, for purposes of discussion, the instrument application diagrams depict practically all measurements as being transmitted to remote panel-mounted instruments except where the instrument is always locally mounted in the vicinity of the point of measurement.

Pulping 2

GROUNDWOOD

Discontinuous Grinder

In a groundwood mill various types of grinders are used to reduce logs into fiber components. Instrumentation is required to maintain a steady production of pulp of the desired quality. The temperature of grinding has a very great effect on the type of pulp produced. Grinding temperature, when steadily maintained at any predetermined point, gives better control of consistency and freeness; this results in increased operating efficiency and maximum production output with less power required per ton to produce it. Temperature control also prevents any sudden deluge of cold water on a hot grinding stone, thus minimizing temperature strains and increasing length of service. The need for frequent burring of the stone is also reduced. Figure 2-1 shows the instrumentation involved in controlling temperature on a typical discontinuous grinder.

Figure 2-1. Temperature control on a discontinuous-type pulp grinder.

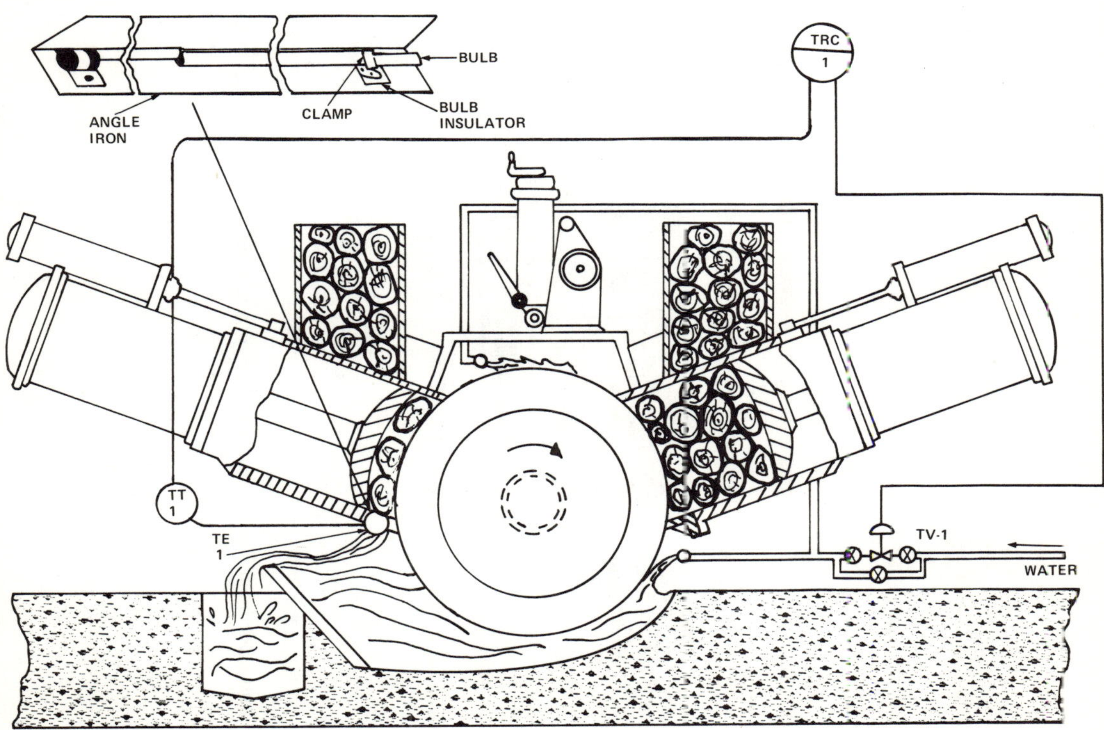

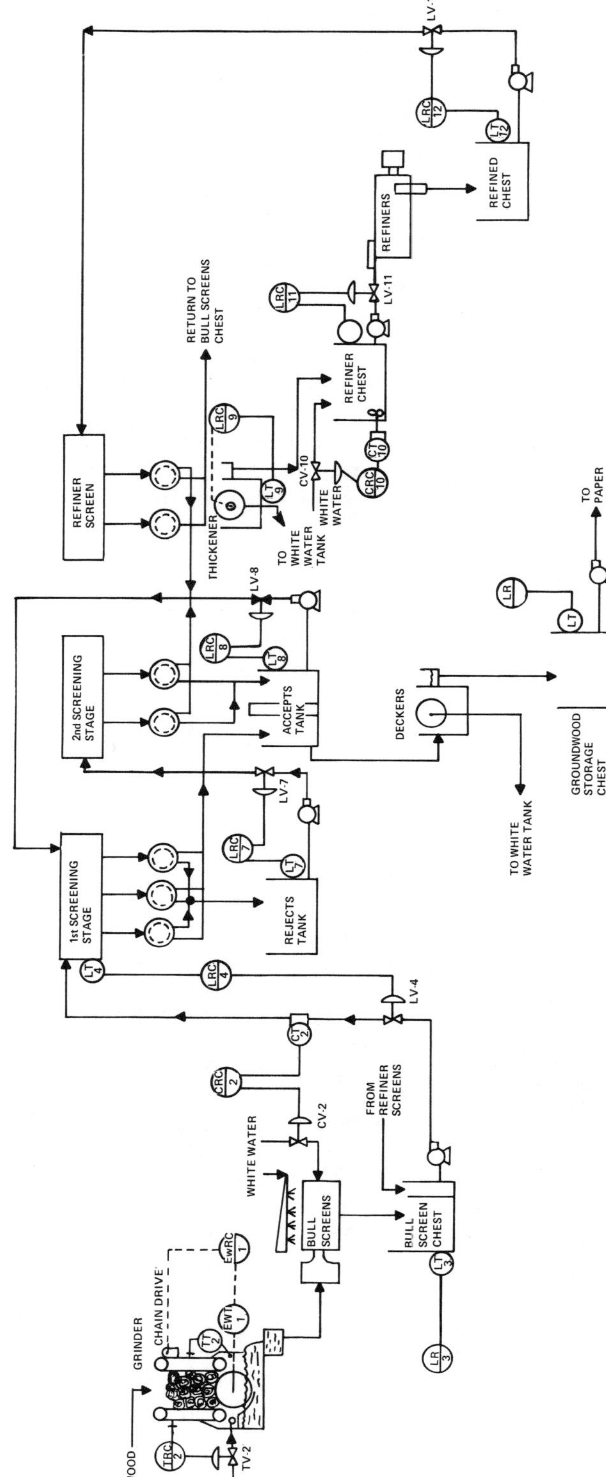

Figure 2-2. Instrumentation of a typical groundwood mill with continuous grinder.

The temperature is sensed by thermal element TE-1, installed as close to the stone as possible. The exact location of the bulb depends on the design of the grinder, but the usual location is on the underside of the finger plate or the bottom of the bridge tee. The length of the bulb is put across the finger plate close to the edge, above the stock level in the pit and across the face of the stone. The temperature measured is the average grinding temperature. The temperature measurement is then transmitted as a signal by temperature transmitter TT-1 to recorder controller TRC-1.

The temperature controller operates an automatic valve, TV-1, in the water spray line to the stone to maintain the grinding temperature at a desired value.

Continuous Grinder

Additional instrumentation associated with the treatment and handling of the pulp after grinding is shown in Figure 2-2. A continuous-type chain grinder is used for the purpose of discussion, although other types of grinders can be used in this operation.

The chain speed of the grinder is controlled by power recorder controller EwRC-1 which receives its signal from power transmitter EwT-1. This maintains a uniform quality of pulp from the grinders. The grinder pit temperature is controlled by a measurement of the vapor temperature over the stone face with an averaging-type bulb whose signal is transmitted by TT-2 to controller TRC-2. The output of controller TRC-2 regulates valve TV-2 on the shower water line. Some installations measure temperature with thermocouples mounted in a metal plate extending across the width of the grinder pit, or located below the finger bars where the stock will strike. When there is good circulation in the pit, these measurement locations provide more accurate representation of the temperature. Consistency of stock supplied to the bull screens is regulated by controller CRC-2 which adjusts valve CV-2 on the white water line, based on consistency measurement CT-2, located on the pulp line to the screens. The bull screen chest level is recorded by LR-3, with the level measured by LT-3.

Constant flow through the cleaning and treatment system is maintained by controlling the level in the screen feed by controller LRC-4 which operates valve LV-4 in order to obtain a constant head to the screens. The head to the screens is measured by LT-4.

Level controller LRC-7 is provided on the chest for the rejects of the primary screens and, in turn, regulates the feed to the secondary screens. Level in the accept tanks is maintained by controlling the level in the vat or trough of the thickeners prior to the refiners with controller LRC-9, which adjusts the speed of the thickener drum. The consistency of the stock in the refiner chest is controlled through the regulation of dilution white water by CRC-10. CRC-10 operates on a consistency signal correlated to a wattage measurement of the circulating motor by CT-10. The level in the refiner feed chest is controlled by a level recording controller, LRC-11, which acts on valve LV-11, controlling the rate of feed to the refiners.

The refined stock chest level is maintained constant by a level recording controller, LRC-12, regulating the rate of feed to the refiner screens.

In general, level measurements are made by transmitters fitted with air-purged bubble systems or diaphragm-type sensing elements. If fibers exist in solutions, the latter type is preferred. Diaphragm-type and ball-type control valves are also the preferred valves for use in applications involving the flow control of fiber-containing liquids.

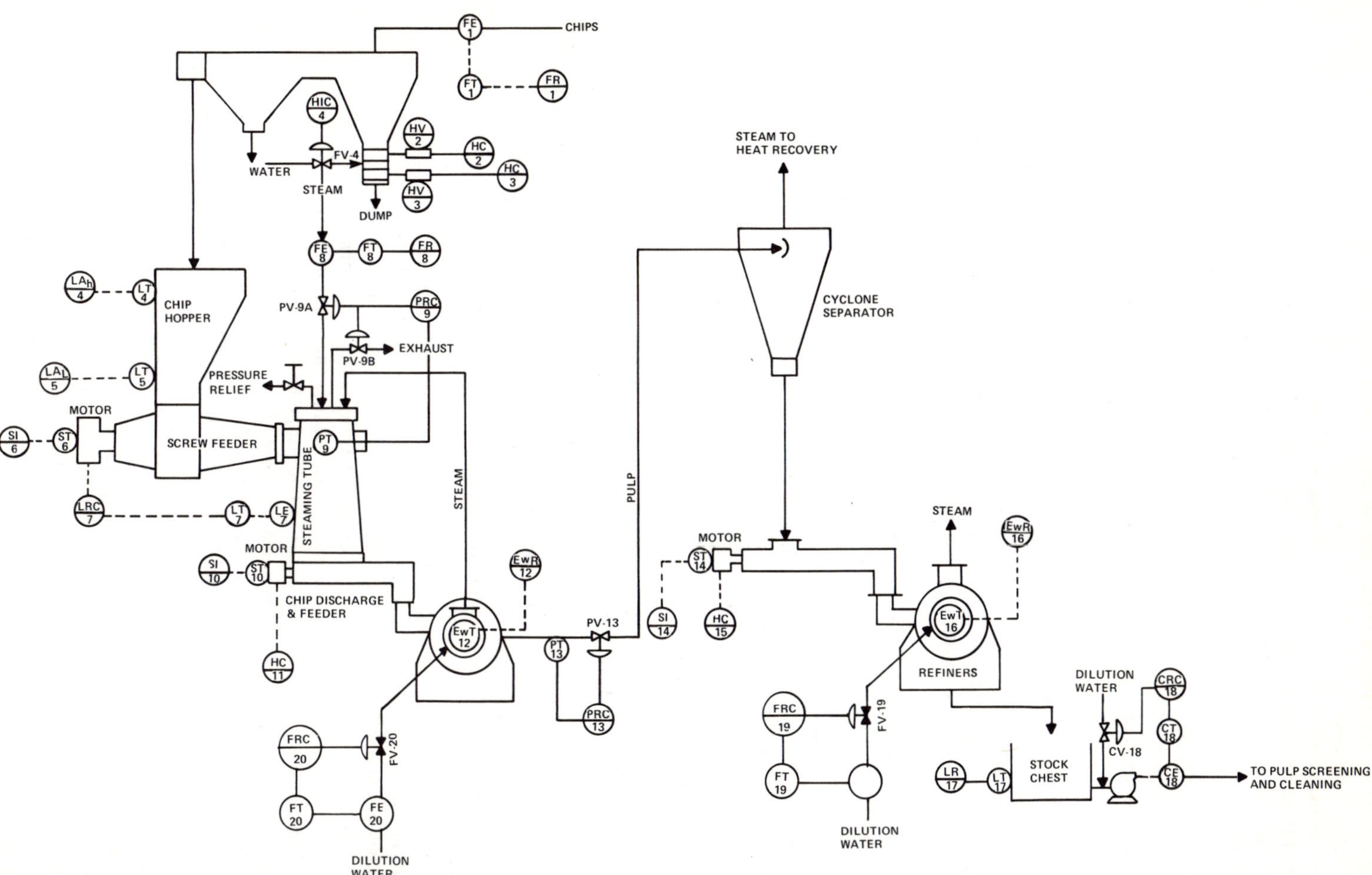

Figure 2-3. Thermomechanical pulping instrumentation.

THERMOMECHANICAL

In view of the increasing cost of raw materials, labor, and power, along with the demand for decreasing polluting effluents, the paper industry has recently turned its attention to thermomechanical pulping, which promises to increase yield from wood, decrease pollution load, and eliminate costly upgrading of old pulping facilities. Use of thermomechanical pulp in newsprint can go a long way in reducing the amount of chemical pulp used. Many other grades of paper might incorporate varying amounts of thermomechanical pulp in their furnish; these include printing and writing papers, tissues, and boards.

The difference between thermomechanical pulp and other mechanical pulp is that thermomechanical pulp is produced by thermally softening the lignin binding material in the wood structure before liberating the individual fibers by the use of pressurized refiners instead of atmospheric refiners. Thermomechanical pulp characteristics are more like chemical pulp than any other mechanical pulp, thereby enabling increased substitution of thermomechanical pulp for chemical pulp in grades of paper where mechanical pulp is mixed with higher priced chemical pulps. This means better wood utilization since the process yield of 95 percent is nearly twice that of chemical pulp.

The process does not require whole logs or chips. Thermomechanical can be produced from wastes of woodworking industries—such as slabs, sawdust, shavings, veneer refuse, as well as wood chips—furthering wood utilization.

In the thermomechanical pulping process, temperature, flow-through, and consistency of refining are the most important variables that affect pulp characteristics. Chips or sawdust are preheated and subjected to mechanical forces in a refiner under pressure. The chips are presteamed at a temperature above 100°C so that the temperature of refining is between 120° and 130°C (15 to 25 psi). It has been theorized that if heated above 140°C, the fibers separate at the middle lamella layer, which is rich in lignin. The lignin goes through a glassy transition to thermal softening, resulting in a thick coating of softened lignin around the separated fibers, which reverts to a glassy state on cooling. This condition hinders further refining to produce fibrillation for strength development.

At a temperature of 120° to 130°C (15 to 25 psi), most of the lignin, while softened, is still in a glassy state. Under these conditions, the fiber separation occurs in the outer layers of the secondary fiber wall, thereby rendering the fibers accessible to fibrillation in subsequent refining, with no heavy coating of surface lignin.

A number of systems have been developed for the manufacture of thermomechanical pulp. Figure 2-3 is a simplified diagram of one of them and illustrates the typical instrumentation found in such a mill.

Basically, the thermomechanical pulping system consists of a chip metering system, FE-1, FT-1, and FR-1. The chips are washed in a conditioner which consists of a water chamber with a low volume upflow of water across which the chips must float and allow heavy debris to drop out. This debris is dumped periodically through the double-valve system, HC-2, HV-2 and HC-3, HV-3. Water flow to the conditioner is manually adjusted with HIC-4, FV-4. After washing and draining, chips drop into the chip hopper which is fitted with high and low alarms, LA_h-4 LA_l-5, used to warn the operator when chips drop below a minimum or rise above a maximum height so that he can take appropriate actions. Chips are fed from the hopper to the steaming tube by a tapered screw feeder whose speed is indicated on SI-6 and controlled to maintain level in the steaming tube by LRC-7. The steaming tube level measurement is made with a nuclear radiation type of sensor, LE-7. Steam to the steaming tube is measured and recorded on FR-8 and the flow is controlled by steaming vessel pressure controller PRC-9 through valve PV-9A. The chips are moved from here to the refiner by a screw-type conveyor whose speed is indicated on SI-10 and manually set by HC-11. Chips are fed to the refiner by a ribbon-type feeder. Steam produced in processing the chips in the refiner is recycled back to the steaming vessel. This steam is sufficient to maintain the required pressure in the vessel which closes valve PV-9A by recording controller PRC-9 on the main steam supply. If the pressure tends to rise over the desired pressure in the vessel, the controller will then commence opening valve PV-9B on an exhaust line.

The refiner power loading is recorded on EwR-12.

Pressure in the refiner is controlled by the throttling valve, PV-13, from PRC-13 whose measurement, PT-13, is made in the blowline to the cyclone separator. The pulp, which is separated from the steam, drops into a conveyor system whose speed is indicated on SI-14 and manually adjusted by HC-15. The conveyor carries the chips to the second-stage refiner whose motor load is recorded on EwR-16. The refined stock then drops down into a stock storage chest whose level is recorded on LR-17. As the stock is pumped from the chest to the screening and cleaning system, dilution water is added by controller CRC-18 in order to regulate the consistency of the pulp at a value for efficient operation.

CHEMICAL PULPING

As previously mentioned, the two most prominent full chemical pulping processes are the sulfate, or kraft, and the sulfite processes. Wood chips are conveyed to either process by belt-type conveyors on which they are weighed on some type of weightometer. As the chips are loaded into the digester, moisture content of the chips is measured by radiation- or capacity-type instruments. This permits the calculation of the bone dry weight of the wood chips from the weightometer reading. Of the two chemical processes, the sulfate is the most widely used.

The wood chips are then cooked in either batch- or continuous-type digesters.

Sulfate Batch Digester Liquor Charging

In preparing a batch pulp digester for cooking chips, the digester is charged with a predetermined amount of wood chips and chemical cooking liquors (white and black liquor).

The amount of white and black liquor charged into the digester must be closely measured to attain uniform cooking. This is accomplished by a drawdown-type system using measuring tanks or an integrated flow measurement system using magnetic-type flowmeters.

Measuring Tank Type. The batch digester liquor measuring system, shown in Figure 2-4, describes a method by which prescribed quantities of liquor may be measured and delivered to the digester being charged.

In operation, the measured amounts of each type of liquor are determined by setting a level set point on controllers LRC-1 and LRC-2 of each measuring tank. The liquor supply pumps are then started, allowing the tanks to be filled. When the desired levels are reached, the controller automatically shuts down the liquor supply pumps. When a digester requires filling, a dump button is activated in proper sequence to obtain the desired addition of white and black liquors by opening drawdown valves LV-1 and LV-2 through solenoid valves on the air supply lines to the valve operators. After drawdown is completed from each tank, the level controllers will automatically start the liquor transfer pumps to refill the tanks to levels preset on the controllers.

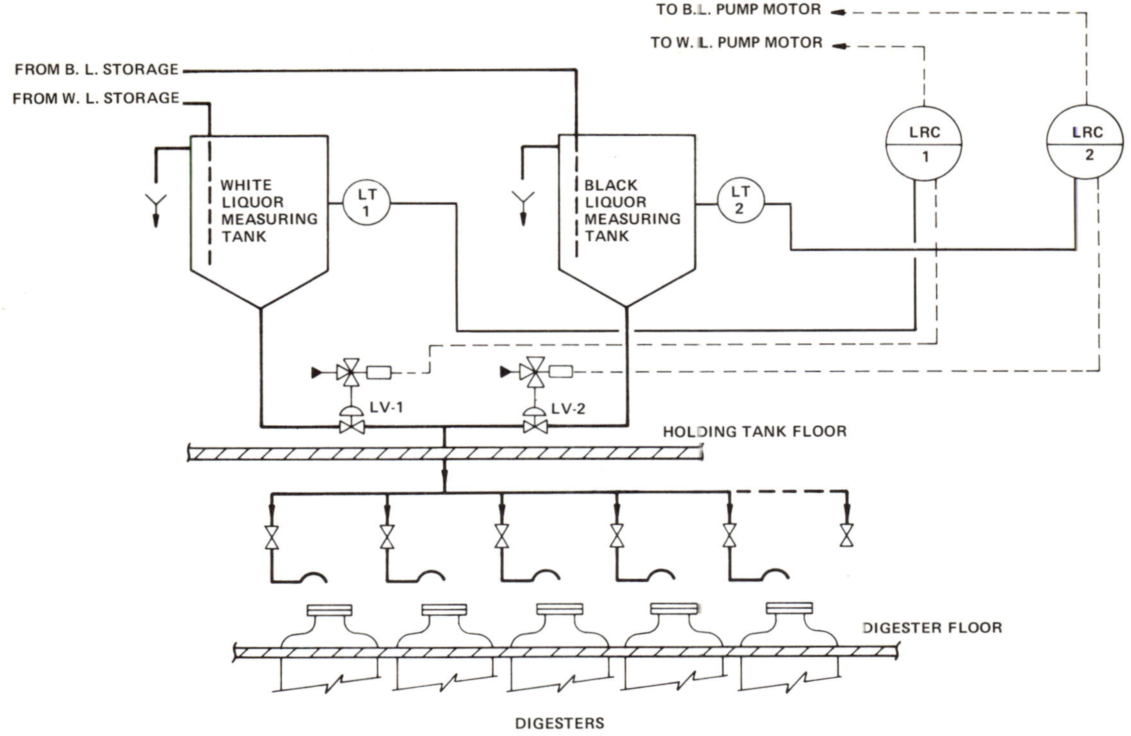

Figure 2-4. Digester liquor charging system—measuring tank type.

Often such systems incorporate various logic circuits designed for specific requirements. Such functions as start-stop, open-close, and on-off may be sequenced, as well as the interlocking of any number of devices which may be provided. Manual overrides are also provided where desired.

Level transmitters LT-1 and LT-2 may use air bubble purge systems to detect level in the measuring tanks. However, the flanged-diaphragm type transmitter is more widely used today.

Integrated Flow Measurement Type. The digital batch control system shown in Figure 2-5 permits batching of the digester cooking liquors directly from storage into the digesters.

This batching system uses integrated flow measurement in place of measuring tanks. The primary flow element, FE-1, is usually of the magnetic type. The meter, coupled with a digital counting system through transmitter FT-1 and converter I/P-1, measures predetermined quantities of cooking liquor. The operation is fairly simple. The required volume of liquor is set on a six-digit counter dial and a pushbutton is activated to start it. The flow rate signal from the magnetic flow transmitter is converted into a train of pulses proportional to the flow rate by converter I/P-1. Each pulse represents a predetermined volume. The pulses go to predetermining impulse counters FQ-1 and FQ-1A, preset to the desired volume. On reaching this preset quantity, the counter operates contacts to stop the flow.

As shown, a single system may be used to measure the black and white liquor components sequentially. However, many users prefer to have two separate, independent systems on black and white liquor flows so that both components can be charged into the digester simultaneously.

The optimum rate of flow of relief gases, which is dependent on the rate of steam flow to the digester, will prevent excessive liquor pullover. A high steam flow requires a low relief flow, and a low steam flow will allow a relatively high relief flow. Relief flow transmitter FT-3 measures flow of the relief gases and transmits a signal to relief flow indicating controller FIC-3. The output from the steam flow transmitter, FT-1, is inversed by reversing relay RR-1. Digester circulation depends largely on the steaming rate and relief rate and is measured by the temperature difference between the top and bottom of the digester. The schedule for the rate of relief gases on each digester is determined by experience. The record of top digester temperature and pressure difference is essential for establishing the relief flow operation. This is just one of several relief systems designed for batch digesters. Other systems depend on time-scheduled relief valve opening, temperature difference, or temperature-pressure relationship, of the relief gases for the basis of their operation.

The plugging of the screen located in the neck of the digester is controlled by blowback controller BBC-4. Blowback consists of opening blowback control valve BBV-4 and closing relief flow control valve FV-3 for a fixed period. The blowback controller measures the difference between the set point signal and the measurement signal to the relief flow controller. When the measurement signal is sufficiently less than the set point signal, the relief valve is wide open and the screen is plugging up. The blowback control will close the relief valve and open the blowback valve, energizing a timer which reverses the operation after a set time has elapsed. If measurement and set point are still not satisfactory for proper relief, the operation will be repeated until the proper relationship is obtained. Other blowback systems have been designed to operate on a pressure difference measurement between the digester and the relief line.

Indirect Steaming. When the cooking liquor in a batch digester is heated without direct injection of steam, it is equipped with a tubular-type heat exchanger as shown in Figure 2-7. Liquor is withdrawn from the digester through a strainer section by the action of a centrifugal pump and passed through the heater tubes, then returned to both the top and bottom of the digester.

Condensate level indicating controller LIC-8 holds the level in the condensate outlet line between a minimum (to preserve the condensate seal) and a maximum (to keep the condensate from covering the heater tubes). When the condensate rises to cover the heater tubes, heater capacity is proportionately decreased with consequent loss of production time. When condensate level drops so low that the seal is broken, steam blow-through results in significant heat loss and consequent increase in operating costs. LT-8 is a differential-type pressure transmitter applied to the shell or condensate leg of the heat exchanger. This measures the condensate level and sends a signal to the level controller which strokes control valve LV-8 on the condensate line.

The condensate, if not contaminated, is returned to the boiler house for reconversion into steam, a feature not available with direct-steamed batch digesters. A conductivity cell and transmitter, ConT-9, provide a means of immediate detection and disposal of condensate by-liquor from defective heater tubes. The signal is transmitted to controller ConRC-9 which switches the three-way valve, LV-9, so that liquor is diverted to the mill sewers in the event of contamination, as evidenced by a high-conductivity measurement. A high alarm signal on the controller alerts the operator when this condition exists so that he can initiate corrective action.

Circulation pump efficiency, the condition of the digester strainer and the heater tubes, is determined by making a load measurement on the pump motor drive with transmitter EwT-7 and is recorded on the wattmeter, EwR-7.

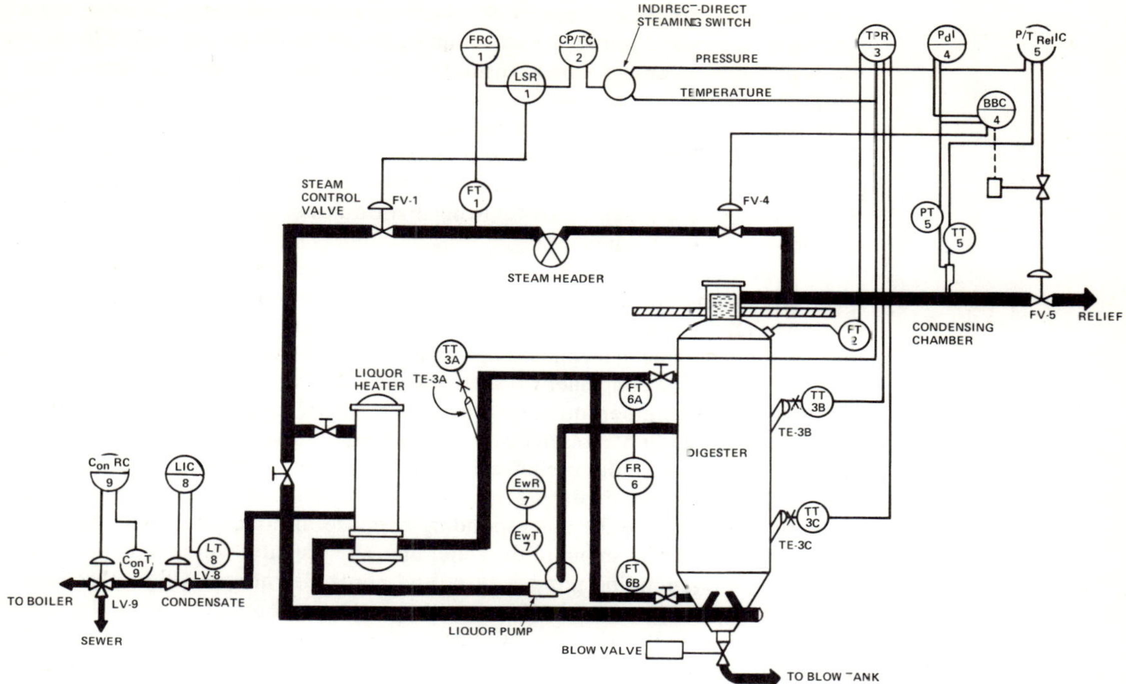

Figure 2-7. Instrumentation for an indirectly heated sulfate batch digester.

Flow through both sections of the liquor circulating line from the liquor heaters is measured with flow nozzle primary elements. Flow transmitters FT-6A and FT-6B on these elements must utilize sealed or purged static impulse lines to guard against possible plugging by fines in the liquor. The flows are recorded on FR-6 from which valves are manually adjusted to obtain correct distribution between top and bottom of the digester. Ratio controllers operating automatic control valves in the circulation lines are also used.

For a given cooking cycle, steaming is initially by steam flow controller FRC-1; subsequently by time-cam controller CP/CT-2. Transfer from steam flow to time-cam control is accomplished by low signal autoselector relay LSR-1. When the digester temperature resulting from the controlled rate of steam flow reaches the temperature corresponding to the position of the cam follower on the moving cam in the cam controller, the cam controller takes over by selective action of the low signal autoselector relay, LSR-1. The cam rotates from the start of the cook and its rise segment represents the total time to reach *hold* temperature and pressure. The actual temperature at which the transfer occurs is determined by the set point of the steam flow controller, the time-schedule cam, the initial heat content of the liquor and chips, and the indirect heater efficiency. Although considerable latitude in the actual point of transfer is tolerable, it always occurs before the hold temperature is reached, thus ensuring uniform rise times and hold temperatures. The combination of steam flow and cam-temperature control protects against boiler overload during heavy steaming at the start of the cook, while maintaining the cook cycle repeatability.

The piping arrangement is such that, with proper valving, the digester can be operated as a direct-steaming type in case of heater outage due to malfunction or repair. In this case, the cam controller can be switched from temperature to pressure by a three-way indirect-direct steaming switch in the output signal lines of the heater outlet temperature transmitter, TT-3A, and digester pressure transmitter PT-2.

Digester temperatures and pressures are recorded on multipoint recorder TPR-3. These measurements can be scaled so that the temperature-pressure relationship can be compared with saturated steam tables and used as a guide to the amount of noncondensable gases existing in the digester and the degree of "false" overpressure it is creating.

To illustrate another method of relieving noncondensable gases formed in the digester, the temperature-pressure relationship type is shown. This procedure involves the measurement of the temperature and pressure of the gases as they pass through the relief line in a condensing chamber. The relief controller, P/T_{Rel}IC-5, is an indicating-type controller with two pointers and two scales. One scale is calibrated in degrees temperature and the other in psig. The temperature element is calibrated so that it indicates in terms of corresponding steam pressure. The controller action is to regulate relief valve FV-5 to maintain the pressure and temperature at the same point during the cook. If all steam exists in the relief line, the temperature will equal the corresponding pressure. If noncondensable gases or air are mixed with the steam in the relief line, the pressure will tend to increase above the corresponding temperature; and controller action will increase the relief valve opening to bleed off the excess pressure, thereby removing the noncondensables accumulating in the digester and returning the temperature and pressure to the proper relationship. In actual practice, the digester is not relieved to zero excess pressure because steam is also continuously relieved with the undesirable gases and to maintain zero excess pressure would cause a waste of steam. Therefore, some fixed excess pressure is maintained in the digester. The amount of excess pressure is determined by the species of wood being cooked and from experience as to the best setting for optimum relief of noncondensables and turpentine recovery with minimum steam loss. The relief control systems based on measurement of relief flow as described for direct-steaming batch digesters can also be used on indirect-steaming digesters.

The screen-cleaning blowback system shown in Figure 2-7 is based on the pressure difference measurement between the digester, made by PT-2, and the relief line, measured by PT-5. These signals are transmitted to indicator P_dI-4 and blowback controller BBC-4, which opens blowback steam valve FV-4 and closes relief valve FV-5 when pressure difference indicates a plugged up screen in the neck of the digester. Steam is blown back through the screen for a fixed period as set on a timer in the blowback controller, after which the steam valve is closed and the relief valve returned to relief control. If the blowback time is not sufficient to clean the screen and bring the pressure differential back to normal, another cycle is initiated.

Other blowback systems have also been used on indirect-steaming batch digesters.

Sulfate Continuous Digester

In recent years, much progress has been made in the continuous cooking of wood chips. There has been a large increase in the number of continuous systems used, with many mills converting from the batch method. The various designs that have evolved during the development of continuous digesters can be generally classified in one of three categories:

1. Vertical. Types with either downflow, upflow and countercurrent flow of wood chips and cooking liquor.
2. Horizontal. Types with forced movement of chips and cooking liquor.
3. Inclined. Types with forced movement of chips and cooking liquor.

Pulping systems have been built consisting of one single unit or a multiple of units in series, depending on the type of wood being cooked and the final type of product desired.

Vertical Type. Although there have been a number of vertical-type digesters used in the paper industry, one of the more widely used systems utilizes the chip and cooking liquor downflow method of operation. Figure 2-8 is a schematic diagram depicting such a system with the instrumentation normally used.

In operation, wood chips enter from a surge bin discharging via a chip hopper into a chip meter, whose speed determines the production rate of the system. From the meter, the chips pass through a low-pressure feeder into a steaming vessel. After presteaming, the chips enter a high-pressure feeder whose rotor inlet port is totally submerged. The level is controlled in the preceding chip chute. The chips are then flushed into a feeder.

As the feeder rotor turns, the chips enter the digester's high-pressure system and are conveyed to the top of the digester by a circulation system. The liquor required for circulation is extracted through a strainer which is cleaned by a screw equipped with a special torsion-type level indicator that measures the chip level in the top of the digester. The volume of cooking liquor is controlled by use of a magnetic flowmeter control system. As chips and cooking liquor are moved down through the digester, they pass through five consecutive process zones: impregnation, heating, cooking, washing, and cooling. Cold blowback liquor is put into the bottom of the digester to dilute the charge for easy removal and to lower temperature before discharge. The cooked pulp is discharged to the blow tank through an outlet device and blow-unit and is controlled by a magnetic flowmeter control system.

The most important control loops on this type of continuous digester system are those recording and controlling the flow of white liquor, FT-3, FRC-3, FV-3; upper and lower cooking temperature, TT-8H, TRC-8H, TV-8H and TT-9H, TRC-9H, TV-9H; digester pressure, PT-11, PRC-11, PV-11 and PIC-10, PV-10; and pulp flow to the blow tank, FT-12, FRC-12, FV-12. These instruments, along with the chip speed indicator, SI-1, determine the input feed rate, retention time, and cooking conditions in the digester, as well as operation of the washing portion of the digester. Steam for the vessel is supplied from the high-pressure flash tank, and a pressure recording control loop, PT-2, PRC-2, PV-2, controls any necessary makeup. The venting of noncondensables from the steaming vessel is accomplished by an indicating control loop, PT-2, PI-2, PV-2.

The level in the chip chute is controlled by overflowing the excess liquor to a tank whose level is controlled through LT-7 and LRC-7 by throttling LV-7 in the digester liquor makeup pipeline. Since the amount of white liquor is controlled with recording control flow loop FT-3, FRC-3, FV-3, and black liquor is controlled with FT-4, FRC-4, FV-4, and both flow through valve LV-7, the level tank acts as a surge vessel whose level actually is the determining factor of the quantity of fresh cooking liquor going to the top of the digester. This flow is indicated by loop FT-7, FI-7, and the recirculation rate is indicated by loop FT-6, FI-6. Temperature of the circulation flow is recorded by loop TT-6, TR-6.

The upper and lower cooking zone temperatures are controlled by control loops at the outlet liquor lines from the upper and lower heaters. Inlet heater liquor temperatures are recorded by loops TT-8C, TR-8C and TT-9H, TRC-9H,

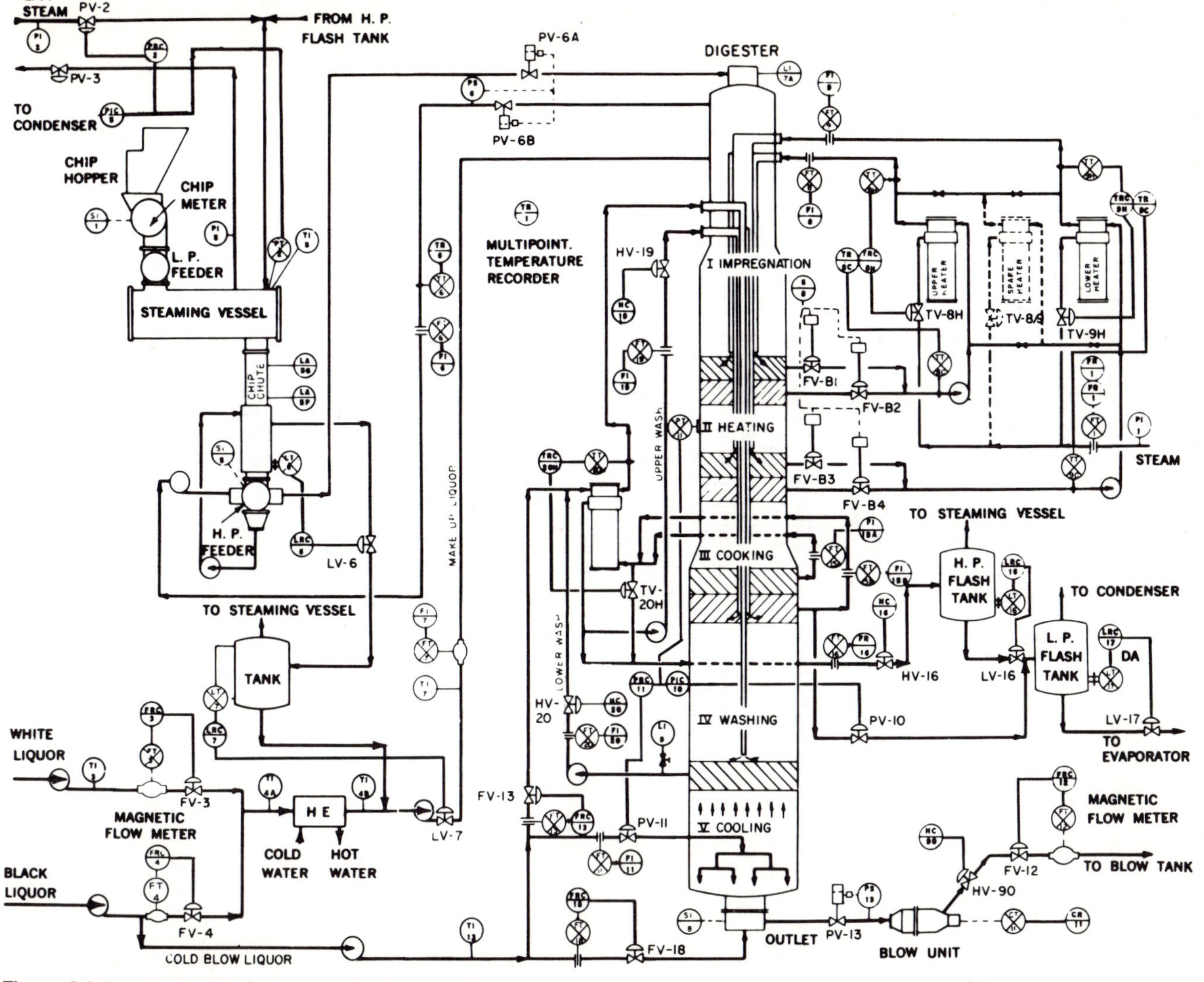

Figure 2-8. Instrumentation for a vertical continuous digester.

and their flows are indicated by FT-8, FI-8 and FT-9, TI-9. There is a spare heater whose valve, TV-8/9, may be operated from either of the two existing temperature controllers. Valves FV-B1, FV-B2, FV-B3, and FV-B4 are butterfly valves on the upper and lower cooking zone screen outlets which are alternately opened and closed by sequence timer S-8, thereby maintaining screens in a clean condition and providing free flow through them.

The liquor flow from the upper washing zone to the heat exchanger is indicated by FT-15A, FI-15A and FT-15B, FI-15B. Washing zone recirculated liquor is indicated by loop FT-19, FI-19 and controlled manually by HC-19, HV-19. The extracted flow to the high-pressure flash tank is recorded by FT-16, FR-16. The temperature of the recirculated upper washing liquor is controlled by TT-20H, TRC-20H, TV-20H by locating the valve at the heater bypass. Recirculated liquor flow in the lower washer zone is measured by FT-20, FI-20 and set manually by HC-20, HV-20 and mixed with the cold blow liquor whose flow is controlled by loop FT-18, FRC-13, FV-13. Digester pressure control loop PT-11, PRC-11, PV-11 regulates dilution liquor to the bottom of the digester, which is indicated by FT-11, FI-11. Flow control loop FT-18, FRC-18, FV-18 controls the counterwash liquor flow to the outlet device. A second digester indicating pressure controller, PIC-10, operating from the same pressure transmitter output, is set to act as a safety valve with PV-10 venting excess liquor to a low-pressure flash tank. Levels on both the high- and low-pressure flash tanks are controlled and recorded by loops LT-17, LRC-17, LV-17 and LT-16, LRC-16, LV-16, with the control valves located in the tank outlets.

A tachometer generator, SI-8, is used to measure and indicate the speed of the outlet device on the discharge system. Consistency of the pulp is measured, recorded, and controlled by a current control system, CT-11, CRC-11, on the blow-unit. Consistency variations are corrected by adjusting the counterwash flow control system. The blowdown valve, HV-90, is manually operated by HC-90 with the pulp flow to the blow tank being measured, recorded, and controlled by FT-12, FRC-12, FV-12.

On temporary shutdowns, cylinder-operated valves PV-6A, PV-6B, and PV-13 are used to "shut-in" the digester. Pressure switches PS-6 and PS-13 prevent opening of these valves until pressure in the top circulating line and blow-unit is high enough to provide the necessary system pressure balance. Level indicator LI-9 is used primarily during initial filling of the digester.

Several local temperature, pressure, and level gauges are used, some of which are shown in the flow diagram. TR-1 is a multipoint temperature recorder used to record selected temperatures throughout the system. Orifice plates with differential pressure transmitters are used quite extensively for liquor flow measurement, and the flanged, flush-diaphragm type level transmitters are recommended for liquor and flash tanks. Magnetic flowmeters are normally used on white and black liquor and pulp flow measurements.

Horizontal Type. Although the horizontal-type continuous digesters have been used primarily for semichemical pulping operations, they can also be used to cook wood chips by the full chemical sulfate process. A number of designs have been developed utilizing this concept. Figure 2-9 illustrates one of them. With this configuration, the chips are conveyed from the chip bin to the inlet valve by a screw-type feeder. A magnetic separator precedes the inlet valve to remove any metallic debris carried along with the chips.

The inlet valve is a *revolving door* type and also provides a pressure seal to the steaming or evacuation chamber. The chips are carried through the evacuation chamber, impregnation chamber, and cooking chamber by a conveyor made up of a series of inclined blades on a rotating shaft. The chips are then transferred

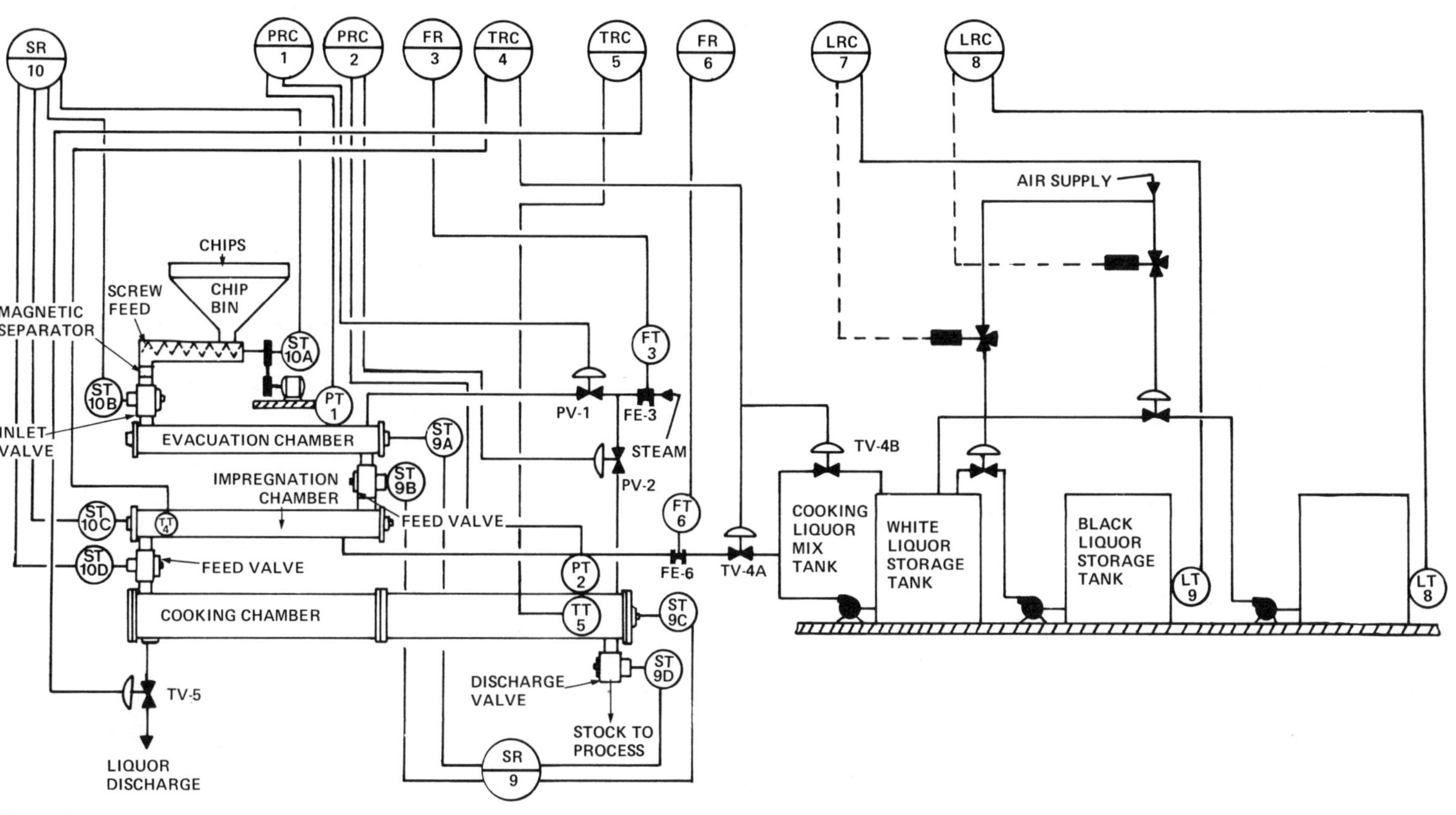

Figure 2-9. Instrumentation for a horizontal continuous digester.

from one chamber to the other by rotary-type feed valves. The pulp is discharged through another revolving-door type valve, the same as the inlet feeder.

Pressure recording controller PRC-1 controls the pressure in the evacuation chamber by automatically adjusting control valve PV-1 on the steam supply. The pressure in the cooking chamber is controlled in the same manner by PRC-2. The total steam being consumed is recorded on FR-3. The liquor level of the impregnation chamber is controlled by sensing the temperature of the liquor and operating control valves TV-4A and TV-4B on the liquor feed lines through temperature recording controller TRC-4. As the temperature sensing element becomes immersed in the liquor, valve TV-4A will close as TV-4B opens. When liquor level drops below the temperature probe, the opposite action occurs, thus providing recirculation to aid mixing in the cooking liquor mix tank.

In the same manner, TRC-5 controls level in the cooking chamber by regulating valve TV-5 in the liquor discharge line. FR-6 records the flow of cooking liquor to the digester. The amount of white and black liquor added to the cooking liquor mix tank is proportioned by drawdown level recording controllers LRC-7 and LRC-8. Speed records of screw feeds, feed valves, cooking chamber rotating conveyors, and discharge valves are made on multipoint recorders SR-9 and SR-10 from signals transmitted from speed transmitters mounted on their driving mechanisms.

Inclined Type. The inclined-type digesters also were first used for the manufacture of semichemical pulps and were primarily suitable for cooking sawdust and other sawmill residues. They were later adopted and used for the production of full chemical pulps from a wide range and combination of wood chips. The inclined digester is another type which has become readily accepted in the paper industry for this purpose.

The digester is essentially a cylindrical pressure vessel usually mounted at a 45 degree angle. Figure 2-10 is a schematic drawing showing a typical basic unit used to cook wood chips. As shown, wood chips from the preparation and storage areas are fed in metered amounts through a feeder presteamer into a specially designed inlet rotary valve.

The inlet rotary valve is provided with a special exhaust vent to relieve vapor pressure in the rotor pocket after it has discharged its chips and before it reaches the low-pressure inlet side of the valve. When a closed vent chip charging system is used (shown in the drawing as the feeder presteamer and tower), the exhaust from this vent, together with blow steam from the blow tank, is used to preheat the incoming chips in a feeder presteamer. The discharge rotary valve does not have such a vent because the pressure is needed to blow the pulp into the blow tank and is relieved accordingly.

The wood chips from the inlet rotary valve drop into a series of traveling compartments in the digester. These compartments, formed by conveyor flights, are in series and are chain driven. A variable-speed reducer controls the conveyor movement. The chip-laden flights travel downward on the top side of an internal dividing plate, which separates the tube in halves, and returns on the opposite or bottom side. Heated cooking liquor is added just below the inlet rotary valve. The lower section of the entire inclined tube contains liquor which can be maintained at or varied from any predetermined level. As the flights move downward through the liquor, the chips are steamed. Impregnation takes place as the chips are dragged by the conveyor through the section of the inclined tube containing liquor. Vapor phase cooking takes place while the chips travel from the liquor to the specially designed discharge rotary valve.

The time for each phase of the process is controlled by: (1) the level of the liquor, (2) location of the inlet and discharge nozzles, and (3) the digester

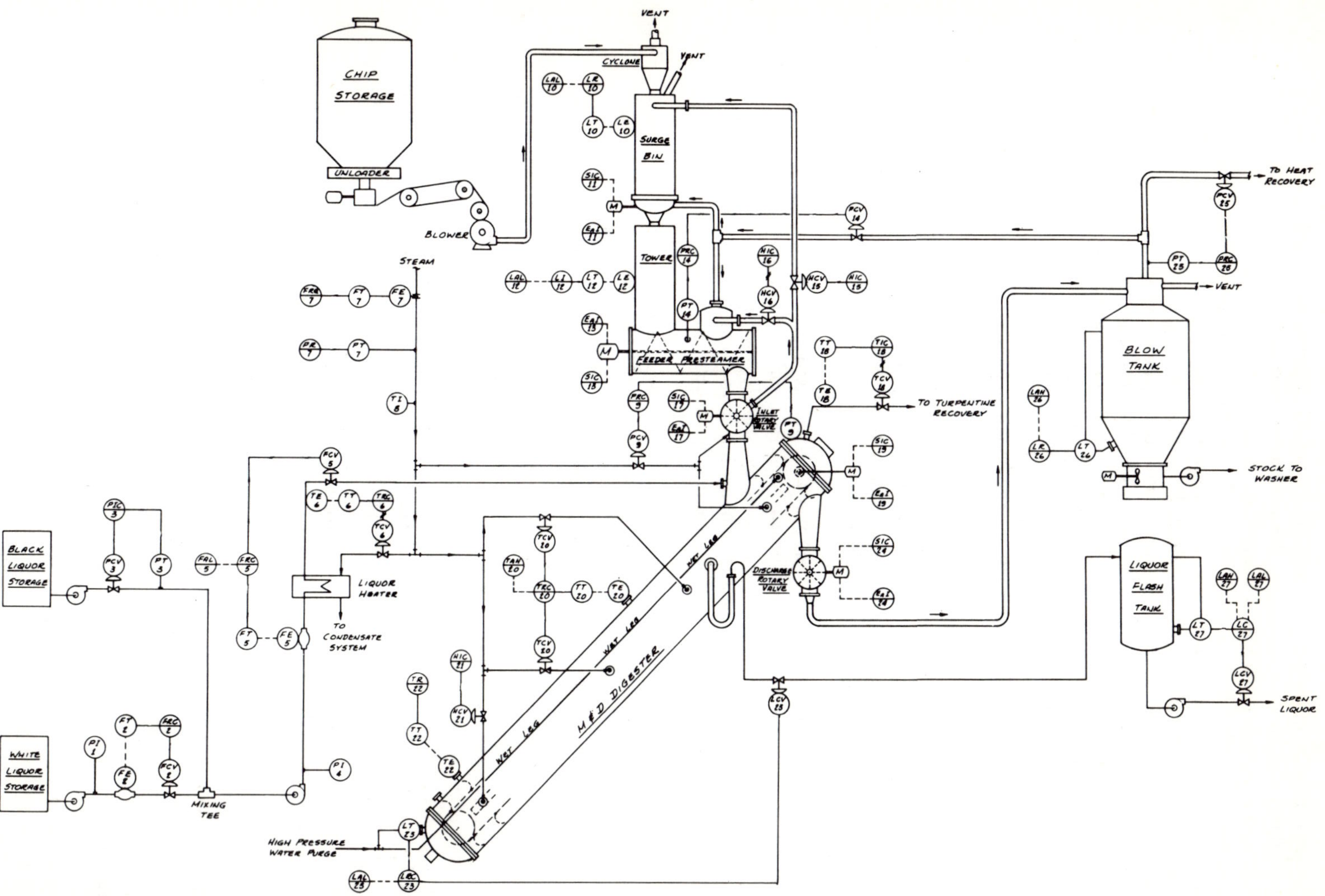

Figure 2-10. Instrumentation for an inclined continuous digester.

conveyor drive system's speed control. All chips proceed through the digester at a uniform rate to achieve uniform cooking.

Several cooking zones can be established in the digester as desired; they include steaming, impregnating, and cooking (either submerged or vapor phase). The type of cook, yield, conditions, and tonnage governs the specifications for a particular unit. Also, the single unit can be coupled with other complete tubes, each having a specific kind of zone varying the temperature, pressure, or liquid as desired.

The cooking features include such things as the use of a variety of cooking methods utilizing any type of liquor, with temperature maintained through direct or indirect heating. Liquor is constantly agitated by the movement of chips through the unit. If desired, liquor can be circulated and heated at various points to achieve the desired temperatures at various stages in the liquid cooking phase.

Installation features include the location of both inlet and discharge on the same floor level. Building requirements and installation costs are lessened by locating 90 percent of the equipment outside of the building and varying the angle of the tube to meet structural restrictions.

Figure 2-10 shows the white and black liquor being mixed in a mixing tee to make up the cooking liquor. The white liquor is passed to the tee under flow control. The flow measurement, FE-2, is usually made with a magnetic-type flowmeter. This control loop is set to maintain the proper chemical/wood ratio based on wood chip conditions. The black liquor flow to the tee is under pressure control with pressure transmitter PT-3 having a suitable seal to prevent plugging of the measurement connection. This loop automatically adds black liquor to maintain the proper liquor/wood ratio. If a mixing tank were used instead of a tee to prepare the cooking liquor, the black liquor pressure control loop would be replaced with a level control loop on the tank.

Total cooking liquor to the digester is added under flow control, also using a magnetic-type flowmeter, FE-5. Cooking liquor is heated in a surface-type heat exchanger with the flow of steam regulated by a temperature control loop using a resistance-type temperature detecting element, TE-6. Cooking liquor is sometimes heated by direct injection of steam. The steam supply flow is measured by an orifice plate and a differential pressure transmitter and this measurement is recorded and totalized. It is also necessary to measure and record the pressure of the steam supply.

The pressure in the digester is maintained at the desired value with a pressure control loop regulating the steam supply. A sealed connection at the point of measurement on the dome of the digester is also required on this transmitter, PT-9. Temperatures along the digester zones are controlled by temperature control loops regulating steam flow to optional steam inlets along the tube. Resistance-type temperature detecting elements, TE-20 and TE-22, are used here. Liquor level in the digester is adjusted and controlled, regulating the flow of excess liquor to the liquor flash with a level control loop utilizing a diaphragm-sealed, flange-mounted differential-type level measurement, LT-23. This transmitter is provided with a wet leg seal on the low-pressure side to obtain proper liquor level measurement. Chip level measurements LE-10 and LE-12 on the surge bin and feeder presteamer tower are usually made with radioactive detecting devices. Pressure in the feeder presteamer is controlled by a pressure controller regulating the flow of exhaust steam from the blow tank. The speed control on the feeder presteamer determines the rate at which chips are fed to the inlet rotary valve and digester unit. Speed controllers are also used to adjust and regulate operation of the rotary valves and internal conveyor in relation to the speed of the feeder presteamer.

Level measurements on the blow tank, LT-26, and liquor flash tank, LT-27, are diaphragm-sealed, flange-mounted differential types with wet leg seals on the low pressure, also.

Sulfite Pulping

Sulfite pulping presents the same general problems that were described for sulfate pulping. The primary difference is that the cooking liquor is an acid solution of calcium, magnesium, or ammonia bisulfite and sulfurous acid, depending on the process used. Sulfur dioxide gas is used in the manufacture of sulfite cooking liquor. Sulfur dioxide is produced by burning sulfur in either a rotary or spray burner. The hot sulfur dioxide gas is then cooled in gas coolers and reacted with the specific hydroxide base in an absorption system to produce the acid cooking liquor desired.

Rotary Sulfur Burner. No matter what type of acid cooking liquor is being manufactured, the first step is the formation of sulfur dioxide. This has been accomplished by roasting sulfur ores, but more commonly by the burning of elemental sulfur. Figure 2-11 shows a system used for burning sulfur in a rotary burner and the instrumentation involved.

Temperature indicating controller PIC-1 controls the steam pressure for jacketed molten sulfur lines to maintain desired fluidity of the sulfur. Overheated sulfur becomes very viscous with the risk of "freeze up." The amount of dry sulfur from the hopper required to maintain the desired level in the molten sulfur storage tank is controlled by applying steam to a tubular melter when the level drops below the temperature sensing element of transmitter TT-2. The signal from transmitter TT-2 is used by temperature controller TIC-2 to regulate valve TV-2 in the steam supply line. At the same time, electrical control contacts are actuated, starting the conveyor which refills the hopper and melter. The reverse occurs when the temperature sensor becomes covered. Temperature controller TIC-3 maintains the molten sulfur storage temperature at the proper value. Level in the burner is maintained by level recording controller LRC-4, which throttles the molten sulfur flow through control valve LV-4. If this level fluctuates, combustion will be erratic due to varying surface areas of molten sulfur, with sulfur

Figure 2-11. Rotary sulfur burner.

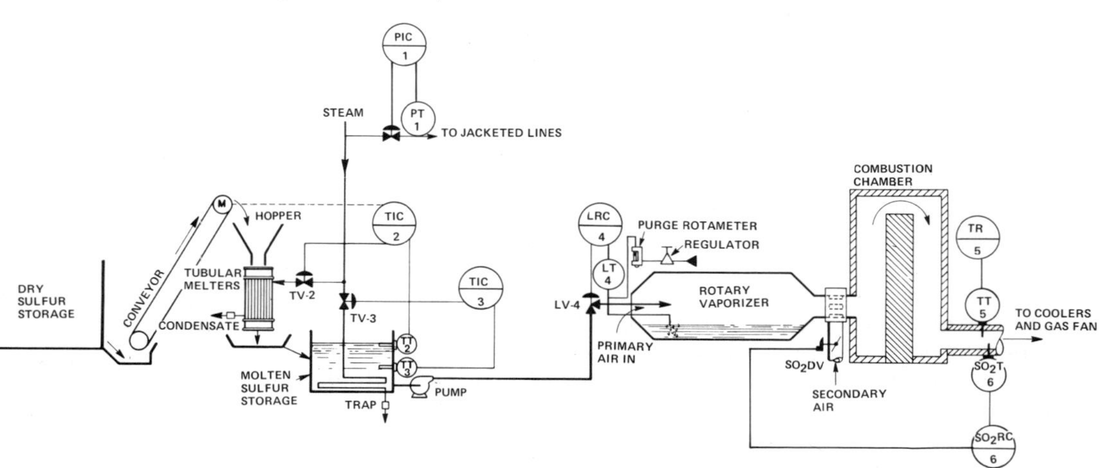

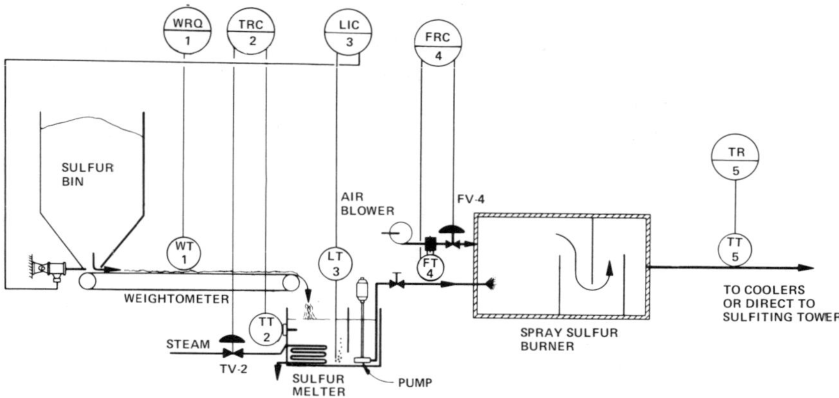

Figure 2-12. Spray sulfur burner.

loss at times through sublimation and sulfur trioxide formation. Temperature recorder TR-5 is used to monitor temperature at the exit of the combustion chamber, which should be held within definite limits for efficient combustion.

Most efficient operation of the burner is obtained when the primary and secondary air are correctly adjusted. Improper burning produces sulfur trioxide which forms a sludge in absorption towers, or it may sublime the sulfur and clog the coolers. In addition to controlling the level of molten sulfur in the burner, some mills control the secondary air damper with a percent sulfur dioxide controller operating on gas samples as shown by sulfur dioxide controller SO_2RC-6.

Spray Sulfur Burner. Figure 2-12 illustrates a sulfur burner system using a spray-type burner and the typical instrumentation employed, this being another method of handling the dry sulfur. The weight of dry sulfur is recorded and totalized by WRQ-1. The temperature of the sulfur melter is maintained by temperature controller TRC-2, throttling the valve, TV-2, on the steam supply line to the heating coil. Level in the melter is controlled by adjustment of the sulfur bin discharge gate valve from level controller LIC-3. The flow of combustion air is measured and controlled at an optimum value by flow controller FRC-4. Exit gas to the coolers is recorded on TR-5 for monitoring purposes.

Gas Cooler. The hot sulfur dioxide gases coming from the burner must be cooled quickly under carefully controlled conditions to prevent chemical losses or formation of undesirable sulfur trioxide. This is done in a variety of cooling systems (see Figure 2-13 for a typical example).

Excess sulfuric acid in the effluent from the primary cooler is recorded on pHR-1. It guides the acid maker in balancing the system variables to prevent chemical losses. The flow of spray water through the primary cooler must be sufficient to cool the gas, but not excessive enough to form waste acid which is lost through the effluent. A weir meter-type recording flow controller, FRC-3, controls the rate of spray water flow.

If the normal water supply to the cooler should fail, cooling would not take place and excessive loss of sulfur dioxide would occur in the absorption towers. Serious equipment damage might be caused by the high gas temperature; a secondary source of emergency water prevents this. Pressure switch PS-6, connected to the primary water line, instantly detects the loss of water supply and an alarm warns the operator.

For balanced operation and most effective cooling in the secondary tower, an indicating liquid level controller, LIC-4, is used. This instrument adds just enough makeup water to counteract losses by maintaining the necessary level in the bottom of the secondary cooler. Temperatures should be known at various critical points to maintain proper balance throughout the system so that the operator can get the desired cooled gas temperature. A multirecord temperature recorder, TR-5, provides continuous temperature records from temperature transmitters TT-5A-E. To obtain the highest degree of gas cooling, the circulating water must be cooled in a heat exchanger. To stabilize the heat extracted in this exchanger, the flow of cooling water is regulated in accordance with its temperature by recording temperature controller TRC-6.

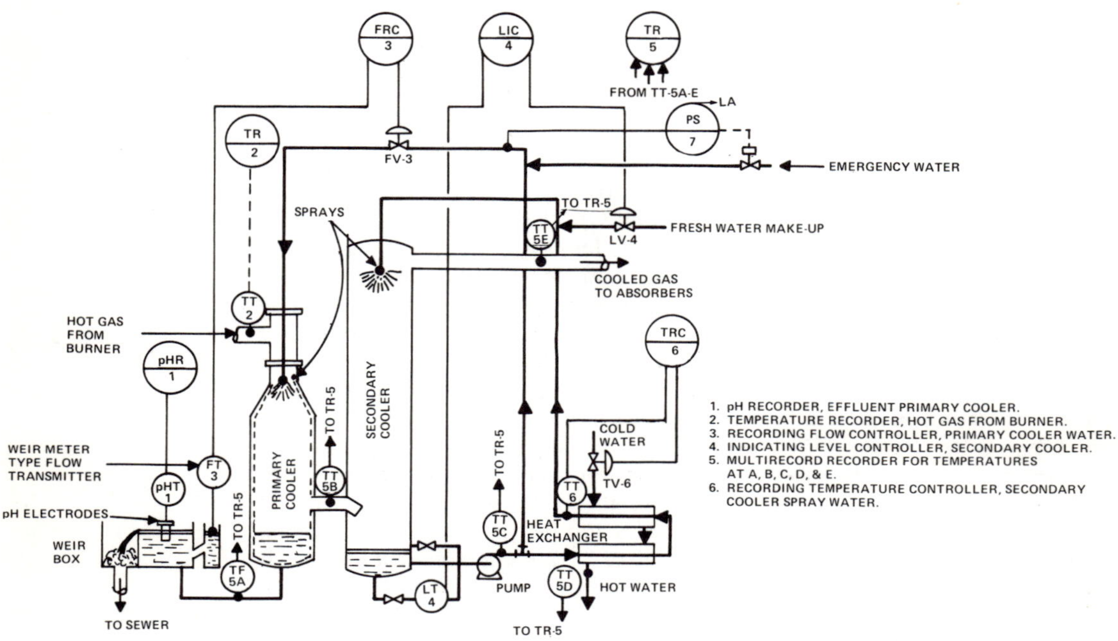

Figure 2-13. Sulfur dioxide gas cooler.

Gas Absorption. If the acid cooking liquor being manufactured is calcium based, the next step is to allow the cooled sulfur dioxide to react with limestone and water in absorption towers, as shown by a typical system in Figure 2-14.

Sulfur dioxide gas flows through the absorption towers packed with limestone, with flow controlled by the gas fan. Water flows countercurrent to gas flow and it is imperative that the water flow be controlled proportionally. Recording flow recorder FRC-5 provides positive control of this variable regardless of variations in supply pressure.

The temperature of water to the weak tower must be controlled to effect proper combustion with the sulfur dioxide gas. It may be necessary to heat the water if it is too cold, or cool it if it is too warm. TR-4, a temperature recorder, is used as a guide to this temperature. A temperature controller and heating-cooling exchanger are used to maintain temperature of this water automatically. A pressure tower may be used to build up the "free" sulfur dioxide strength during periods of warm weather. Part of the sulfur dioxide is diverted through a compressor to the tower where it flows countercurrent to the flow of raw acid.

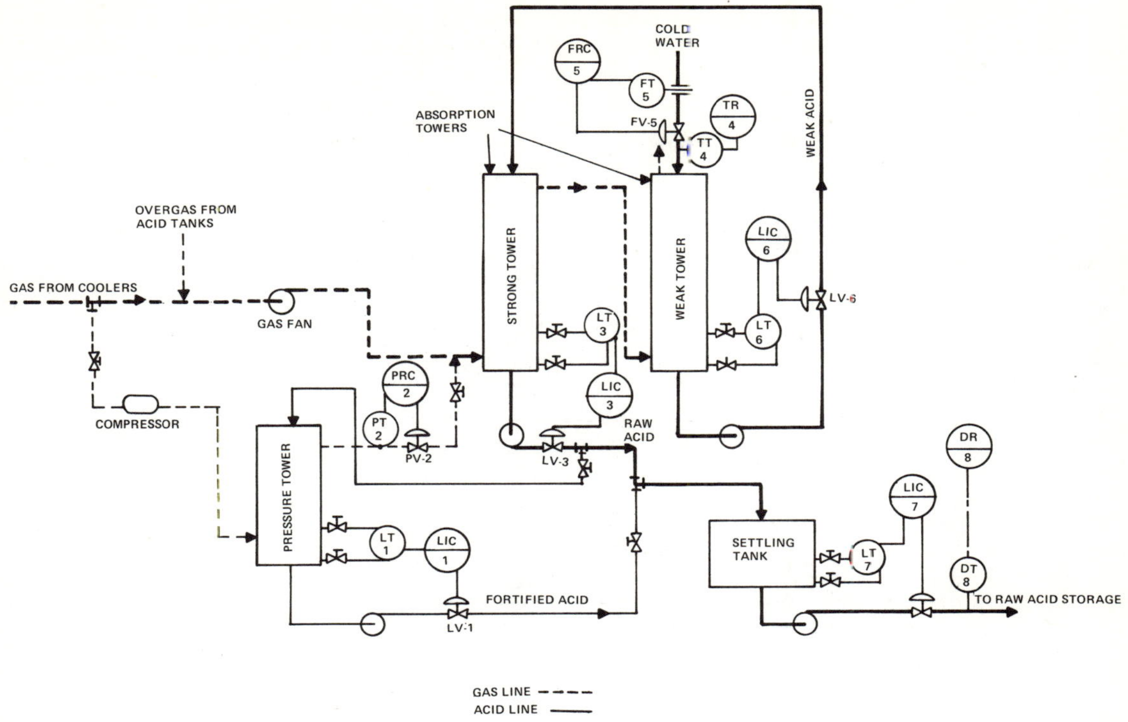

Figure 2-14. Sulfur dioxide gas absorption system.

Since the amount of gas absorbed is proportional to the pressure in the tower, a recording pressure recorder, PRC-2, is used to control the tower pressure by bleeding off excess pressure.

The liquor supply must be maintained in the storage and weak absorption towers, the pressure tower, and the settling tank. Indicating liquid level controllers LIC-1, LIC-3, LIC-6, and LIC-7 are used for this purpose.

DR-8 is a density recorder. It provides a continuous record of the specific gravity of the acid leaving the system as a check on the uniformity of its operation. The density measurement is made in the acid line to storage and transmitted to the recorder by transmitter DT-8.

Hot Acid Accumulator. The cold acid coming from the absorption system is usually fortified before using it as a cooking liquor in the digester. Figure 2-15 shows a typical system used on calcium-based cooking liquor. Sulfur dioxide gas is recovered and the acid is heated by the digester relief gases to produce a uniform hot cooking liquor.

The hot acid system consists of a series of pressure vessels or accumulators to collect gas and entrained liquor relieved from the digester. The sulfur dioxide in this gas is absorbed by the acid in the accumulators and its heat elevates the temperature of the acid. The system is divided into zones so that gases can be reclaimed at various pressures encountered in the cooking cycle.

High-pressure digester relief gas and liquor are directed through a high-pressure eductor where they mix with warm acid being pumped to high-pressure accumulators which are vented to the low-pressure accumulator under control of pressure controller PRC-11. When the digester pressure falls

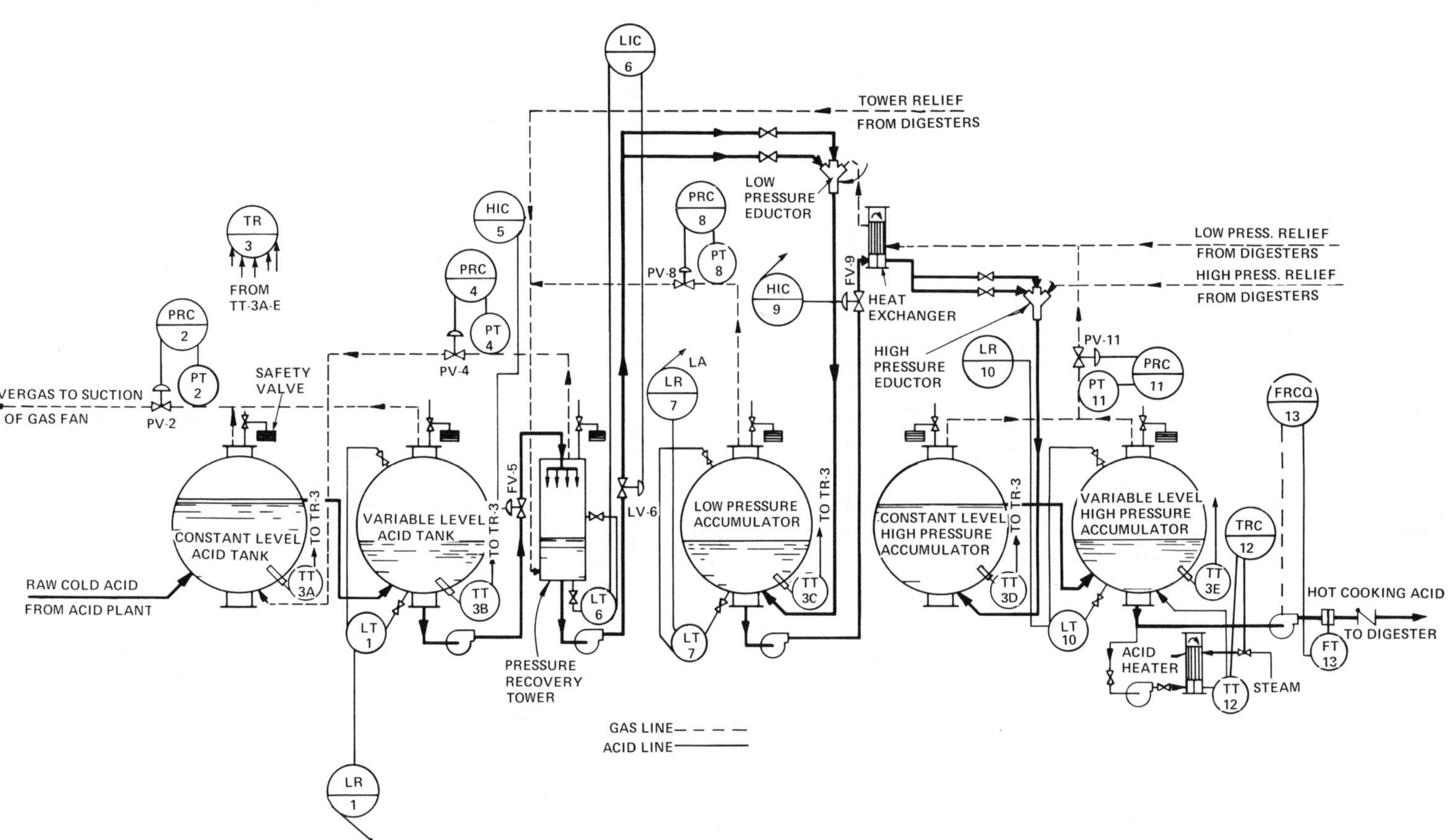

Figure 2-15. Hot sulfur dioxide accumulator system.

off, relief gas is diverted to the low-pressure eductor, mixing with acid being pumped to the low-pressure accumulator. The low-pressure accumulator is controlled by recording pressure controller PRC-8 by venting excess pressure to the pressure recovery tower.

In the same manner, excess pressure in the pressure tower is vented to the cold acid and from there to the absorption system. Pressures are controlled by recording pressure controllers PRC-2 and PRC-4.

The several supply tanks involved use level recorders LR-1 and LR-7 with alarms or are automatically controlled by controller LIC-6. Acid transfer is controlled manually by the operator with remote manual stations HIC-5 and HIC-9. Multirecord temperature recorder TR-3 provides continuous records of the acid in each of the tanks. Recording temperature controller TRC-12 controls the final temperature of the acid by regulating steam to the acid heater. FRCQ-13 is a flow recording controller with totalizing integrator on the hot acid flow to the digesters; it operates the transfer pump and records the rate of acid flow, digester charging time, and the number of digester changes.

Sulfite Batch Digester. The sulfite batch digester's control system shown in Figure 2-16 is typical for direct steaming with natural or forced circulation, and for indirect-steaming operation. Control can be from measurement of either digester temperature or steam flow. By replacing the temperature cam with a steam flow cam in time schedule cam controller KT/FC-1, turning switch SW-1 and operating two hand valves, control can be changed from temperature to steam flow control.

A temperature controller can respond to variations in temperature only at the point of measurement. Unless that point is representative of the changes in steam input, satisfactory control based on temperature is impossible. Forced circulation with indirect-heating digesters is best controlled from a temperature measurement at the transmitting point, TT-1, because the circulation rate assures a fast temperature response to changes in steam to the heater. A temperature schedule cam is used in KT/FC-1, allowing steam to flow through valve T/FV-1 to the heater. Valves A, B, D, F, and G are open and valves C, E, and H are closed.

Forced circulation with direct steaming may be used regularly or when the indirect heater is temporarily out of service. Because the liquor is circulated at a relatively high rate, a temperature measurement, TT-1, is representative of the average temperature in the digester. Control is from a temperature schedule cam in KT/FC-1 and steam enters at the bottom of the digester. Valves A, B, E, and H are open and valves C, D, F, and G are closed.

Natural circulation, direct-steaming digesters are best controlled from a predetermined schedule of the rate of the steam flow. Circulation in the digester is established by convection and steam velocity, and the temperatures at various locations in the digester are neither rapidly nor consistently responsive to changes in steam flow. However, if the heat input is accurately controlled from a predetermined rate of steam flow, the average digester temperature will follow the prescribed curve with surprising accuracy. In operation, the switch, SW-1, is turned to *steam flow* and KT/FC-1 controls the steaming from a steam flow schedule cam rather than from a temperature cam. If a digester's forced circulation pump is temporarily out of service, the operation can be changed to steam flow control. Should the mill be using natural circulation and direct steaming but expect to change to forced circulation at a later date, it is possible to start off with steam flow control and change over to temperature control with the same basic instrumentation.

Temperature is the most important variable in the cooking process inasmuch as it directly affects the acid penetration, cooking time, yield, and pulp quality.

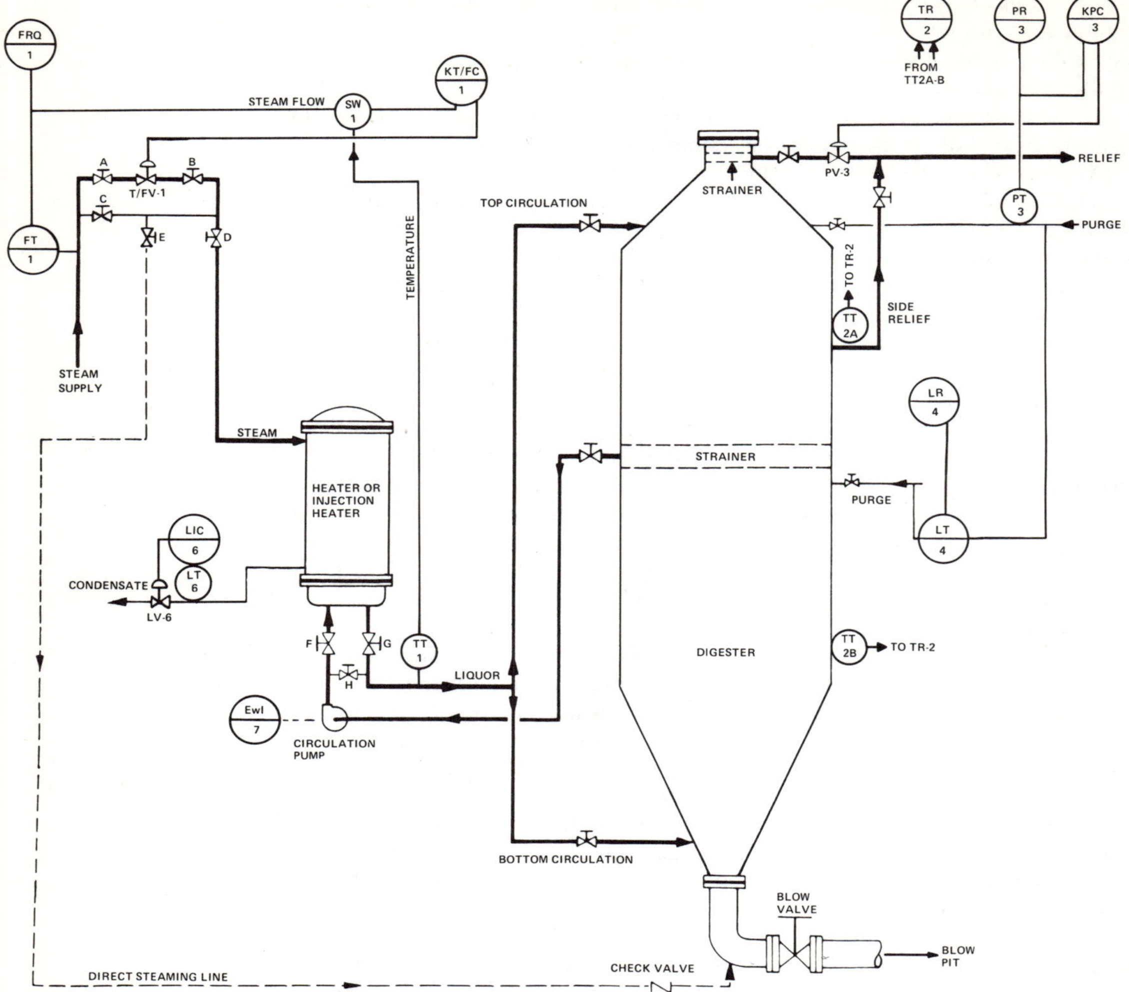

Figure 2-16. Sulfite batch digester.

Excessive temperature reduces potential yield by breaking down the cellulose chemically and physically and partly decomposing it into soluble products that are subsequently washed out of the pulp; yet the temperature must be high enough to assure a satisfactory product.

The precooking segment of the schedule—the heating of the digester to cooking temperature—must also be controlled according to a predetermined schedule to prevent excessive screening losses and steam consumption, and promote natural circulation within the digester. Normally, the controller mechanism in KT/FC-1 is actuated by a signal transmitted from TT-1. The system reproduces the temperature curve for which the cam is cut.

The pressure of the gas liberated from the acid during the cook is directly related to the strength of the acid and its rate of penetration into the chips. It should be high enough, consistent with insurance and safety regulations, for maximum pulp production and should be controlled to assure uniform penetration and product with minimum sulfur consumption. Top relief pressure control is essential throughout the cooking cycle, especially during the final gas-off period at the end of the cook.

The relief pressure controller, KPC-3, has a pressure schedule cam. In starting a cook, it is necessary only to engage the friction clutch and rotate the cam to the starting point. The cam rotates until the cycle is completed; then it is stopped automatically.

More uniform pulp with higher yields results when the temperatures in various portions of the digester are essentially the same. It is generally accepted that the top and bottom zones, TT-2A and TT-2B, represent the extremes of potential temperature difference. Therefore, a continuous record of these temperatures is an indication of circulation within the digester and serves as a guide in maintaining optimum circulation at all times. The digester temperatures are recorded on temperature recorder TR-2. By means of an automatic electric switching device, the temperatures are recorded in such a manner that the record of control is essentially continuous and the other temperatures are shown periodically.

The records can alert the cook to impending circulation difficulties and corrective action can be taken before off-quality pulp is produced.

Many mills meter the steam to each digester with flow integrating recorders, FRQ-1 (see Figure 2-16). This procedure is generally recognized as excellent production control practice and provides figures for comparing cooks, digesters, and mills. Various cooking factors, such as temperature schedule, bleachability, quality and kind of wood used, and chip packing, can be studied by means of this procedure.

Digester liquor level is important. It must be high enough to cover all the chips when the cooking starts, yet must allow sufficient gas space during part of the cycle to insure maximum dry gas recovery and prevent a "wet cook." Digester level is measured by a steam-purged differential-type level transmitter, LT-4, and level recorder LR-4.

To maintain high efficiency of operation in the tubular heat exchangers, it is essential that condensate be removed as rapidly as it is formed. A differential-type level transmitter, LT-6, measures this level and transmits a signal to controller LIC-6 which operates valve LV-6 in the condensate discharge line to maintain the desired level.

Circulation pump efficiency, the condition of the digester strainers, and the heater tubes are indicated by a load measurement on the pump motor with wattmeter EwI-7.

Other Digesters

There are a number of other types of digesters designed primarily for use in pulping processes, such as neutral sulfite semichemical, although they have been used for sulfate cooks.

Rotary Digester. This type of digester has been in such forms as horizontal cylindrical, vertical cylindrical, and spherical. Their operation is similar with the most common shape being the spherical, shown in Figure 2-17.

Typical instrumentation used is a time-cam pressure controller, KPC-2, which regulates the steam flow to raise the digester pressure on a desired cam schedule. The relief temperature-pressure relation controller, $T/P_{Rel}C$-3, gets its measurement signal from temperature transmitter TT-3, whose thermal element extends into the digester through the trunnion. The pressure measurement is received from the digester pressure transmitter, PT-2. Based on the temperature-pressure relationship, the controller operates a relief valve to remove air and other non-condensable gases to maintain temperature-pressure at the proper value for good

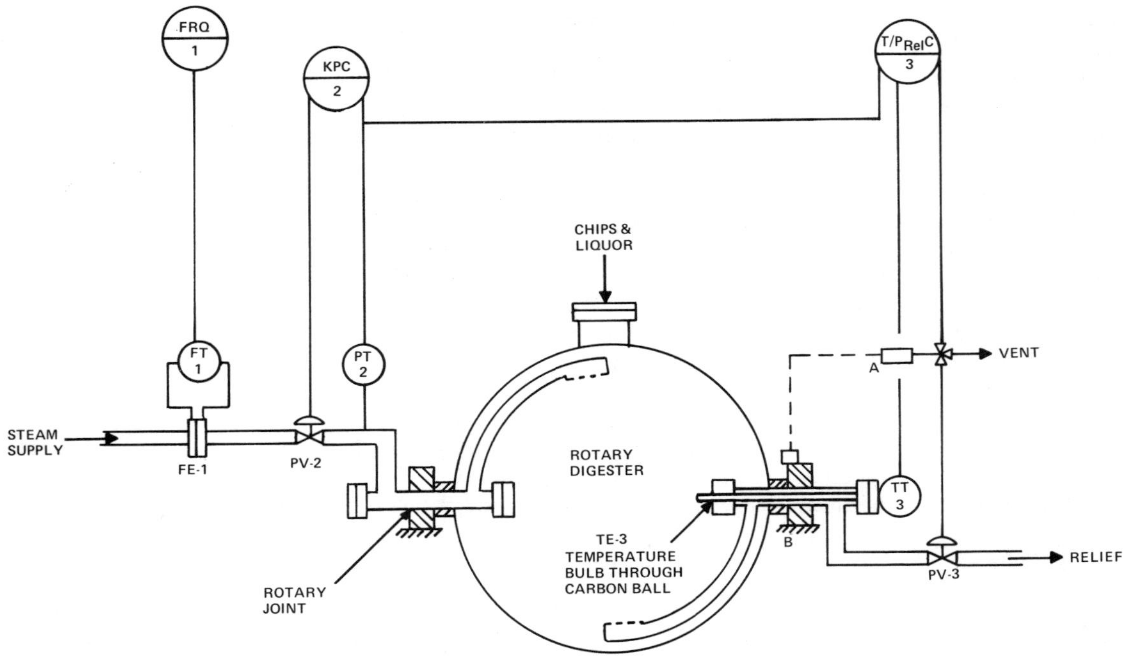

Figure 2-17. Rotary batch digester.

relief. An integrating steam flow recorder, FRQ-1, is usually provided as a check on the quantity of steam usage for each cook.

TURPENTINE RECOVERY

Turpentine is a by-product of the pulp mill which is obtained by the recovery of the volatiles from the sulfate pulping of the softwoods. Crude turpentine production varies depending on the wood species, the season of the species, growing condition, age of the wood cut, pulping conditions, and the turpentine recovery system that is used. The yields can vary from 1 to 10 gallons per ton of pulp produced.

Crude turpentine recovery is relatively simple. The procedure consists of condensing cooking relief gases and decanting off the crude turpentine oil fractions as shown by the simplified diagram of a typical turpentine recovery system (Figure 2-18). Gas relieved from the digester during the sulfate pulping of wood is separated from any liquor carryover in a gas-off separator or cyclone. The vapors are then condensed in a heat exchanger. Temperature controller TIC-1 maintains the temperature between 90° and 100°F by throttling the flow of water through the heat exchanger with valve TV-1 from a temperature measurement made by TT-1 on the turpentine and steam condensate leaving the heat exchanger. The noncondensable gases are vented to atmosphere or are treated to minimize pollution. The vented noncondensables are monitored by flow alarm FA-2. The mixture of turpentine and steam condensate flows by gravity from the condenser to a decanter tank where the turpentine floats to the surface and is skimmed off. Excess steam condensate is usually sewered. Level in the tank is measured and indicated by LT-4 and LI-4. The vent of this tank is also monitored by a flow alarm, FA-3.

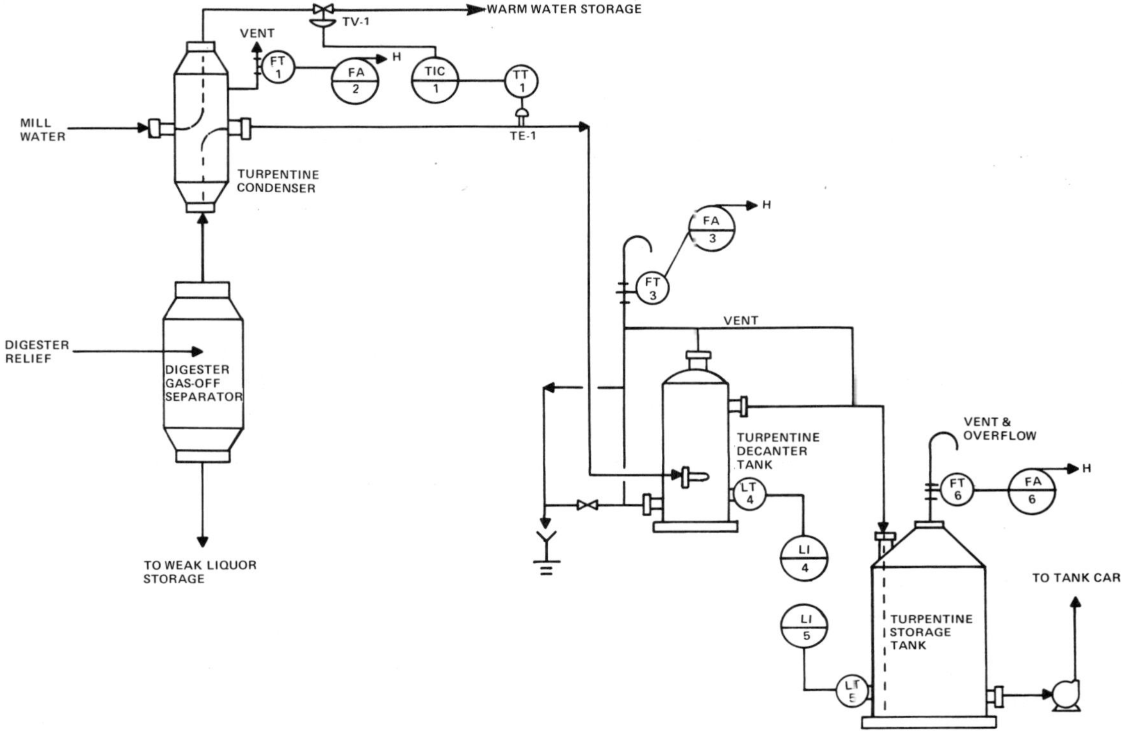

Figure 2-18. Turpentine recovery system.

The decanted turpentine is then stored in a closed tank and pumped intermittently to tank cars for shipment to refineries for production of commercial sulfate turpentine by fractional distillation. About three gallons per ton yield of refined product, which is used in paint thinners and as intermediates in the manufacture of insecticides, is considered economical.

3 Blow Heat Recovery, Screening, and Washing

GENERAL

After the wood chips are chemically converted into pulp in the digester, the resulting slurry must be passed through a screening and washing process. While transporting the pulp and cooking liquor from the digester by *blowing* it into the storage blow tank, it is separated from the steam in a cyclone and the heat released is recovered for use in the mill. This is accomplished by a blow heat recovery system.

BLOW HEAT RECOVERY

The function of the blow heat recovery system is to recover maximum heat from each digester blow. A typical system is shown in Figure 3-1. The cyclone is usually located above the blow tank. The temperature sensing element of transmitter TT-1 is located in the outlet of the blow steam jet-type condenser. When a digester is blown into the cyclone, the separated steam enters the condenser. Transmitter TT-1 senses the heat from the blow steam and sends a signal to controller TRC-1 which, in turn, rapidly opens valve TV-1 in the water line to the condenser. This controller is usually furnished with a batch control function which prevents "wind-up" of the instrument between digester blows, thus eliminating overshooting the control point on startup. This effects a steam saving.

Figure 3-1. A typical blow heat recovery system.

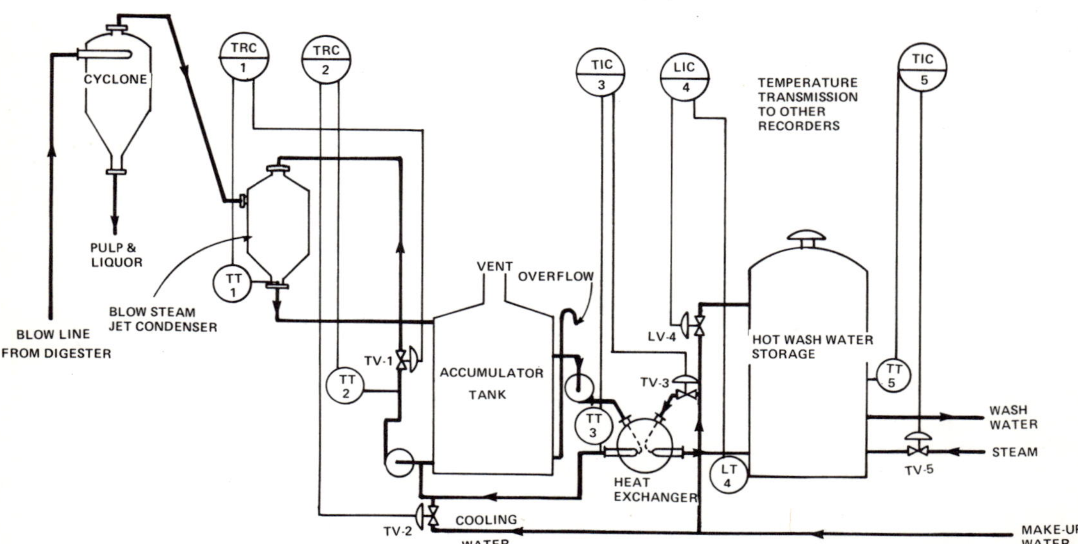

The water pump suppling water from the accumulator tank to the jet condenser runs continuously. If the water from the accumulator tank is too hot, as measured by the temperature sensing element of transmitter TT-2, a signal is transmitted to controller TRC-2. The signal from controller TRC-2 opens valve TV-2, located in the makeup water line to the pump inlet.

Temperature transmitter TT-3 measures the temperature of the accumulator tank outlet from the heat exchanger. It sends a signal to controller TIC-3 which regulates the flow of clean, cold makeup water to attain maximum heat recovery. The liquid level controller, LIC-4, maintains the storage level at a prescribed setting regardless of the wash water demand. Temperature controller TIC-5 provides auxiliary heating for the wash water if the demand exceeds the capacity of the blow heat recovery system.

SCREENING

Screening is essentially the removal of oversized foreign materials such as knots, uncooked wood particles, large slivers, extraneous dirt and other debris from cooked pulp, either before or after washing, by the use of mechanical separation from the acceptable pulp fibers. This is normally accomplished by coarse or fine screens located between the blow tank and pulp washing operation or after the washing operation. There are three general types of coarse screens: basket, inclined plate, and cylinder screens. The fine screens are available as a flat, perforated type, or as a rotary type.

Hot Stock Screening

The more favorable economics of screening at higher consistencies in an environment of relatively high temperature and alkali concentration has led to the development and use of pressure-type rotary screens. The process is commonly referred to today as hot brown stock screening because it is located after the blow tank and prior to washing.

Figure 3-2 depicts the operation of typical hot stock screens and Figure 3-3, their use in a system. Cooked pulp is drawn from a blow tank and passed through the screens which, in this case, operate on a pressure differential between the inlet and outlet sides of the screen. This is accomplished by developing pulsations across the screen by a foil sweeping over the screen face at close tolerances. The unscreened stock is introduced in a cylindrical basket-type screen and in a tangential direction. The screening action, as well as the movement of the rejects, is carried out by the rotating foils. The number, arrangement, clearance, and form of the foils are matters of design, which vary from manufacturer to manufacturer. Reject removal and the addition and control of both the dilution and purge water are usually designed to suit the characteristics of the specific screen unit. Numerous arrangements are possible and exist in practice. Such combinations and the recycling of accepted pulp from the various stages are dependent on individual mill situations and cleanliness requirements.

The major objectives of the instrumentation involved in a hot stock screening operation are to control:

1. The consistency of unscreened stock to the screens.
2. The flow of unscreened stock to the screens.
3. The flow of dilution water to the screens.
4. The level of the rejects storage tank.

Instruments LT-1, LR-1 measure and record the level in the blow tank. The recommended level transmitter used here is of the flush diaphragm, flanged type which is backloaded by a water purge connection to the top of the tank to compensate for pressure variations in the blow tank. An indicating hand control station, HIC-2, is used to regulate the recirculation of unscreened pulp from the fibrillizer back to the blow tank by adjusting control valve HV-2. A pressurized, in-line type of consistency transmitter, CT-3, detects variations in consistency of the unscreened pulp going to the screens and transmits a signal representing this change to consistency recording controller CRC-3. The output from the controller is used to operate secondary dilution valve CV-3b. This output is also recorded as representing the valve position and as the measurement to consistency indicating controller CIC-3. Controller CIC-3 operates primary dilution control valve CV-3a in relation to secondary dilution valve CV-3b. The total flow of unscreened pulp is measured with an electromagnetic flow transmitter, FT-4, and is recorded on flow recorder FR-4. Load conditions on the fibrillizer and screen motors are monitored by current indicators AI-5, AI-6, AI-9. A flush-diaphragm, flanged-type level transmitter, LT-7, measures the level in the primary brown stock washer and transmits the signal to level recording controller LRC-7.

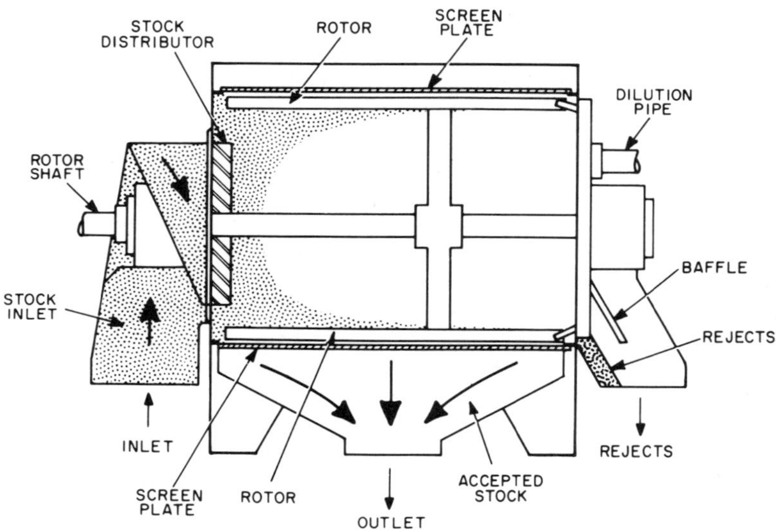

Figure 3-2. A typical centrifugal hot stock screen.

Controller LRC-7 controls the level in the washer vat by throttling level control valve LV-7 in the black liquor dilution line from the brown stock washer vat to the screens. The vacuums in No. 1 and No. 2 screens are measured by flush-diaphragm, flanged-type vacuum transmitters. A high selection relay automatically selects and transmits the higher of the two output signals from the vacuum transmitters to vacuum recording controller PRC-8. The output from this controller adjusts control valve PV-8 in the supply line to the screens in order to maintain optimum vacuum and screening conditions in the screens. Level transmitter LT-10 measures the level in the screen rejects tank and sends a signal to level recording controller LRC-10, which controls the tank level by modulating a control valve in the discharge side of the rejects pump.

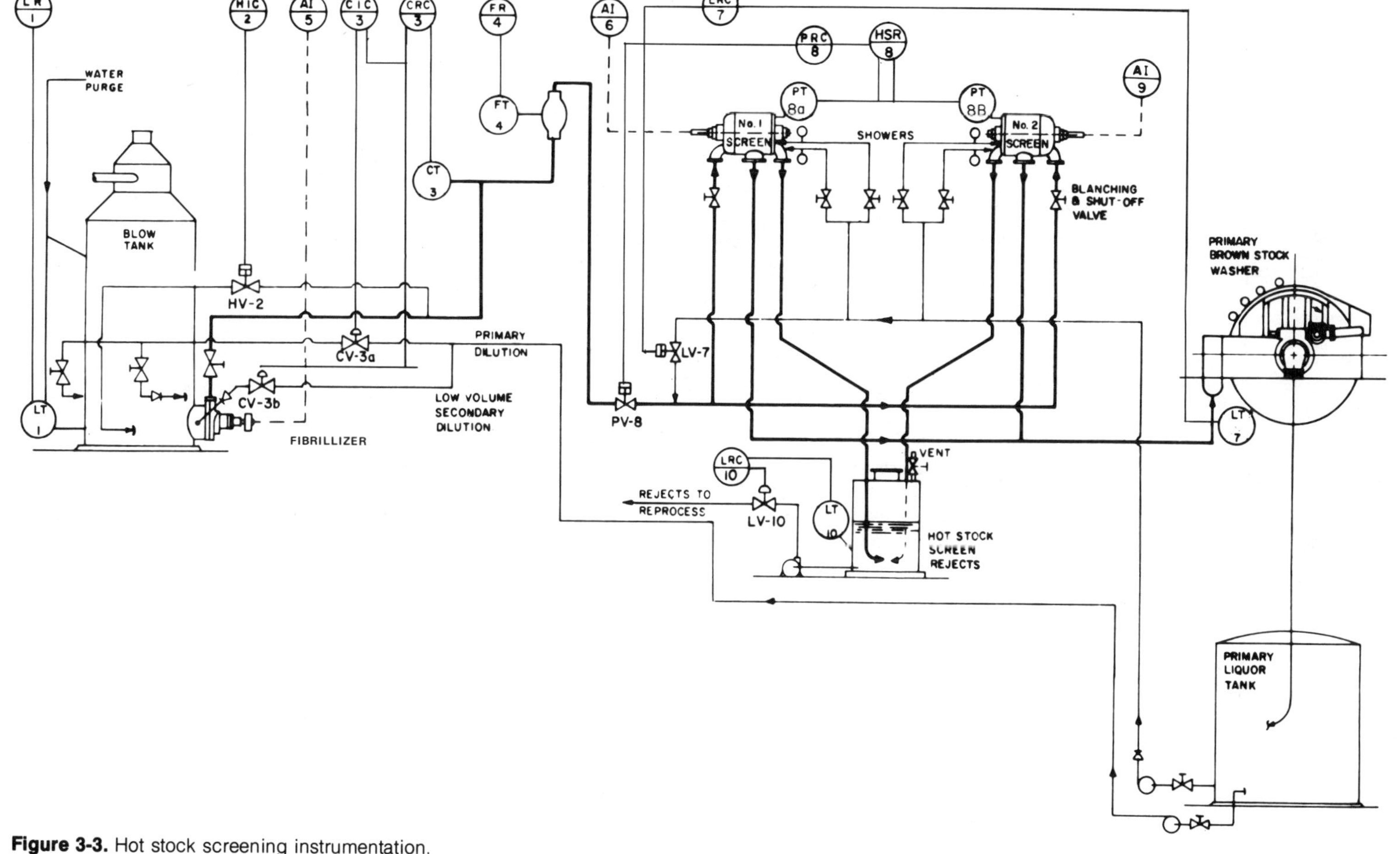

Figure 3-3. Hot stock screening instrumentation.

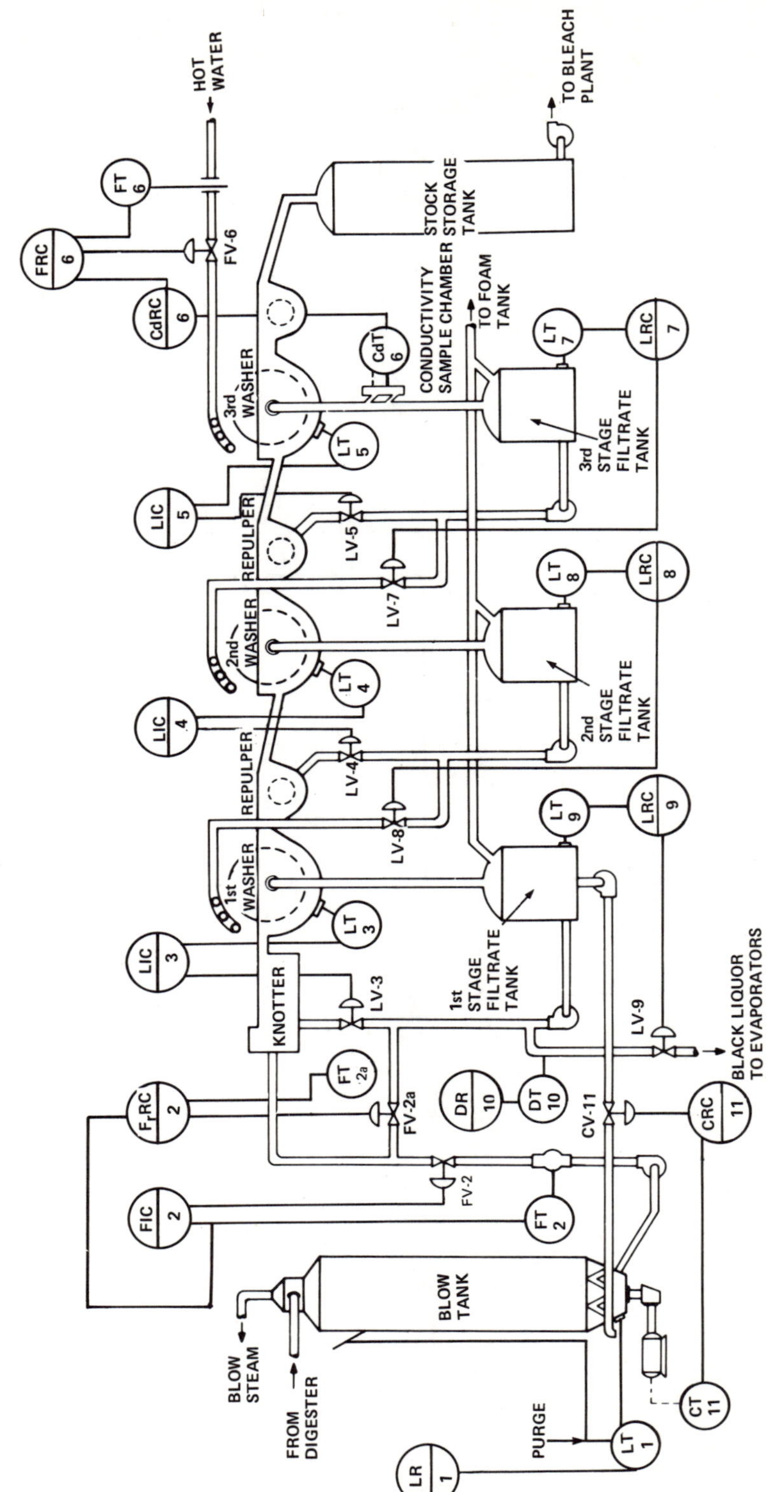

Figure 3-4. A rotary filter brownstock washing system.

WASHING

Pulp washing is the removal of washable, spent cooking liquor and intercellular matter dissolved by the cooking liquors during the cooking process. This must be accomplished without diluting the wash liquors any more than is absolutely necessary, as all these liquors must be evaporated later in the process of chemical recovery. Incomplete washing renders subsequent bleaching of the pulp difficult and impairs the value of the pulp. Excessive washing renders recovery of chemicals too costly. Obviously, the maximum amount of wash liquors must be conserved and none should be allowed to escape as waste.

Washing has been accomplished in the past by hosing down the pulp in drain tanks, by the use of screw presses, by the use of rotary vacuum washers, and lately, by the use of continuous diffusion washers. The use of rotary vacuum washers is the most common method.

Rotary Vacuum Washers

There are several well-known and widely used brown stock washing systems based on the use of rotary vacuum washers. A typical countercurrent multistage system is schematically shown in Figure 3-4.

In this system, pulp from the blow tank is discharged at regulated consistency to a pump which delivers the pulp to a series of vacuum washers. These vacuum washers consist of wire-screen-covered cylinders suspended in vats and provided with a means of applying vacuum to the inside. The slurry in the vats is drawn against the wire screen by the siphon action of filtrate falling through the down legs, causing the pulp fiber to form a mat on the surface. Three stages of washing are shown, but systems can consist of any other number of stages depending on the production it is to handle and the type of pulp to be washed. Hot water is used in the last stage of the wash, with the effluent or filtrate from this stage used as wash water in the preceding stage, and so on to the first stage. Three definite mechanical actions take place during this operation:

1. Dilution of the pulp slurry with wash liquors.
2. Filtering of the pulp fiber from the slurry over the wire-covered drums.
3. Displacing of cooking liquors from the pulp mass by progressively weaker black liquor and, eventually, hot water.

The filtrate from the first stage contains approximately 14 to 20 percent solids, and is pumped to a storage tank prior to subsequent processing for the recovery of chemicals.

The major objectives of the instrumentation involved in the pulp washing operation are to control:

1. The flow rate of unwashed pulp per process requirements.
2. The flow rate of recycled filtrate in fixed ratio to stock flow.
3. The flow rate of wash water input in accordance with washing demand.
4. The stage-to-stage balancing of filtrate counterflow.

Level recorder LR-1 provides a continuous record of blow tank level. This shows the availability of stock for delivery to the brown stock washers, as well as the availability of space for blows from the digesters. It receives its measurement from a flanged-type, diaphragm differential transmitter, LT-1, mounted directly on the blow tank, thus eliminating any intermediate piping which would be susceptible to plugging or clogging. To eliminate the effect of blow tank pressure variations above the stock level, a pressure line is run from the top of the blow tank to the low-pressure connection of the differential pressure transmitter. This line is air-purged to prevent blow tank carryover from collecting in the line and producing a measurement error.

Consistency control of the brown stock leaving the blow tank is based on the fact that the apparent viscosity of the pulp suspension changes with any change in consistency. This viscosity change is reflected in the power required by the electric motor to drive the agitator.

Consistency of the stock being pumped from the blow tank is detected by an agitator motor load measurement, CT-11, using a thermal watt converter. It is controlled by consistency recording controller CRC-11, modulating control valve CV-11 in the black liquor dilution line from the first-stage filtrate tank. Inflow rate is measured by magnetic flow transmitter FT-2 and is controlled by flow indicating controller FIC-2 and flow control valve FV-2 in the pulp line to the washers. The pulp flow signal, together with the recirculated filtrate flow measured by FT-2a, is also transmitted to flow ratio recording controller FrRC-2, which maintains the proper inflow consistency by operating a flow control valve, FV-2a, on the recirculated filtrate flow line.

Level control for each washer is accomplished by level indicating controllers LIC-3, LIC-4, LIC-5 from signals transmitted by diaphragm, flanged-type level transmitters LT-3, LT-4, LT-5, and by operating level control valves LV-3, LV-4, LV-5 in filtrate recirculation lines. These controls (combined with filtrate tank level controllers LRC-7, LRC-8, LRC-9, throttling control valves LV-7, LV-8, LV-9, using signals from the same type of transmitters, LT-7, LT-8, LT-9, on the filtrate tanks) maintain balanced stage-to-stage counterflow operation while the pulp is being washed.

The concentration of recoverable chemicals in the filtrate from the third-stage washer is measured by a conductivity measuring cell in a sampling chamber in the filtrate downleg. A thermal element mounted in the sampling chamber measures filtrate temperature and automatically corrects for temperature variations. This compensated signal is transmitted by CdT-6 to conductivity recording controller CdRC-6, which is the primary instrument of a cascade loop. The output of the cascade loop sets the control point of the hot wash water, secondary flow recording controller, FRC-6. A differential flow transmitter, FT-6, is used to measure this flow, and flow control valve FV-6 in the hot water line is throttled to maintain the final stage filtrate at a constant concentration.

If the filtrate chemical concentration increases, the primary (conductivity) controller raises the set point of the secondary (flow) controller. The water valve is opened wider to increase water flow which, in turn, decreases the filtrate concentration and returns the conductivity to the desired set value. The opposite action occurs on a decrease in filtrate conductivity. The desired set value is determined by the individual characteristics of the washing and recovery equipment and the type of pulp being washed.

Density recorder DR-10 provides a continuous record of recovery liquor density (normally in terms of Baumé) as an indication of reclaimable solids from a signal received from density transmitter DT-10 in the liquor line from the system. This density transmitter is usually of the nuclear radiation or refractive index type.

Screw Press Washers

The brown stock washing system, using screw press washers, operates on the countercurrent principle similarly to the rotary vacuum washer system. Its basic difference is in the use of screw press washers instead of the rotary vacuum washer. Figure 3-5 shows sectional and exploded views of the essential components of a typical screw press.

This screw press consists of a spline-grooved barrel housing and a worm of progressively diminishing volume between flutes. The worm fits in the barrel made up of thin, cylindrical, mild steel perforated plates. The perforations are tapered outward. Pulp enters the chute on the left. It is conveyed into the barrel by the worm. As the volume of the worm decreases, liquor is squeezed out through the perforated plates.

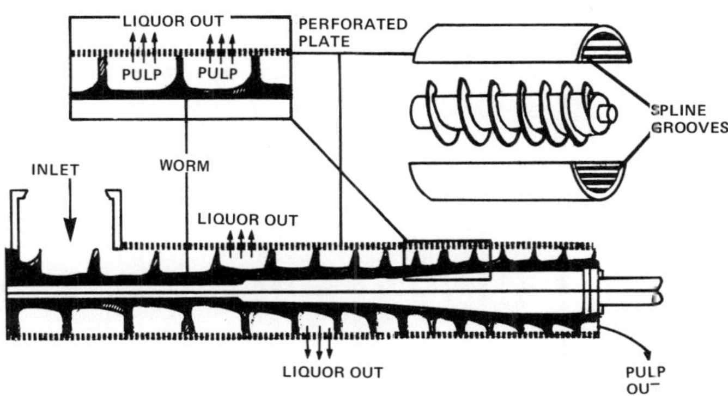

Figure 3-5. A typical screw press.

The instrumentation associated with a basic screw press washing system is shown in Figure 3-6. Level recorder LR-1 provides a continuous record of the blow tank level; its installation and operation are exactly the same as in rotary washer systems. Consistency of the brown stock leaving the blow tank is controlled by CRC-2 which detects changes in the agitator motor load measurement made by CT-2 and modulates control valve CV-2 in the black liquor dilution line. The quantity of unwashed pulp is controlled by thickener level controller LRC-3 which regulates valve LV-3. The control point of this controller is remotely set in a cascade fashion from the output of the stock chute level controller, LRC-4, which maintains the proper head of stock for satisfactory operation of the first press in the system. The load of the motor driving the screw is measured by EwT-5 and is controlled at a value for optimum operation by EwRC-5 by adjusting valve EwV-5 to regulate the amount of wash liquor fed to the press. Recorder EwR-6 records the loads of the motors driving the screws in the other presses in the system, assisting the operator in keeping track of their operation. The levels in the presses are maintained at best operating values in each press by level recording controllers LRC-12, LRC-13, LRC-15, and LRC-17 which adjust the flow of liquor filtrate extracted with valves LV-12, LV-13, LV-15, and LV-17. Intermediate and weak liquor storage tank levels are controlled by recording level controllers LRC-14 and LRC-16, respectively, by operating three-way valves LV-14 and LV-16 which direct liquor filtrate into its respective storage tank, if needed, to maintain the desired level. If the desired level is satisfied, a portion or all of the filtrate is diverted to the previous liquor storage tank, upstream in the system. LRC-11 controls the level in the strong liquor tank by regulating the flow of filtrate from the thickener. What is not needed to maintain the level is sent to the evaporators for further processing and recovery of chemicals. The total flow of strong liquor pumped to the evaporators is totalized and recorded on FRQ-7.

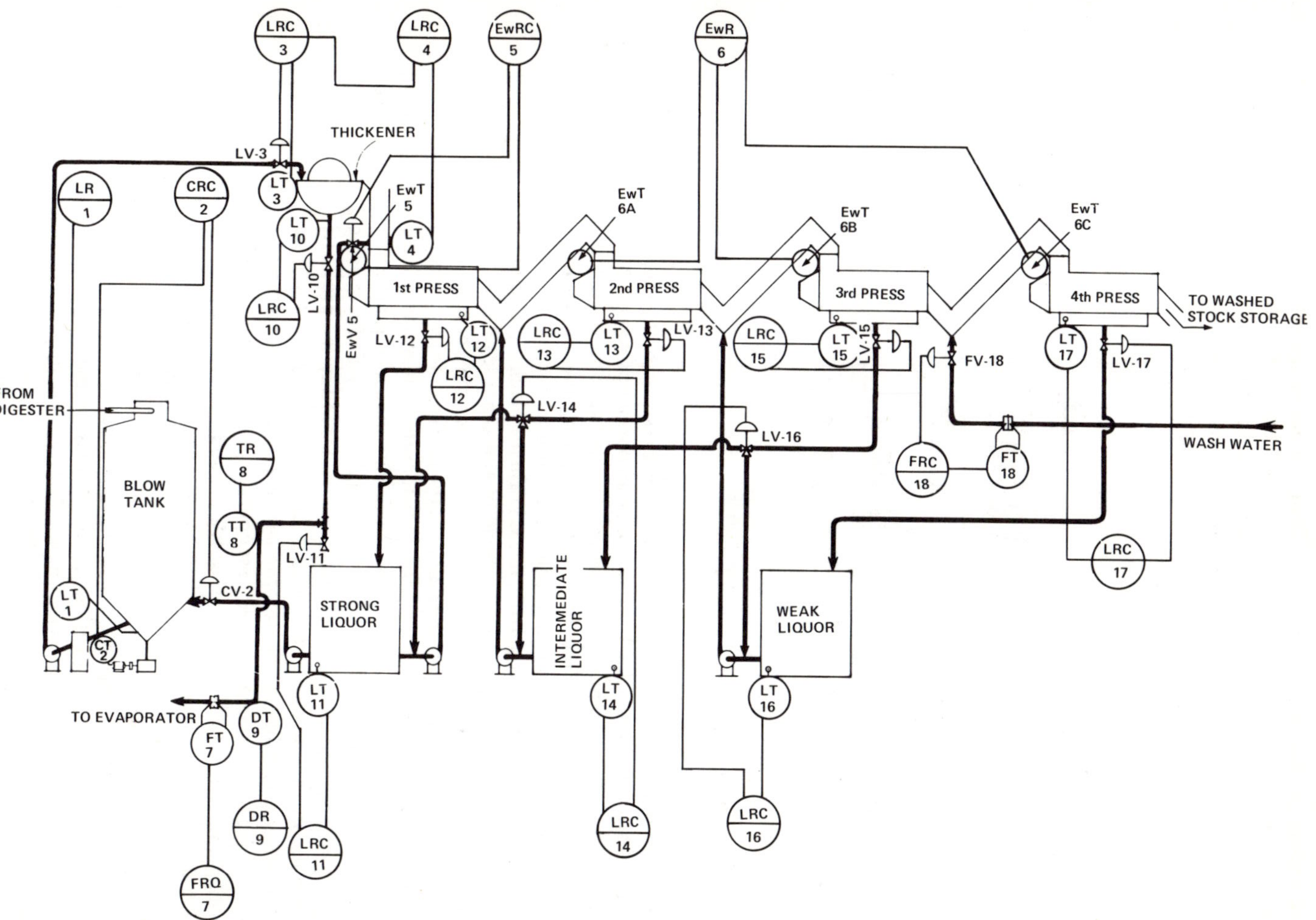

Figure 3-6. Instrumentation for a basic screw press washing system.

The temperature of this liquor is recorded on TR-8 and the density is measured by transmitter DT-9 and recorded on DR-9. This information is used to make adjustments to the process to maintain optimum washing efficiency.

WASHED STOCK SCREENING

After brown stock washing, certain grades of paper require further removal of *slivers* or *shives* and other undesirable residual debris before they are bleached and processed further. A screening and cleaning system, as depicted in Figure 3-7, has been used to accomplished this operation.

Washed stock chest level is recorded on LR-1 which is used as a guide to keep the screening and cleaning operation in pace with the washing operation. Consistency of the stock pumped to the primary headbox is controlled at the proper value by CRC-2 by adding decker filtrate with valve CV-2, just ahead of the pump, and measuring it with CT-2, just after the pump. Stock flow to the headbox is measured by an electromagnetic flowmeter, FE-3, and regulated by recording controller FRC-3 through valve FV-3. Additional dilution is made by ratioing the decker filtrate to the stock flow by flow ratio recording controller FrRC-16. Square root converter $\sqrt{16}$ is used in the output of flow transmitter FT-16 to linearize the differential pressure flow signal so that proper ratioing can be accomplished by the controller, which receives a uniform primary signal from transmitter FT-3.

Headbox level recorder LR-4 is used to obtain the proper balance of flow to the screens by maintaining the level just below the overflow dam. Remotely operated valves LV-6 and LV-7 are adjusted to obtain uniform flow to the screens and are regulated to the screen drive motor load as measured by EwT-8 and indicated on EwI-8, together with the inspection of the reject rate and quality of the screened stock. The second primary screen pressure is indicated on PI-7.

Accepted stock is dewatered on the decker. Vat dilution is adjusted by HIC-14 and CV-14 from the inspection of mat formation. The decker speed is automatically adjusted by LIC-9 in order to control it at a predetermined vat level, as measured by LT-9. The mat on the decker is washed with fresh water showers to reduce the liquor content further, and also to heat or cool the stock, if necessary, for later bleaching or processing. This flow of hot or cold water is measured by an orifice plate and differential-type transmitter FR-10, and is controlled by FRC-10.

The screened stock chest level is recorded by LR-12 as a guide to the adjustments made to the system to provide a constant supply to the bleaching operation. If additional dilution water is necessary, it can be added by remote manual operator HIC-11 and valve FV-11.

Decker filtrate is recycled back for dilution and for various level controls in order to conserve water. Level recorder LR-13 is provided to keep a record when surplus water overflows the filtrate tank. This situation can occur when more water enters the system from the washed stock chest and the showers than leaves with the decker stock mat. Where different volumes and pressures are involved, separate filtrate pumps are used. This eliminates interaction of some of the instrumentation for the system.

Primary screen rejects go to a tank for dilution and then are pumped to the secondary screen headbox. The operation is similar to the primary screens, with screen drive motor load measured by EwT-19 and indicated by EwI-19. Accepts return to the primary headbox for another screening; rejects go to a vibrating dewatering screen. Stock feed rate is self-regulating out of the primary rejects tank. The pump to the secondary headbox is sized to handle primary rejects and some dilution water added by level indicating controller LIC-15 to control the level

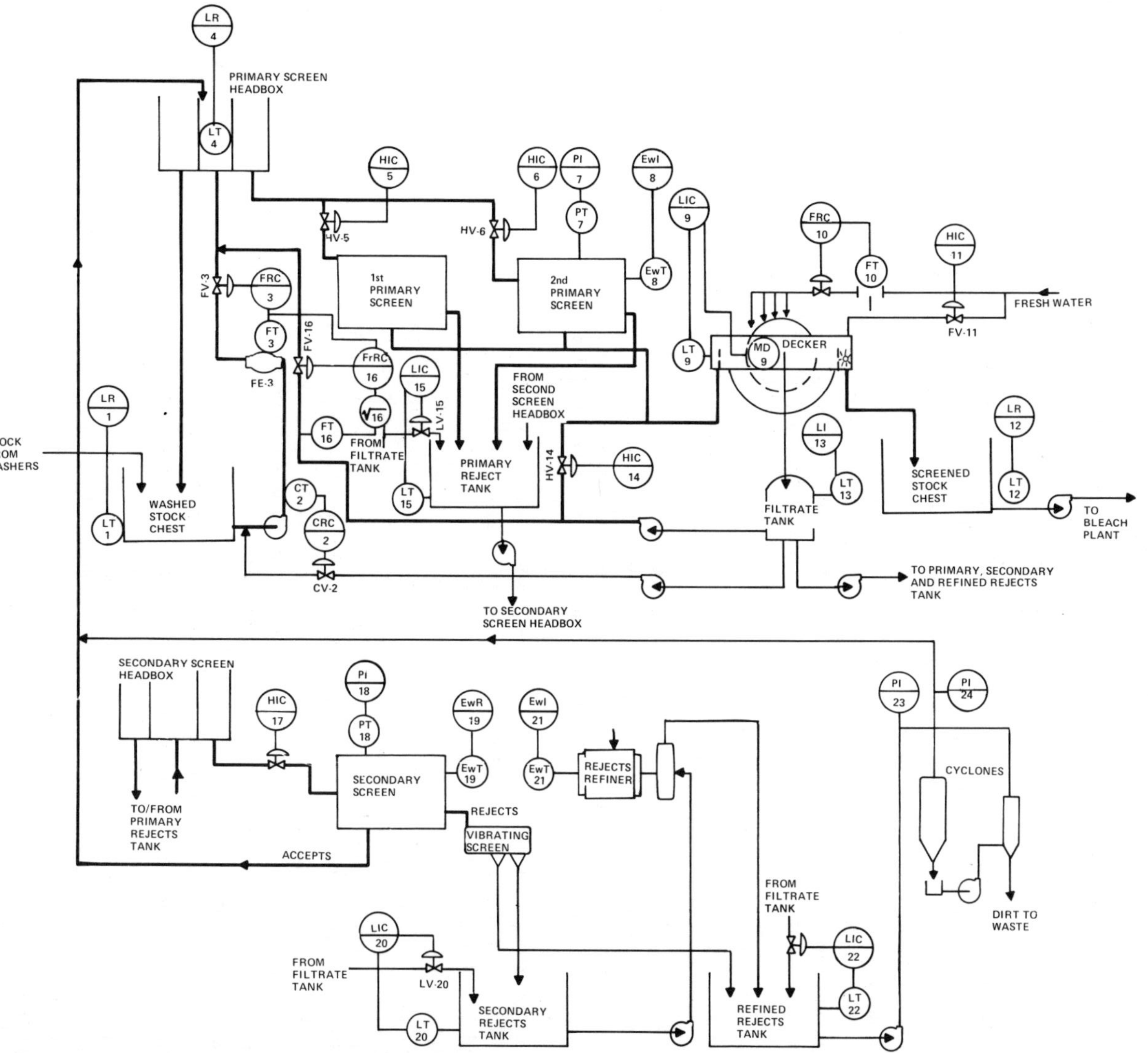

Figure 3-7. Washed stock screening for further removal of residual debris.

measured by LT-15. If the primary reject rate increases, the tank level tends to rise, the level control reduces dilution in the flow, and the consistency to the secondary screen increases enough for the secondary screen to handle this load.

Rejects in the secondary reject tank are diluted by the addition of decker filtrate through level indicating controller LIC-20. These diluted rejects are pumped to the rejects refiner and dropped into the refined rejects tank where additional filtrate is added by level controller LIC-22. Refiner drive motor load is measured by EwT-21 and indicated on EwI-21. The refined rejects are then pumped to two stages of hydraulic cyclones for grit and dirt removal.

In operation, throughput rate and screening quality are closely monitored so that proper corrective action can be made if the operation becomes abnormal. Consistency is reduced if throughput decreases due to screen plugging. If the decrease is due to too much water coming in with the feed, then throughput must be increased. Motor loads, vat and chest levels, stock discharge lines, and cyclone discharge cones are periodically inspected to insure proper operation of the screening and cleaning system.

4 Evaporation

In the economics of the pulping process, it is essential that the spent chemicals from the cooking process be recovered, regenerated, and reused. Therefore, the first step in the chemical recovery cycle is to concentrate the spent cooking liquor which has been separated from the pulp fiber in the washing stage by evaporation. This removes large quantities of water so that the liquor can support combustion in the following burning stage in the recovery furnace and maximize heat recovery. The design and configuration of the evaporator used is determined by the production and type of pulping process used. Because of the nature of sulfate liquor, the evaporators are usually of the multiple effect type and constructed of mild steel, although stainless steel tubes are also used. The corrosive nature of sulfite liquors makes it impractical to use anything less than stainless steel. In the earlier types of sulfite evaporators, the general arrangements were similar to conventional multiple-effect evaporators used on the sulfate process. Due to the serious scaling problems caused by conventional sulfite liquors, special and unique arrangements of evaporators were devised.

SULFATE EVAPORATOR SYSTEMS

As shown in Figure 4-1, a typical multiple-effect evaporator used on sulfate liquor is actually a series of individual evaporator bodies arranged so that the vapor generated from one evaporator body becomes the steam supply to the next evaporator in the series. The multiple-effect principle allows a given quantity of heat to be reused for evaporation a number of times by boiling a series of less dilute liquids with successively lower boiling points at decreasing vapor pressures and with the final effect usually operating under vacuum.

A countercurrent backward feed system is used with weak liquor entering the sixth effect at 15 to 20 percent solids and leaving the first effect at 50 to 55 percent solids.

Steam is introduced to the steam chest of the first effect; it then condenses and boils off a proportional amount of water. The vapor formed is then introduced in the steam chest of the second effect which is set at a lower pressure. Because the liquor is more dilute, its boiling point is lower than the condensing temperature leaving the first effect. This difference in temperature is the driving force which allows vapor from the first effect to boil off a proportional amount of water from the second effect and so on. In this way, it is possible to remove 4.0 to 4.5 pounds of water per pound of steam.

As the thin black liquor is concentrated, the solubility of the organics present, called soaps, is exceeded. These must be removed to prevent their precipitation and subsequent fouling of the downstream effects, thereby decreasing the efficiency of evaporation. A soap separation tank is usually located between the third and fourth effects; the soap is allowed to float to the top of this tank and can be removed by skimming. In the mills using wood species producing appreciable amounts of soap, it is used to produce tall oil which is sold as a by-product.

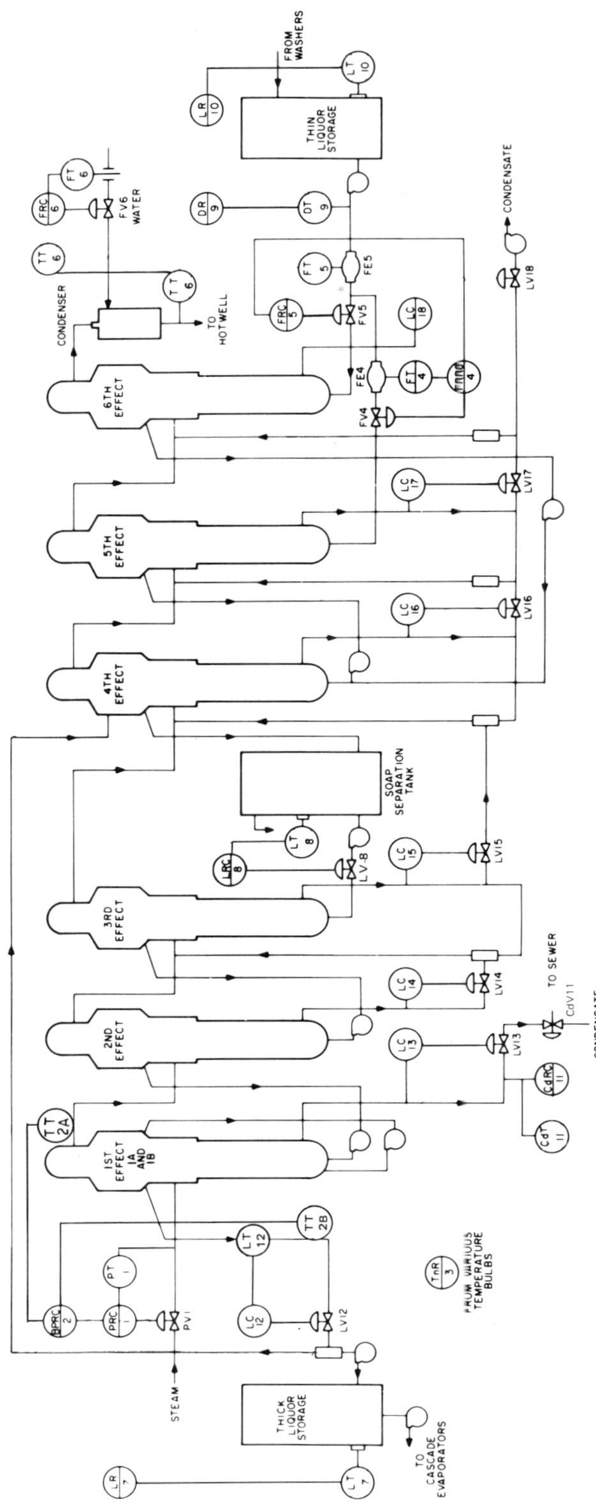

Figure 4-1. Sulfate black liquor multiple-effect evaporator.

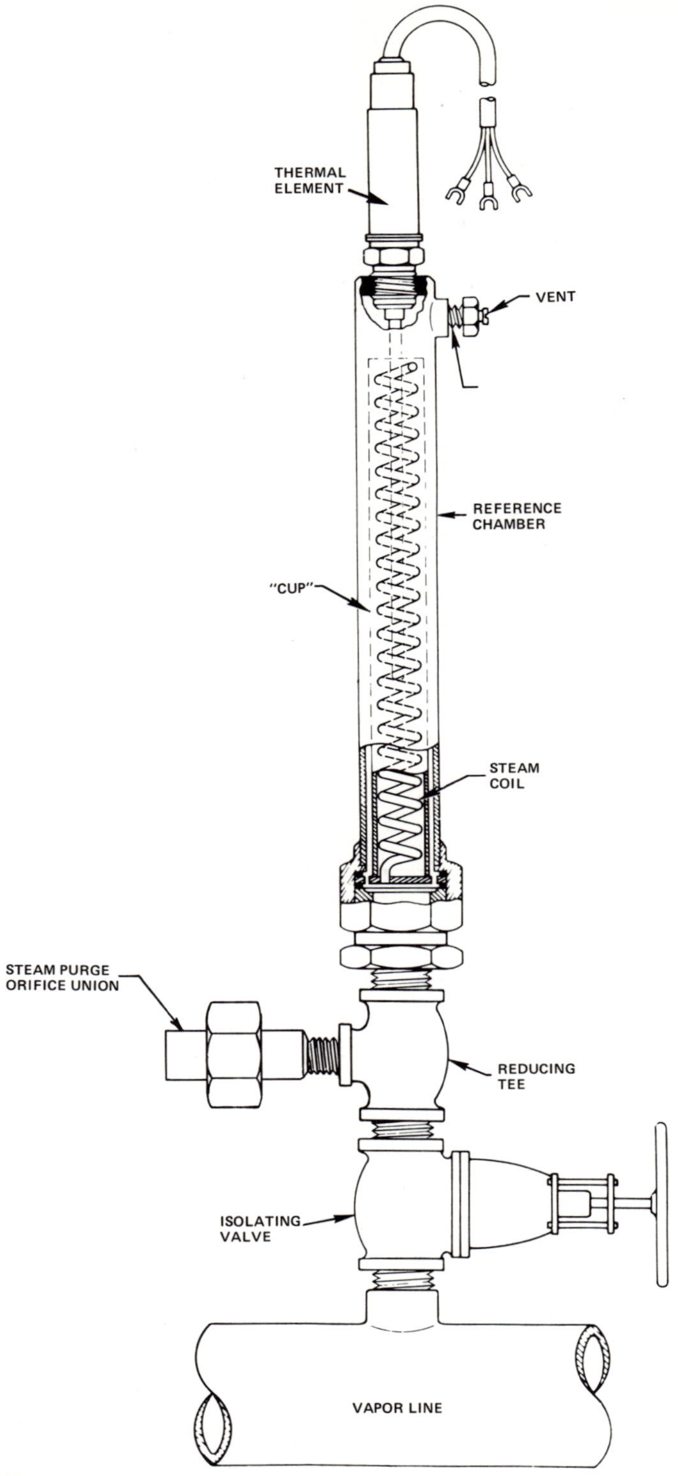

Figure 4-2. Condensing chamber for boiling point rise temperature measurement.

Instrumentation

The objective of the typical instrumentation illustrated in the diagram (Figure 4-1) is to attain the necessary balance between heat input and product flow and to produce a uniform solids-to-liquid ratio in the thickened liquor.

Steam at about 60 psig and 300°F is admitted to the first effect under pressure control by PRC-1, whose set point is automatically and remotely set in cascade control fashion from the output of density recording controller BPRC-2. The measurement of density for this controller is based on the boiling point rise principle. Temperature measurements are made and transmitted to the controller by TT-2A and TT-2B. The controller then adjusts the control point of the steam pressure controller in accordance with the difference in temperature measurements; this is essentially the difference between the boiling liquor and saturated steam at the same pressure. A known relationship exists between the boiling point elevation and the density of specific types of liquors. Where this relationship is not known, it can be easily established by each mill. The thermal element for TT-2B is located so as to ensure continual submergence in the thickened liquor by level loop LT-12, LC-12, LV-12, while the thermal element of TT-2A is installed in a condensing chamber tied into the vapor line which serves to desuperheat the vapor, thus providing a saturated steam temperature at evaporator pressure.

A typical condensing chamber is shown in Figure 4-2. This device operates to provide a cup of water boiled by the circulation of purging steam up the spiral coil. This circulation is assured by the radiation of heat from the outer wall of the chamber, condensing the surplus steam that is continually drawn up the coil. This condensate flows down between the cup and outer wall, draining into the vapor line through four holes in the plate supporting the coil and cup. As the steam passes up the spiral, enough condenses to maintain boiling in the cup surrounding it. If no noncondensable gases are present in the vapor line, a condensing chamber without steam purging is sometimes used. In this case, the superheat is sufficiently removed from the vapor by radiation to the atmosphere.

Density of the thickened liquor is also measured by other methods, based on radioactive nuclear sensors and refractive index detectors. A variation in liquor density to the first effect will cause a pressure change in that effect. Immediately, this condition will be partially corrected by the pressure controller as it modifies the steam flow rate. DRC-2 senses the change in density and gradually alters the set point of the pressure controller, PRC-1, to a new value (see Figure 4-1). This will cause a further change in the steam flow rate to complete the compensation and reestablish the density to the original desired value.

Condensate from the first effect is returned to the boiler house and is controlled by level controller LC-13 to maintain the level of condensate below the evaporator tubes and prevent steam blow through the condensate line. If contamination with black liquor occurs, conductivity recording controller CdRC-11 immediately operates the three-way diversion valve, CdV-11, to divert the condensate flow to the sewer.

Thin liquor feed to the evaporator is divided between the fifth and sixth effects because, in most cases, the sixth effect alone could not handle the total feed rate without flooding. Instruments FRC-5 and FRRC-4 control the feed of this thin black liquor. The feed is usually split as shown, with flows automatically ratioed in desired proportions. Flow measurement using orifice plates and purged differential pressure transmitters have been satisfactorily employed for this function. However, the magnetic-type primary flow elements are now commonly used here because they present no obstruction to entrained solids.

DR-9 is a continuous Baumé density recorder on the liquor feed which receives its measurement signal from transmitter DT-9. It gives a continuous trend of weak liquor gravity changes to the operator so that he may periodically reset flow recording controller FRC-5 (which records the total volume feed rate) to maintain a constant total solids throughput. The density measurement transmitter, DT-9, is usually of the bubble tube type. However, radioactive and refractive index density measuring elements have also been used here.

Liquor leaving the fourth effect has been concentrated to the point at which resin soaps formed during the cooling operation will no longer stay dissolved, but will tend to separate out. If allowed to continue into the following effects, the soaps would cause excessive foaming, scaling, and reduced capacity. The soap separates in the separation tank and is skimmed off the top of the black liquor. Proper liquor level is maintained by level recording controller LRC-8, regulating the feed rate to the third effect with valve LV-8. The level transmitter, LT-8, can be a bubble-tube or a flanged, flush-mounted diaphragm force-balance type.

Level controllers LC-14, LC-15, LC-16, LC-17, and LC-18 hold condensate seals on each effect and allow flash condensate from control valves LV-14, LV-15, LV-16, and LV-17 to add to the overhead vapor of each effect. Flow recording controller FRC-6 regulates the water rate to the barometric condenser in accordance with its set point as cascade set from the discharge temperature.

Level recorders LR-7 and LR-10 provide the evaporator operator with continuous information on thick and thin liquor inventories.

TnR-3 represents a multipoint temperature recorder that is receiving signals from temperature transmitters installed in selected locations throughout the evaporator system, thus providing an aid to efficient operation and an indication when an effect is fouling up and requires cleaning.

Feedforward Control. The large capacity and dead time inherent in multi-effect evaporators contribute to the problems encountered in controlling their operation. An upset must be corrected slowly or the entire system will become unstable. At times, the primary controlled variable, for example product density, cannot be kept under automatic control when large and/or fast changes occur. If feed to the evaporators suddenly increases, the product density is upset and feedback control becomes inadequate. To operate satisfactorily, the steam pressure controller must then be switched from a remote set point (set by the density controller) to a local set point.

The instrumentation shown in Figure 4-1 is made up primarily of feedback control loops. The basic operating principle of feedback control depends on a deviation between measurement and set point before corrective action is taken; therefore when an upset occurs in the system no corrective action is taken until it appears in the output of the process, which is after-the-fact and may be too late.

The feedforward control concept actually prevents the upsets from entering the process by beginning the corrective action at first sensing of the upsets or load changes, thereby preventing their appearance as upsets in the output. Comparing configurations *A* and *B* in Figure 4-3, note that in the feedforward system the load is an input to the control system as well as to the process.

When applied to an evaporator process, the primary loads can be considered as the thin liquor flow; the manipulated variable is the steam pressure or flow as set by the feedback measurement signal which is the thick liquor density (controlled variable). With feedback control, the load changes in liquor flows are transmitted to the feedforward control system before entering the process, and the steam flow is controlled accordingly. Although steam pressure or steam flow can be used as the manipulated variable, steam flow is preferred when implementing the control with the feedforward concept in order to facilitate the

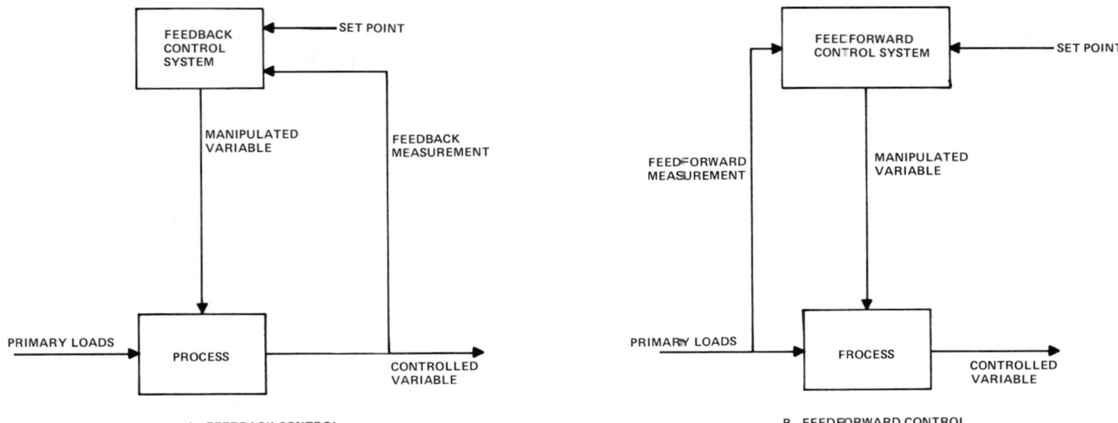

Figure 4-3. Feedback and feedforward control.

matching of the steam flow to liquor flow on an equivalent basis.

There are other factors (minor load changes) which are either unknown or cannot be measured but must be corrected by manipulating the steam flow; otherwise they will appear as minor variations in the output. This is accomplished by adding a feedback loop to trim the feedforward control system as shown in Figure 4-4. A dynamic compensator is also shown because the transient effects of controls and load changes vary from unit to unit and the compensator matches these transient effects, delaying or speeding up the measurement to the feedforward control system. These rates are adjustable to the process characteristics.

In order to illustrate how the feedforward concept can be applied to the multi-effect evaporator system in Figure 4-1, the process can be simplified by grouping the first three effects and the last three effects, and showing the three major loops in Figure 4-5 as:

1. Feed flow control to the last set of evaporator effects.
2. Level control on the soap separator tank.
3. Pressure or flow control of steam to the first set of evaporators with the set point set from the density controller on the thick black liquor.

Figure 4-4. Feedforward control with feedback trim.

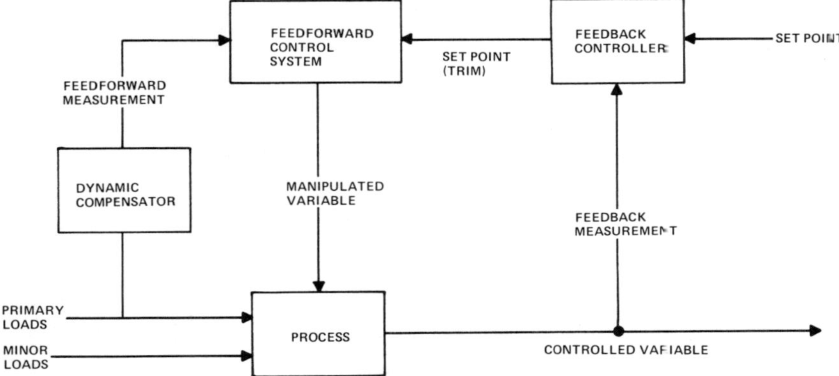

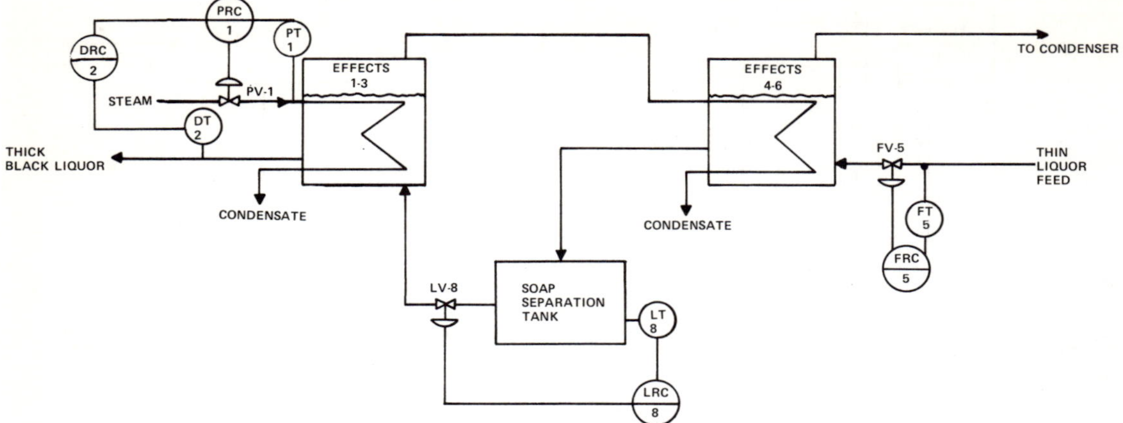

Figure 4-5. Evaporator feedback control.

Figure 4-6 shows the modifications and additions required to accomplish feedforward control as compared to the feedback system. The primary feedforward signal is the feed rate of the liquor into the first three effects from the soap separation tank. This flow signal from FT-8 passes through the dynamic compensator unit to match the transient effects of the process which, in this case, is a lag effect. After dynamic compensation, the measurement signal passes through a multiplier which establishes the desired steam/liquor feed ratio that is used as the set point to steam pressure controller PRC-1. The density signal output from the density controller is the feedback trim signal which alters the ratio between the liquor feed and the steam if the final density is not at the desired value.

If the change in load originates in the thin liquor feed to the last three effects, this change is used to initiate the feedforward action after appropriate dynamic compensation and multiplication to obtain the proper ratio between the feed to

Figure 4-6. Evaporator feedforward control.

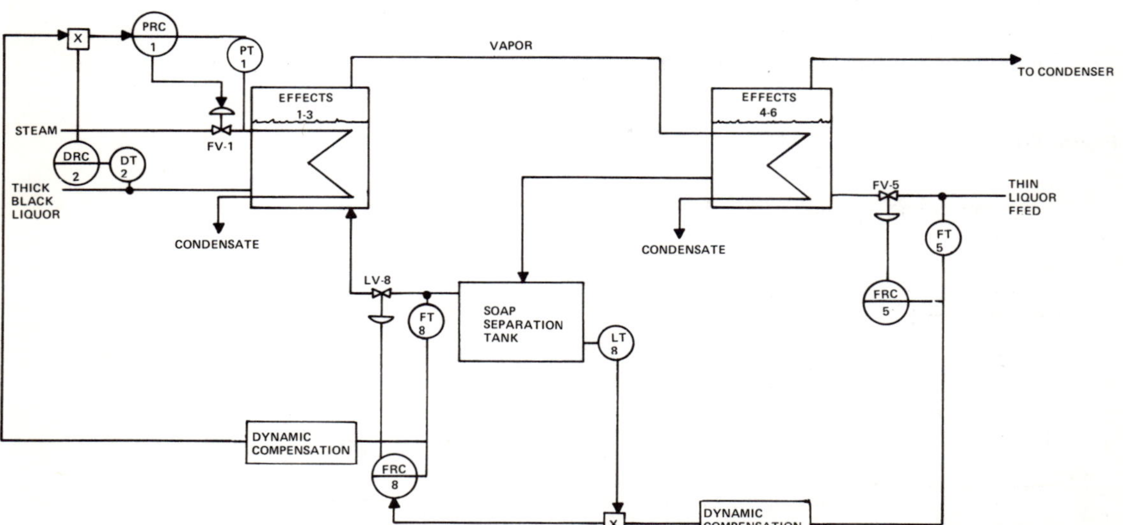

the last effect and the feed to the third effect. This ratio is altered by the output of the soap tank's level controller to maintain proper level. This level also represents the proper ratio desired between these two flows and can be used as the feedback trim correction to the feedforward flow loop.

The dynamic compensation is adjusted so that the flow into the soap tank increases almost simultaneously with the flow out. This increases the steam to the first effect before changes in the thin liquor feed are felt at the soap tank. Thus, the steam rate is changed so that it coincides with the flow as it reaches the first effect, thereby minimizing density swings of the final thick black liquor.

Evaporators with Concentrators

Recent installations of multiple-effect evaporators have incorporated the use of one or two additional effects just ahead of the first effect of the regular set. These additional effects were furnished with a separate supply of steam and evaporated the 55 percent total solids black liquor from the first effect of the regular evaporators up to 60-65 percent total solids. This allows the liquor to be fed directly to the recovery furnace for burning, eliminating the use of direct-contact type evaporation with recovery boiler flue gas. This avoids the stripping of odor-causing sulfur compounds from the liquor by the flue gases, which are one of the major sources of air pollution and sulfur losses from the process.

The instrumentation used is similar to that used for the typical multiple-effect evaporators except that the controls used on the concentrators are the same as those on the first effect, which are shown as loops PT-10, PRC-10, PV-10, DT-11, and DRC-11 in the simplified diagram (Figure 4-7).

When applying feedforward control to multiple-effect evaporators, the feedforward instrument loops normally located around the first three effects are also duplicated around the concentrator effects as shown in Figure 4-8 by loops FT-12, dynamic compensator, a multiplier, PT-10, PRC-10, FV-10, DT-11, and DRC-11. Note that because this is feedforward control, steam flow control is used instead of pressure control.

Figure 4-7. Simplified diagram of evaporators with concentrators.

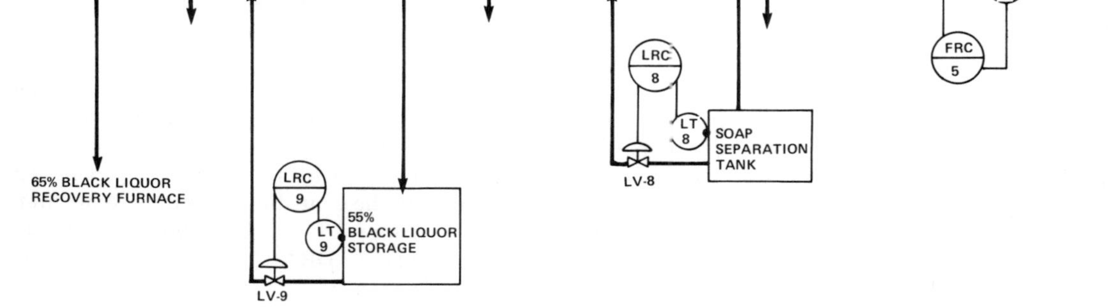

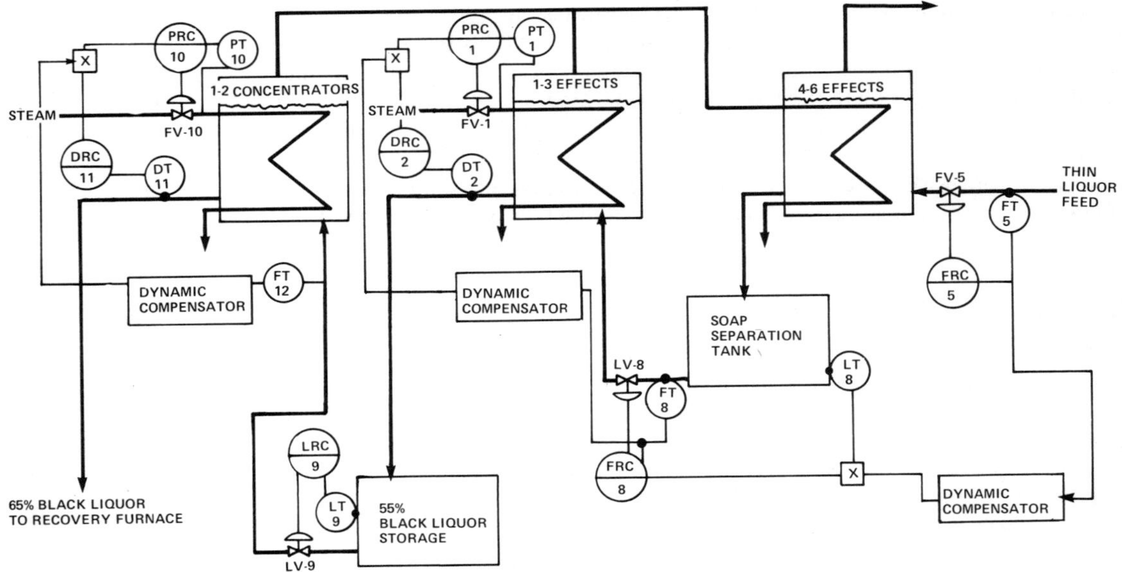

Figure 4-8. Feedforward control on evaporators with concentrators.

Tall Oil

Like turpentine produced from the gaseous relief of sulfate digesters, tall oil is another major by-product of the sulfate pulping process. The primary source of tall oil is from the soap skimmed mechanically from the surface of the black liquor in the evaporator soap tank. Again, depending on the species of wood, the season, the growing condition of the species, the age of the wood, the pulping conditions, the evaporator, and tall oil conversion system used, the tall oil yield per ton of pulp produced can vary from approximately 100 to 165 pounds.

The first step in converting tall oil soap is acidulation of the soap skimming with sulfuric acid to produce a crude tall oil. This is usually done at the pulp mill site. Figure 4-9 depicts a plant process to do this.

Soap from the evaporator's soap separation tank is allowed to settle in a soap storage tank equipped with a level transmitter, LT-1, and a high level alarm indicator, LIA-1. As the soap is pumped from the storage tank to a mixing tee, water heated by reactor temperature controller TRC-4 is continuously added. It is mixed with sulfuric acid being measured by magnetic-type flow element FE-3 and controlled by flow controller FRC-3. A dispersant is sometimes added and mixed before the slurry enters the reactor tank, where temperature is controlled by measurement TT-4 which regulates steam to the dilution water.

The acidified tall oil slurry is screened and sent to the centrifuge feed tank where temperature is maintained by controller TRC-5 by varying the heating steam with valve TV-5. The slurry is passed through strainers on its way to the centrifuge. Its flow is measured by magnetic flowmeter FE-7 and controlled at a value for the best operation of the centrifuge by controller FIC-7. The centrifuge separates the crude tall oil from the spent acid, which drops into the spent acid catch basin. A portion of this is recycled to the centrifuge feed tank to maintain the level under constant control of level loop LT-6, LIC-6, and LV-6. Density transmitter DT-11 and density recording alarm DRA-11 are used to determine when the spent acid becomes too dilute or too strong.

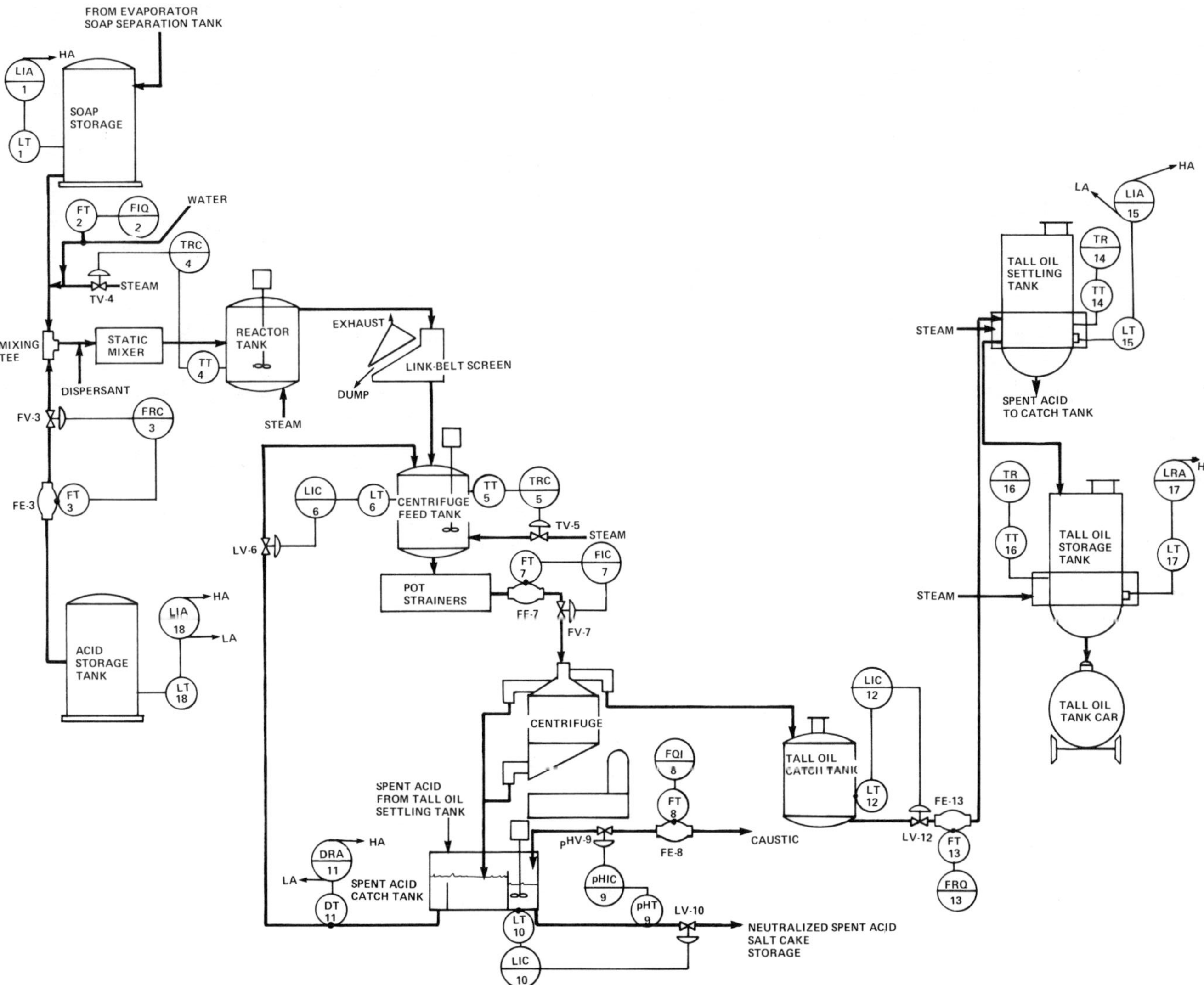

Figure 4-9. Tall oil process. This process is usually done at the pulp mill.

Any excess spent acid in the catch tank is neutralized by the addition of caustic into the overflow section of the catch tank by pH controller pHIC-9. The caustic flow is shown as being measured and totalized by magnetic flowmeter FE-8 and integrator FQI-8. The level in the overflow section is maintained by regulating valve LV-10 in the neutralized spent acid pipeline to the salt cake storage at the recovery furnace area of the mill.

The centrifuged crude tall oil is sent to a catch tank from which it is fed to the tall oil settling tank as level is controlled by indicating controller LIC-12 which operates valve LV-12 in the line to the settling tank. A flow record is made and the crude tall oil produced is totalized on integrating flow recorder FRQ-13.

The crude tall oil is allowed to settle further in the settling tank, and the spent acid removed is sent to the spent acid catch tank. This tank is kept warm by the addition of steam with the temperature being recorded by TR-14. The level is measured and alarmed by LT-15 and LIA-15. After settling, the crude tall oil is sent to storage from which it is pumped intermittently to tank cars for shipment for further refining and distilling to produce fatty acids, rosins, pitch, distilled tall oils and tall oil heads such as palmitic acid. Storage tank temperature is recorded on TR-16 and the level is recorded and alarmed on LRA-17.

SULFITE EVAPORATOR SYSTEMS

In the earlier types of evaporators used to concentrate liquors from the sulfite pulping process, the general configurations were similar to conventional multiple-effect types used on sulfate process black liquor. Because of the corrosive nature of sulfite liquor, it was necessary to use stainless steel as a material of construction. With the capital cost being much higher, the number of effects used was reduced to a minimum. Also, because of serious scaling problems, special and unique arrangements of evaporators and washing systems were devised.

In some of these systems, the flows of liquor and condensate are alternately switched from one side of the heating system to the other. The acid condensate formed washes away scale previously formed from the opposite operation.

The falling film principle has also been applied to multiple-effect evaporators in which the tube blanks are sprayed with liquor on the outside of round tubes which have the condensing steam on the inside of the evaporator tube surface. By arranging the tube bundles in zones which can be made idle, scale washing and dissolving can be accomplished during operation.

Vapor Compression Evaporator

Evaporating by vapor compression consists essentially of recovering vapors from the boiling liquor, compressing the vapors to a pressure high enough to correspond to an increase in temperature greater than the boiling point used in the liquor, and using the compressed vapors as a heating medium to supply further heat to the liquid. In this way, all the energy required, apart from small extraneous heat losses which may be made up by the addition of direct heat, is supplied as mechanical energy, and the addition of a small amount of energy mechanically to the vapor makes possible the recovery of a much larger amount of latent heat in the vapor. It is, thermodynamically speaking, a heat engine in reverse as are refrigerating machines and other types of heat pumps. It is the simplest type of heat pump since the material being heated is used as the heat transfer medium and the use of heat exchangers and throttling devices is avoided. A schematic representation and the instrumentation involved in typical vapor compression evaporators are shown in Figure 4-10.

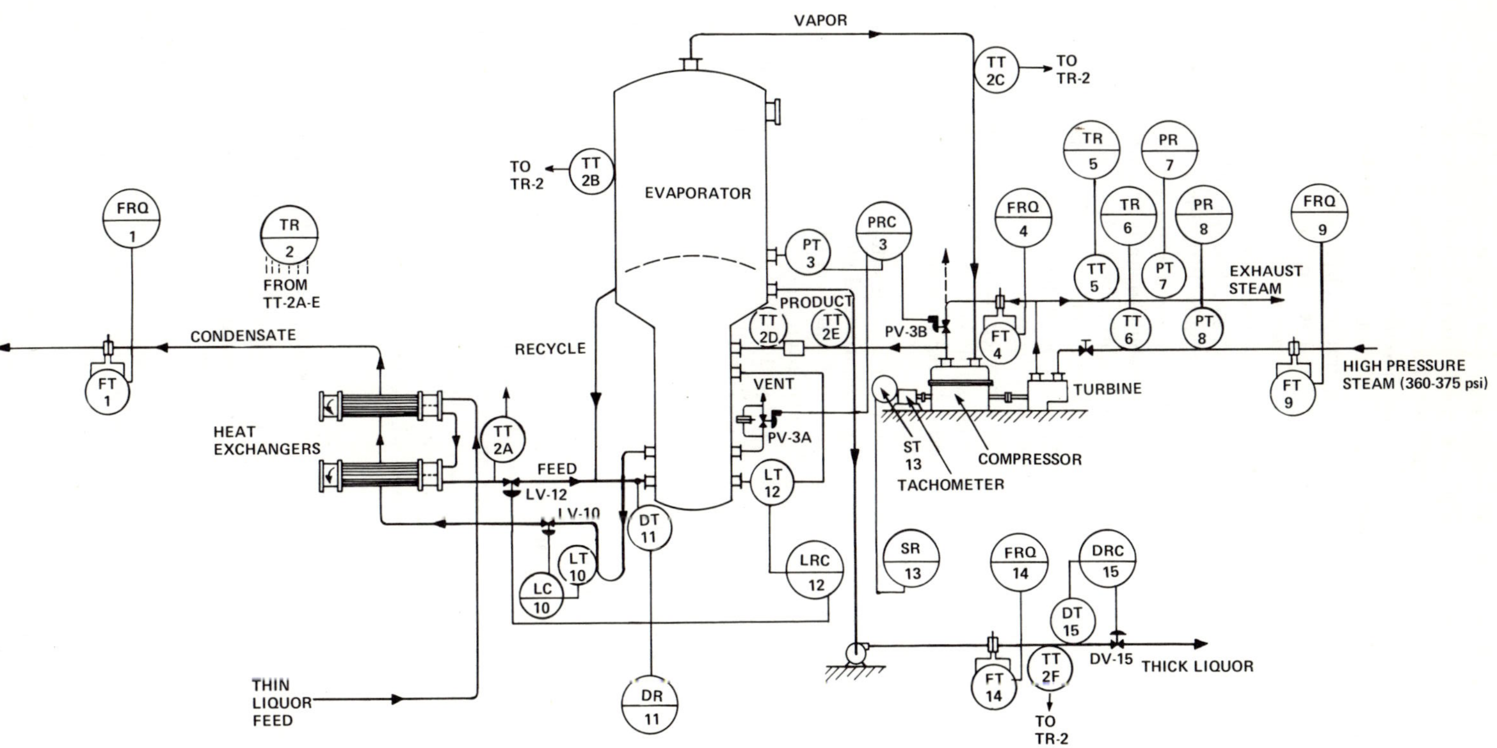

Figure 4-10. Vapor compression evaporator.

In operation, the vapor removed from the boiling liquor being concentrated is compressed a few pounds per square inch and fed back to the outside of the tubes of the evaporator where it condenses and transfers its latent heat to the boiling liquid. Since the required increase in pressure and temperature is only enough to overcome the boiling point rise of the liquor above the boiling point of pure water, together with the temperature difference required to cause heat transfer, the amount of heat recirculated is very large in comparison with the power required to drive the steam compressors.

FRQ-1 records and totalizes the condensate flow from the heat exchanger in which the thin black liquor is preheated by the hot evaporator condensate prior to being fed into the evaporator body. PRC-3 is a recording pressure controller which admits steam or vents the evaporator as required to maintain desired evaporation pressure. The makeup steam is recorded and totalized on FRQ-4 while the exhaust steam temperature record is made on TR-5. A record of the high-pressure steam temperature is made by TR-6. The exhaust steam pressure is recorded on PR-7 and the pressure of the high-pressure steam is recorded on PR-8. High-pressure steam flow is recorded and totalized by FRQ-9.

The level in the condensate leg from the evaporator is maintained by level controller LC-10. Sometimes this is simply an in-line field-mounted, float-type level control unit. A record of the incoming thin liquor feed density variations is made by recorder DR-11.

The level in the evaporator is controlled at the desired value by controller LRC-12, throttling the thin liquor feed with valve LV-12. The operator keeps track of the compressor speed with recorder SR-13, receiving a signal from a tachometer-type speed transmitter, ST-13. The flow of thick black liquor from the evaporator is measured and totalized by integrating flow recorder FRQ-14. Density of the evaporated thick black liquor is controlled by DRC-15, which regulates the liquor outlet valve, DV-15, to increase or decrease residence time of the liquor in the evaporator body as measured by the density measuring element of transmitter DT-15.

Other Evaporators

Black liquor from the multiple-effect evaporators is further concentrated at the recovery furnace by the hot combustion gases leaving the boiler tube section of the furnace after it passes through the feedwater heat exchanger (economizer) or combustion air heater. Types of equipment normally used are cyclone evaporators, cascade evaporators, and venturi scrubbers. The instrumentation used with these evaporators will be reviewed when recovery furnace instrumentation is presented.

BLACK LIQUOR OXIDATION

The more volatile, unstable sulfide compounds such as hydrogen sulfide, mercaptans, and methyl sulfides formed in the sulfate process are likely to escape to the atmosphere, which is a source of sulfur losses and a cause of an odor problem. Oxidation of the black liquor by exposure to air or molecular oxygen converts these unstable sulfide compounds into more stable nonvolatile sulfur compounds, principally thiosulfate, which will not break down into compounds that are normally released and lost to the atmosphere on further processing.

Although the usual point for installation of a black liquor oxidation system is just before the evaporators, some mills oxidize the black liquor at the point on the multiple-effect evaporator system where the liquor is removed for soap separation, or after the evaporators. A serious foaming problem is created with oxidation before the evaporators. Defoamers have been used to minimize this problem.

Air Oxidation

A relatively large black liquor surface area is required to have effective exposure to air for oxidation. This has been accomplished in towers filled with a large number of plates over which the black liquor is circulated countercurrent to an updraft of air. There are various designs of oxidation units, each to suit a particular mill.

Good success has also been achieved by bubbling compressed air from a network of pipes located in the bottom of a black liquor tower. Such a system is illustrated in Figure 4-11. Air from the compressor is blown through a separator where moisture is drained from it. The level in the separator is controlled by level control loop LT-1, LIC-1 and LV-1. The air is then introduced into the bottom of the oxidizing tower. Pressure controller PIC-2 regulates the amount of air to the tower by operating valve PV-2 in an air exhaust line to maintain the desired pressure in the tower. The flow of air is measured and recorded by FT-3 and FR-3. Liquor is sprayed into the tank at the top to help retard excessive foam, and pumped out at the bottom to a stilling tank. The foam is usually broken down by either mechanical or chemical means. The level in the stilling well is controlled at the desired level by controller LIC-4, throttling the discharge flow from the stilling well with valve LV-4.

Oxidation of black liquor efficiencies of 95 to 100 percent are attained with properly operated units. The operation is usually monitored by determining the

Figure 4-11. Black liquor oxidation with compressed air.

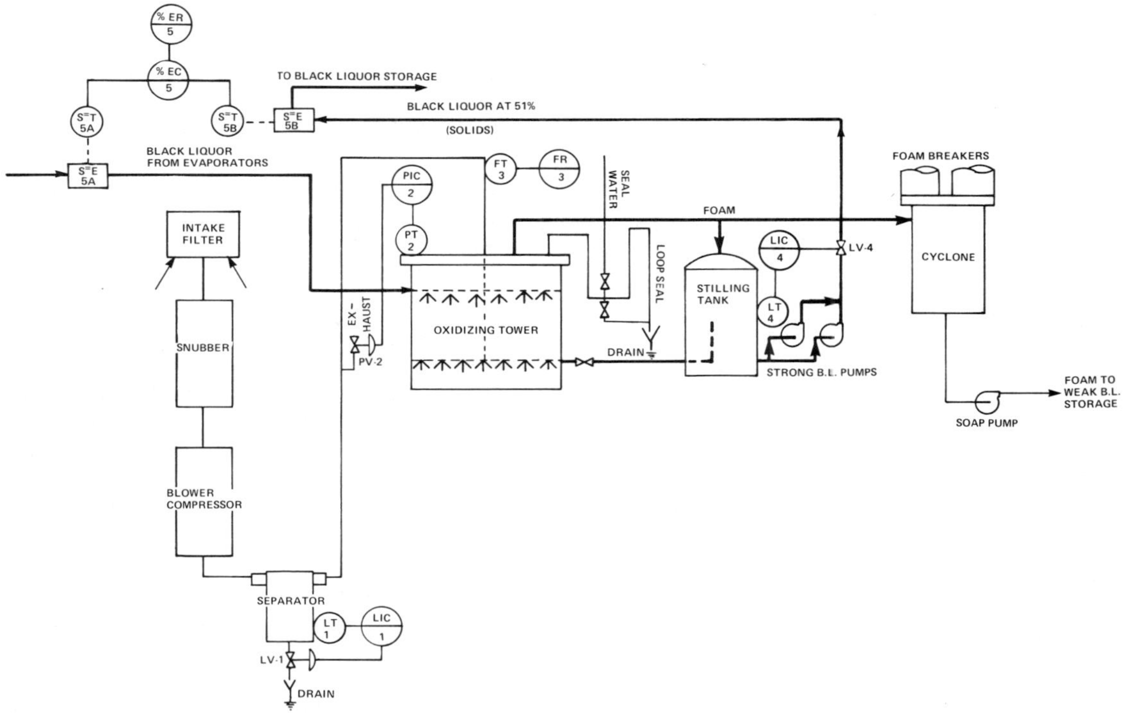

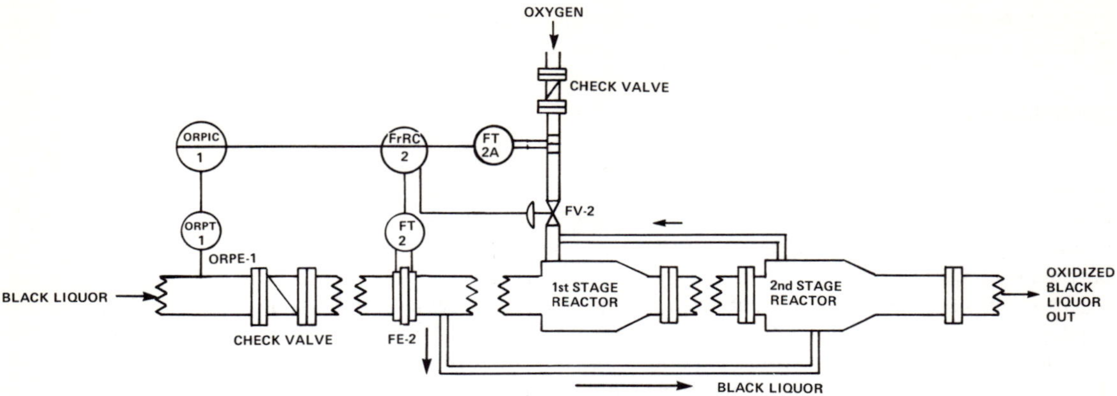

Figure 4-12. Black liquor oxidation with molecular oxygen, using oxidation-reduction potential.

unit efficiency. This is accomplished by taking samples before and after oxidation and testing for the concentration of the unstable sulfide by colorimetric or potentiometric laboratory titration procedures. The difference between the total sulfide in and out of the oxidation unit divided by the total sulfides in is reported as the percent oxidation efficiency.

It has been found that sulfide levels are related to and can be measured by a direct measurement of the sulfide ion. Therefore, continuous measurement of the sulfide ion can be made in and out of the oxidizing unit with sulfide ion selective electrodes, and oxidation efficiency can be computed from these measurements. The sulfide ion selective electrodes are shown as $S^{=}$E-5A and $S^{=}$E-5B in the diagram. These measurements are transmitted by $S^{=}$T-5A and $S^{=}$T-5B to percent efficiency computing relay %EC-5 which performs the following calculation to compute the percent oxidation efficiency.

$$\frac{\text{Total } S^{=} \text{ in} - \text{Total } S^{=} \text{ out}}{\text{Total } S^{=} \text{ in}} = \text{percent oxidation efficiency}$$

This signal is recorded on %ER-5 to be used as a guide to the operation of the system. Where the unit design will allow it, the percent efficiency signal can be used as the basis for automatic control of the flow of air to the oxidizing tower.

Oxygen Oxidation

Successful oxidation of sulfate black liquor with molecular oxygen instead of air has been reported by a number of mills. The procedures tried have involved the use of multistage, tubular-type reactors as shown in Figure 4-12. In this particular trial, the sulfide concentrate of the incoming black liquor was measured as a related value to the oxidation-reduction potential with probe ORPE-1. This signal was then transmitted to controller ORPIC-1 whose output was used to remotely set the ratio of flow ratio controller FrRC-2, regulating the flow of oxygen in accordance with the flow of black liquor to the reactor as measured by FE-2 and FT-2.

Another method of adding molecular oxygen to black liquor is by ratioing the oxygen to a primary and secondary mixer and controlling the total flow from a measurement based on a sulfide ion selective measurement after oxidation. Such a system is shown in Figure 4-13. The oxygen flow system is comprised of a primary flow loop, FT-5, FRC-5, and FV-5, and secondary flow loop FT-6, FRC-6,

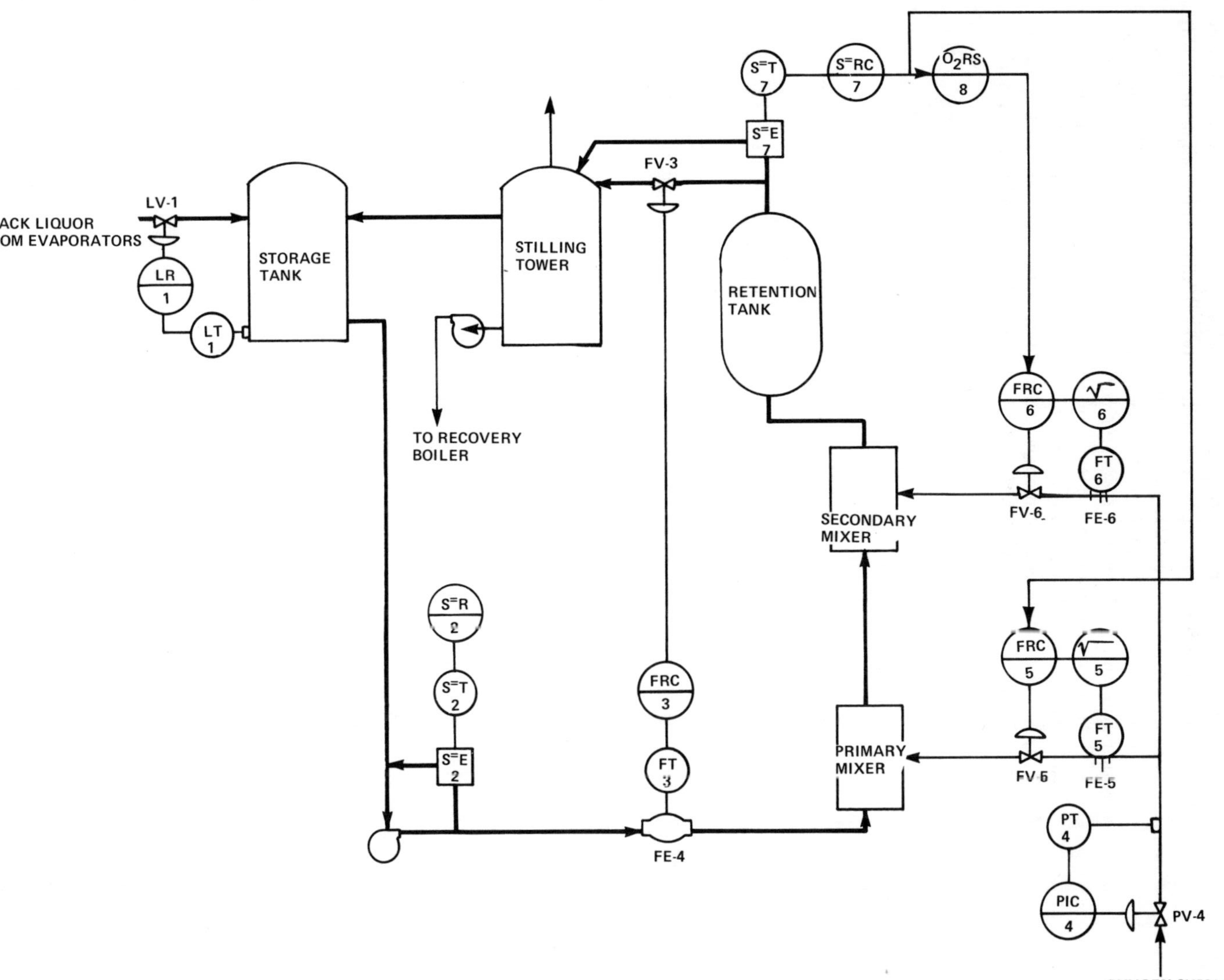

Figure 4-13. Black liquor oxidation with molecular oxygen, using $s^{=}$ measurement and ratio control.

and FV-6. The set point of the flow controller of each loop is adjusted from the output of the sulfide ion recording controller, $S^{=}$RC-7. The primary controller set point is driven directly from the output while the secondary controller set point is driven through oxygen flow ratio station O_2RS-8, whose adjustment will produce a primary/secondary oxygen ratio of flows to the respective mixers.

The flow of black liquor to the mixers is controlled from a magnetic flowmeter measurement, FE-4, by flow recording controller FRC-3 and flow valve FV-3.

The sulfide ion measurement on the untreated black liquor is made with a sulfide ion selective measuring element, $S^{=}$E-2, and recorded on recorder $S^{=}$R-2. By comparing this measurement with that made on the oxidized black liquor by $S^{=}$E-7 the oxidation efficiency of the system can be determined as described under "Air Oxidation."

Power and Recovery Boilers 5

There are as many ways to instrument a power or recovery boiler as there are types of boilers in the paper industry. The following discussions are intended as a review of the types of instrumentation and control used in the steam generation process in the pulp and paper industry. The philosophy and principles which led to the control techniques of boilers will be examined first since they can be related to the instrumentation used on recovery boilers, as discussed later.

In the discussion of power boiler instrumentation, diagram symbols developed for this particular application will be used. These symbols, shown in the standard functional diagram legend (Figure 5-1), differ somewhat from the typical process-type symbols, but they are more compatible with power industry instrumentation.

POWER BOILERS

Combustion Control

The term *combustion control* refers to that part of a boiler control system which performs the two basic functions of energy balance and furnace control. The furnace control system is a slave subloop to the energy balance system. The two systems will be discussed separately to show their different functions.

Energy Balance—Single Boiler. In a steam and/or electric generating process a certain energy system demand (load) must be matched by energy input (fuel). The energy balance system must monitor the relationship between input and output and manipulate energy input via the furnace control subloop. The energy balance system is structured in accordance with process configuration, operating procedure, and expected frequency and rate of load change on the boiler.

The simplest form of an energy balance system is a single-element steam pressure controller which manipulates the furnace control system as shown in Figure 5-2. Steam pressure is an excellent index of energy balance. It responds as the integral between energy input and output, and a proportional-plus-integral type controller is commonly used. Where the steam pressure signal is inherently quiet or can be filtered to eliminate noise, a three-mode proportional-plus-integral-plus-derivative controller would provide added performance.

A single-element pressure control energy balance system is satisfactory:

1. Where load changes are infrequent and/or at relatively slow rates of change (1% to 2% per minute).
2. When load changes are infrequent and/or relatively large deviations from the pressure set point (±5% to ±10%) are tolerable during load changes.
3. When a more complex system cannot be economically justified.

A more complex system must be used where load changes are frequent and/or relatively fast (greater than 5% per minute), and where large steam pressure deviations are tolerated.

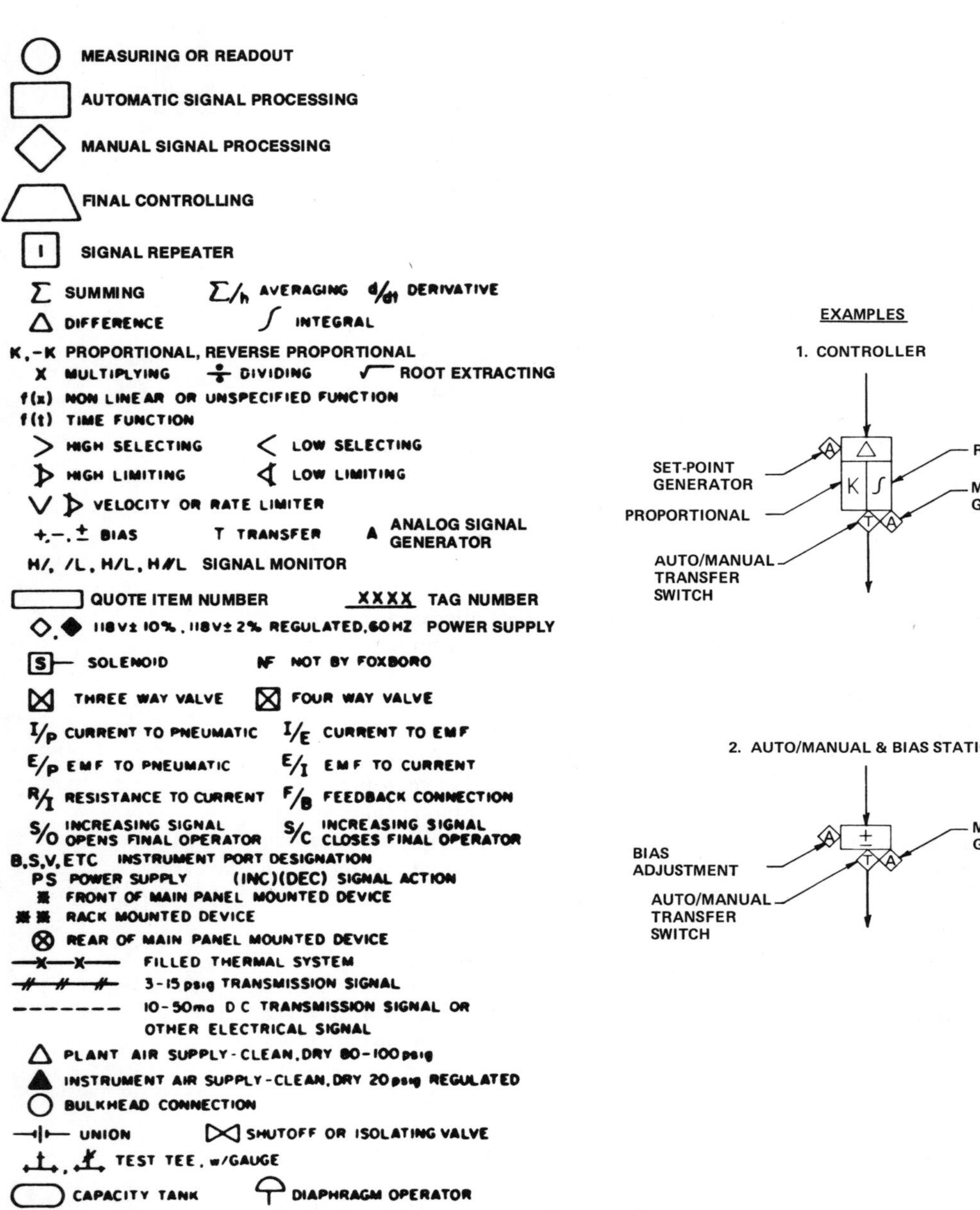

Figure 5-1. Standard functional diagram legend.

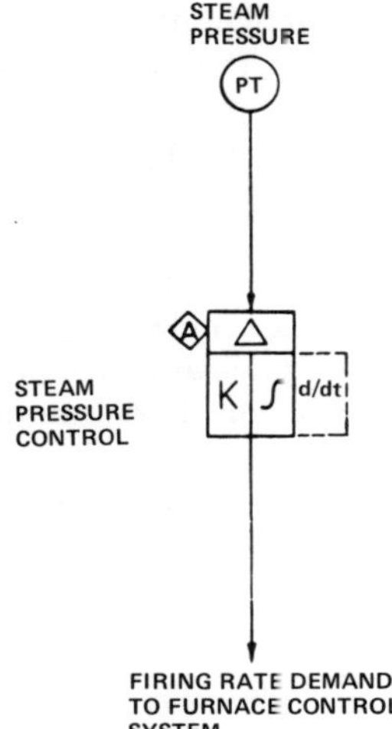

Figure 5-2. Single-element energy balance system—single boiler.

The approach is to calculate the energy demand, match energy input (fuel) to this demand (feedforward), and trim with a steam pressure (energy balance) feedback loop. Appropriate dynamic compensation must be included in the feedforward loop. A simplified version of this configuration is shown in Figure 5-3.

Here, a signal representative of the process energy demand is used to set the boiler firing rate through a multiplier. The second input to the multiplier is from a controller that monitors steam pressure. This feedback loop (steam pressure) adjusts for differences between energy demand and energy supply. A multiplier is

Figure 5-3. Energy balance concept.

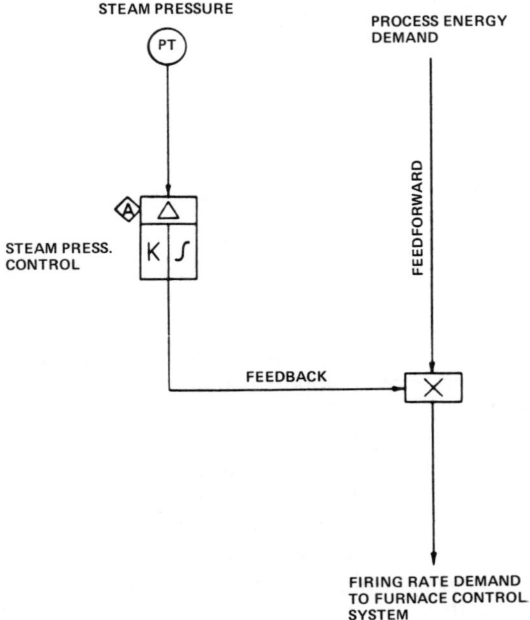

used in this configuration because a fluid property, steam pressure, is being controlled and, thus, the amount of incremental correction required for a given pressure deviation is proportional to the load. In effect, this provides for an "equal percentage" characteristic of the furnace control system.

The signal most commonly available as an indication of process energy demand is steam flow (see Figure 5-4). However, there is a dynamic problem associated with using steam flow as an energy demand index. This practice can act as *positive feedback* under some common operating conditions.

Consider, for instance, a unit being fired manually with a waste fuel, and with steam pressure control by automatic manipulation of a supplemental fuel. If an increase were made manually in the rate of waste fuel, an increase in steam flow would result. The control system would interpret this increase in energy input (*supply*) as an increase in *demand*, and instability would result. This effect is particularly severe on installations where multiple boilers feed a common header.

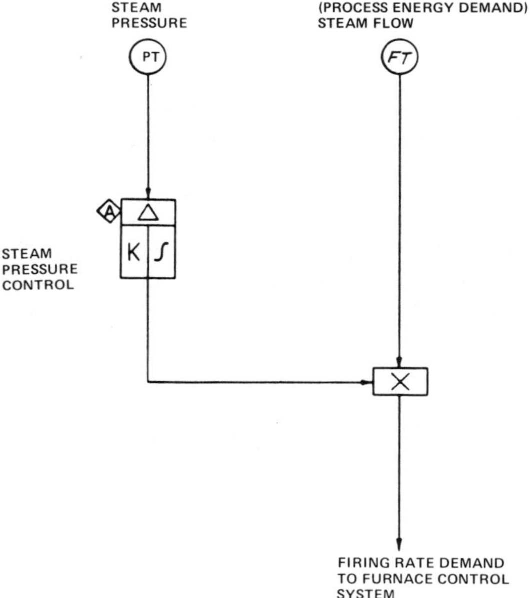

Figure 5-4. Energy balance system—single-boiler installation.

Figure 5-5 illustrates a method of canceling the positive feedback effect for a single-boiler installation.

In this scheme, the deviation of the controlled steam pressure is used as an approximation of change in boiler-stored energy. By subtracting this from the steam flow signal, a signal very closely approximating the actual energy demand is obtained. With this configuration, the previously described upset on the fuel side of the unit would cause no change in the feedback demand.

On a load increase, steam flow will increase and steam pressure will decrease. The steam flow effect will account for the increased firing rate to match the process load. To restore steam pressure, the boiler will have to be overfired temporarily to increase the stored energy in the boiler metal and contained fluid (boiling water). The dynamic compensator added after the derived feedforward signal will accomplish this.

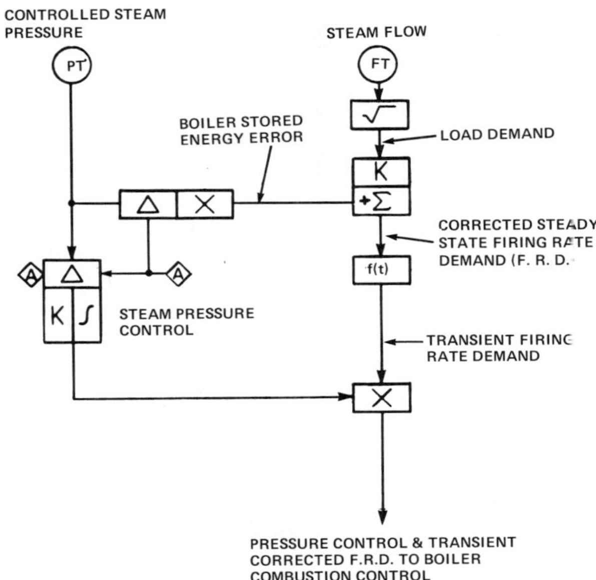

Figure 5-5. Energy balance system—single-boiler installation with transient corrected F.R.D.

On an increase or decrease in a manually introduced fuel, the effect of increased (decreased) flow and pressure will cancel. The pressure controller will adjust the automatically manipulated fuel.

On a pressure set point change, the increase in pressure and flow (due to the controller action after the change) will again cancel each other and eliminate the positive feedback effect.

Energy Balance—Multiple Boilers. Most paper mill installations are equipped with a steam distribution system having several boilers feeding a common header. In this case, the steam header pressure is controlled by regulating the firing rates of two or more boilers in parallel. Provisions must be made for adjusting the load share of each boiler, for base loading any boiler(s), and for startup and shutdown of an individual unit.

Similar techniques are applied to multiple-boiler installations as are applied to single-boiler installations. Also, the same criteria determine the required control system structure.

Figure 5-6 shows the simplest type of system, the single-element steam header pressure controller. Again, the derivative mode is optional. The application criteria for this system are the same as for the single-element, single-boiler system described previously.

The controller output is sent in parallel to each boiler through an auto-manual-plus-bias station. This allows for operator adjustment of the share of the load each boiler carries for manual base loading of any unit(s), and for convenient startup and shutdown of any unit.

Figure 5-7 shows the structure of a feedforward energy balance system for multiple-boiler installations. This system is similar to the single-unit energy balance system (Figure 5-5) but must have an additional compensating loop. A proportional-plus-integral controller (plant master) is incorporated to ensure that the total firing rate demand signal always represents the same energy input. This controller receives the firing rate demand signal as the set point. Its measurement

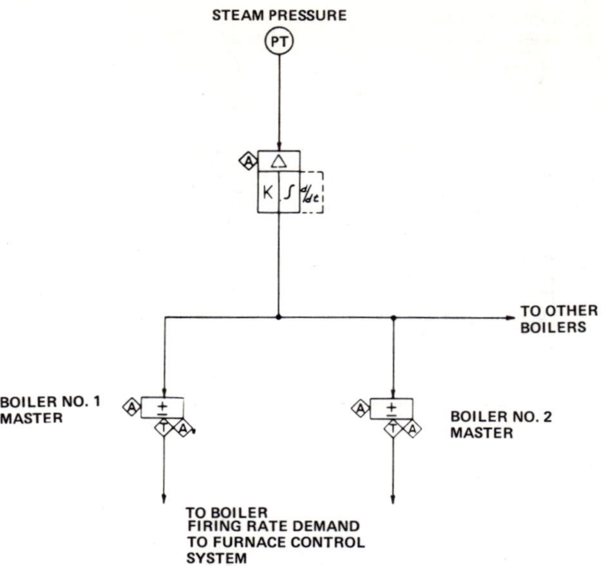

Figure 5-6. Single-element pressure control—multiple boilers.

signal is the sum of the individual boiler firing rate demands. Now, regardless of the number of boilers in automatic service, an incremental change in the firing rate demand is always matched by an equivalent total incremental change to those boilers in automatic service. If boilers of different capacities are installed (often the case, particularly when new units are added to an existing system), then the summing unit must be scaled accordingly.

Figure 5-7. Energy balance system—multiple boilers.

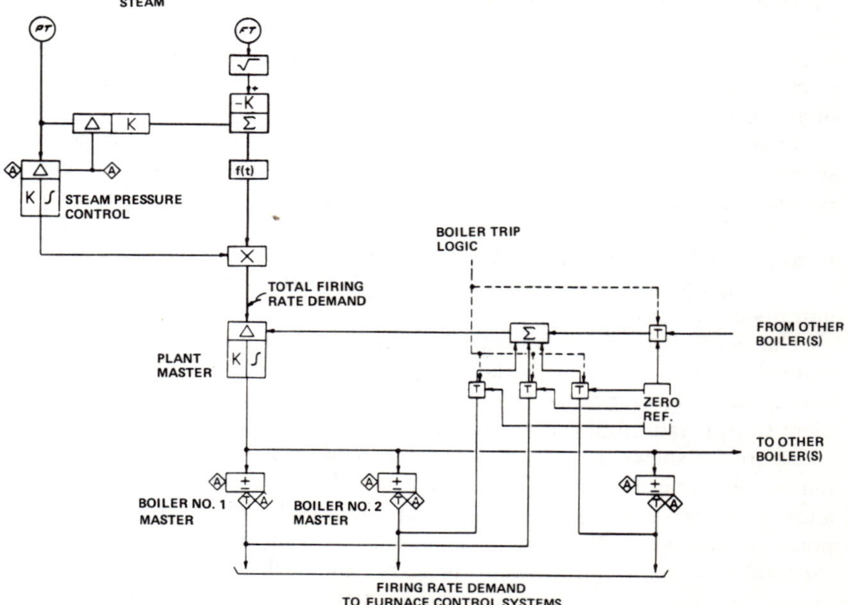

This system should be further compensated to correct the total effective firing rate demand in the event of a boiler trip. Trip logic, in this case, is set up to substitute a zero reference signal for the boiler firing rate demand signal when a trip-out of any boiler occurs.

Furnace Control System

In response to the firing rate demand signal established by the energy balance system, the furnace control system must perform the following functions:

1. Maintain energy input at the level demanded by the energy balance system.
2. Maintain fuel/air ratio. Fuel/air ratio must be held at the proper value to maintain boiler efficiency and to minimize the production of pollutants. Generally, the fuel/air ratio is not held constant over the load range and the control system must be "characterized" to fit the boiler. This is accomplished empirically by field testing on startup.
3. Maintain safe furnace conditions. During both steady-state and transients, as well as during startup and shutdown, furnace conditions must be controlled to prevent the possibility of accumulating excess fuel in the unit. Interlocks, burner management, and furnace control system design must be coordinated properly to ensure safety.
4. Maintain furnace pressure on "balanced draft" units.

Fuel/Airflow Control. Many installations, particularly smaller units or units firing an unmeasurable fuel, utilize what is called a *positioning* type of fuel/airflow control system. In this system, fuel and air are not measured. Their relationship is maintained by holding the position of the final operators in correspondence with each other. (See Figure 5-8.)

The firing rate demand signal directly positions the fuel valve. Every position of the fuel valve is assumed to represent a repeatable value or fuel flow, and a corresponding position of the airflow damper is established. Characterization must be

Figure 5-8. Basic parallel positioning furnace control.

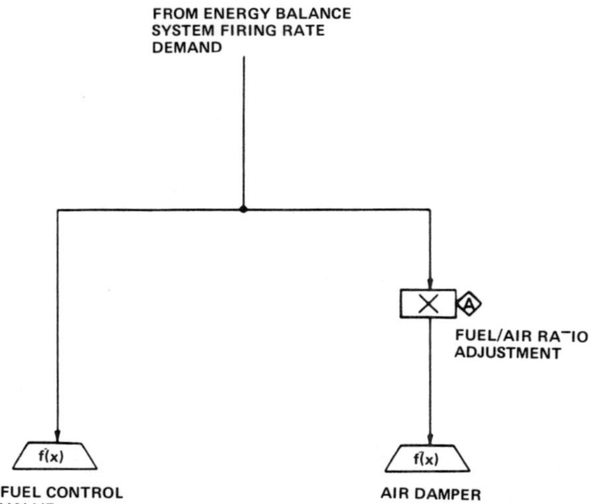

provided in the air damper drive (and sometimes also on the fuel valve) so that the two final operators can be "tracked" in proper correspondence. A limited trim is provided in the fuel/air ratio adjust station to allow some modification of the initially calibrated fuel/air ratio. This may be accomplished automatically but a manual station is used in most cases.

There are advantages and disadvantages to a positioning system. The advantages are:

1. The system is simple and reliable.
2. It is quick in responding. On a pneumatic system with long transmission distances, for instance, positioning systems have been successfully applied in some cases where metered flow loops are undesirably slow.
3. The rangeability of the system is high, limited only by the final operators.
4. This is an inexpensive system.

The disadvantages are:

1. Fuel/air ratio control is not precise since it is not based on direct measurement. It is dependent on other parameters and is affected by varying fuel characteristics, fuel temperature and/or pressure variations, atmospheric conditions, etc. If these conditions are not held constant, the fuel/air ratio will be affected significantly.
2. This system cannot be applied to multiple-burner installations. System flow resistance and, hence, fuel flow rate are functions of the number of burners in service.

To overcome the inherent disadvantages of positioning systems, so-called metering systems are generally applied to furnace control. Fuel and air are measured, and flow control loops are closed on these variables.

Figure 5-9. Series metering—fuel follows air.

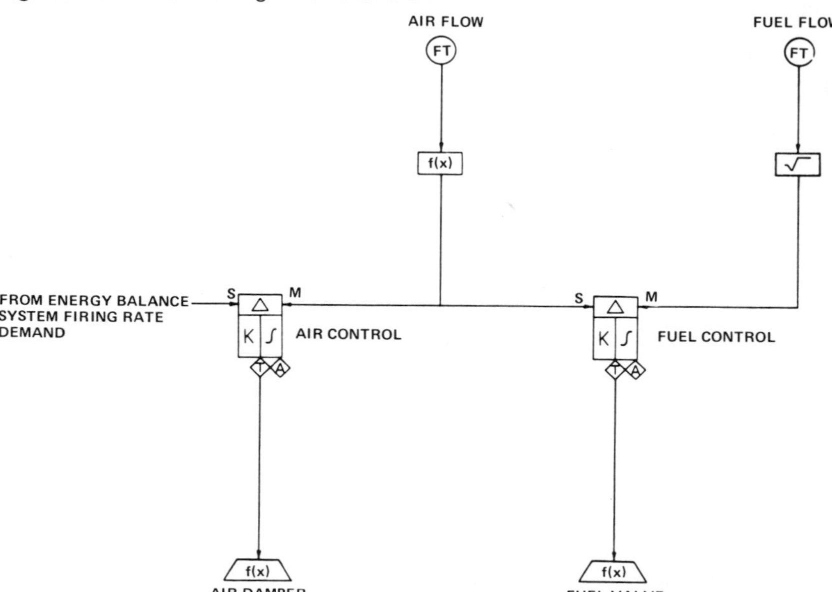

There are two basic approaches to metering systems: series metering and parallel metering.

Figure 5-9 illustrates a series metering system in which *fuel follows air*. In this system, the firing rate demand establishes the set point of the airflow controller, and the airflow measurement establishes the set point of the fuel flow controller. This configuration provides the following features:

1. Maintenance of fuel/air ratio on a measured (metered) basis.
2. Fuel demand that is never greater than the actual measured airflow—an important safety consideration.
3. Fuel follows air on load increase—there is no possibility of unburned fuel accumulation.
4. Since the airflow loop is usually somewhat slower in responding than the fuel flow loop, parallel tracking of the two occurs on load decrease.

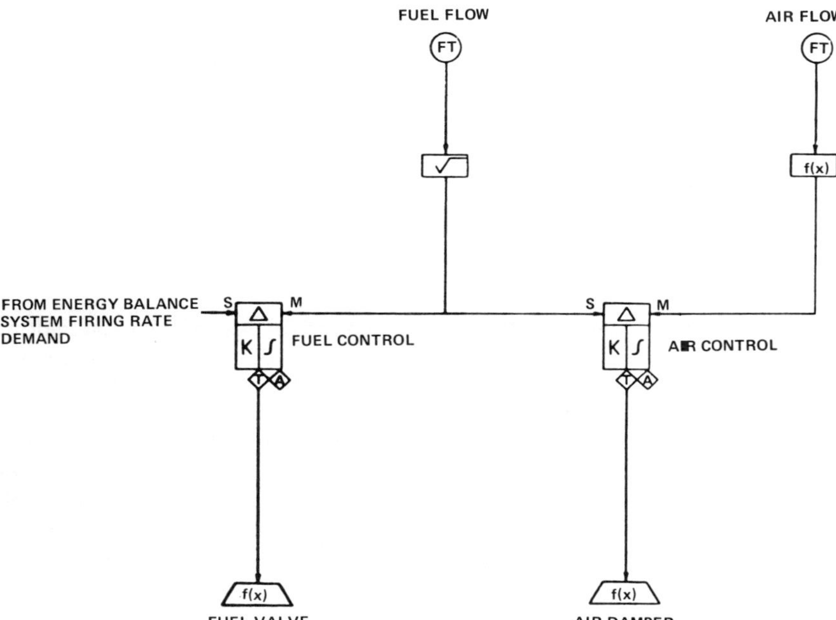

Figure 5-10. Series metering—air follows fuel.

Figure 5-10 shows another series metering system in which *air follows fuel*—structured identically the same as the *fuel follows air* system, except with the fuel and air loops reversed. This system has one advantage. The response of this system is optimum because of the faster response of the fuel loop.

From a safety standpoint, this configuration is undesirable. On load increase, fuel will lead air. Also, fuel demand can be maintained regardless of whether or not air is available for combustion.

The preferred configuration for furnace control is developed as follows. (See Figure 5-11.)

In this basic *parallel metering* system, the firing rate demand signal is applied in parallel as the set point to two slave flow control loops. One flow control loop monitors fuel; the other, air.

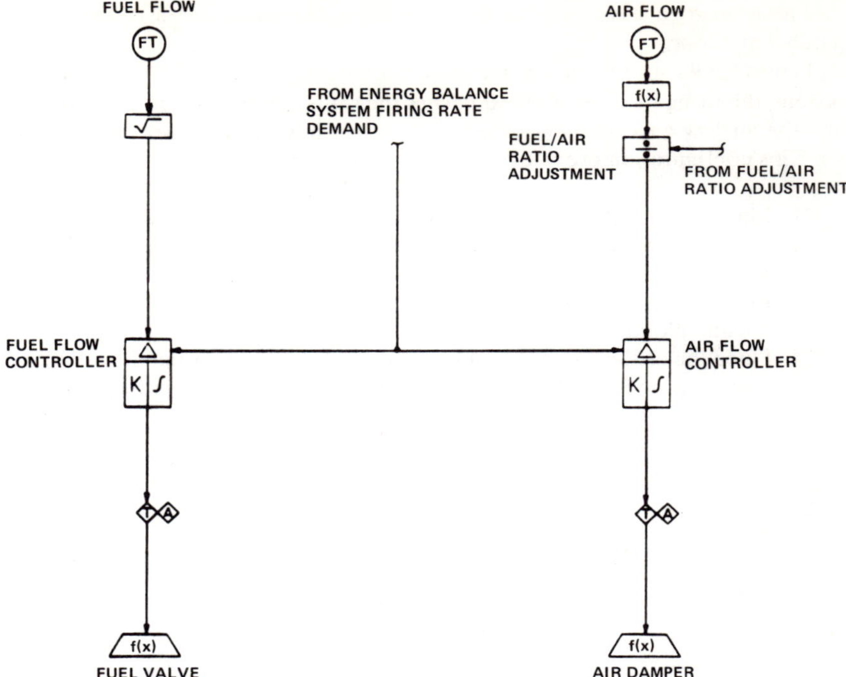

Figure 5-11. Basic parallel metering furnace control.

This system provides for stable, accurate control of energy input and fuel/air ratio during normal steady-state operation, but does not guarantee maintenance of furnace safety during transient conditions or during certain predictable operating irregularities.

A parallel system incorporating a fuel flow interlock is illustrated in Figure 5-12. The interlock is accomplished by inserting a low signal selector in the set point signal to the fuel flow controller. The second input to the selector is controlled airflow. Thus, the demand signal for fuel is either the firing rate demand signal or the measured airflow, whichever is lower. This system provides the following features:

1. Fuel demand can never exceed measured airflow.
2. On a load increase, fuel demand must follow air demand. Fuel cannot be increased until airflow increase is proven through the airflow measurement loop.

A parallel metering system with airflow interlock is shown in Figure 5-13. This interlock is accomplished by inserting a high signal selector in the set point signal line to the airflow controller. The second input to the selector is measured fuel flow. Thus, the demand signal for air is either the firing rate demand signal or the measured flow fuel, whichever is higher. This system provides the following:

1. Air demand can never be less than measured fuel flow.
2. On a load increase, fuel demand must follow fuel. Air cannot be decreased until fuel flow decrease is proven through the fuel flow measurement loop.

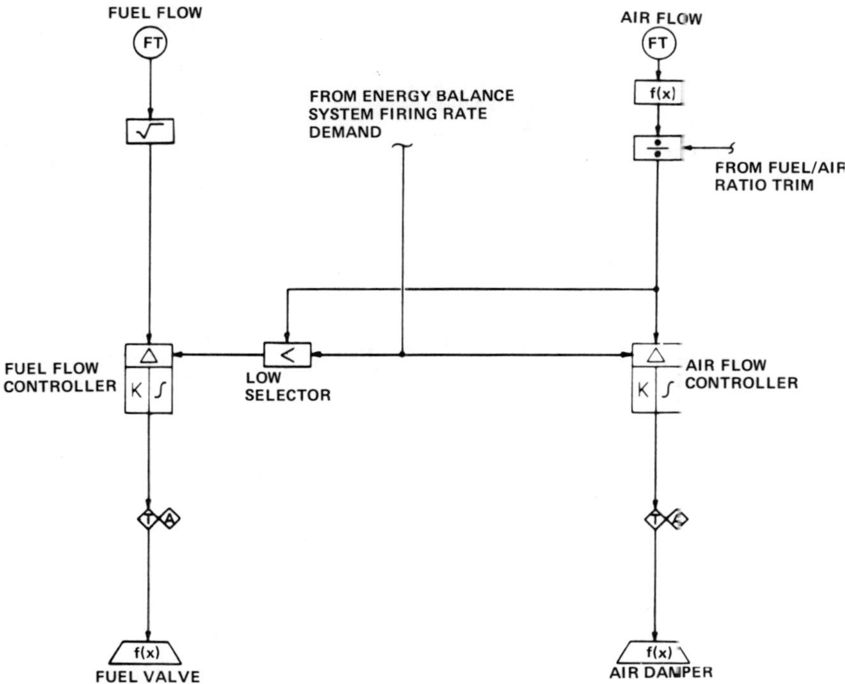

Figure 5-12. Basic metering furnace control with fuel flow interlock.

Figure 5-13. Basic parallel metering furnace control with airflow interlock.

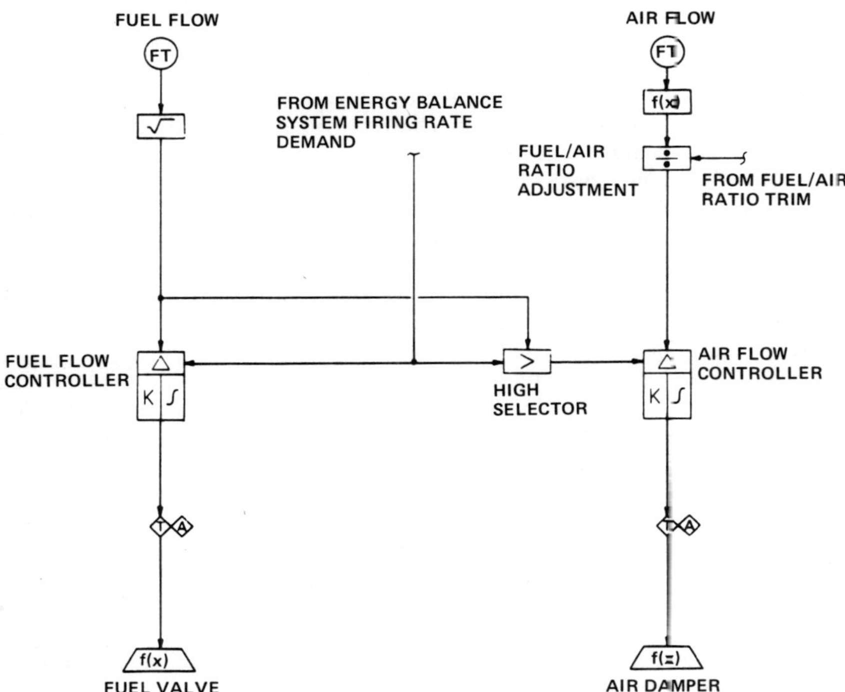

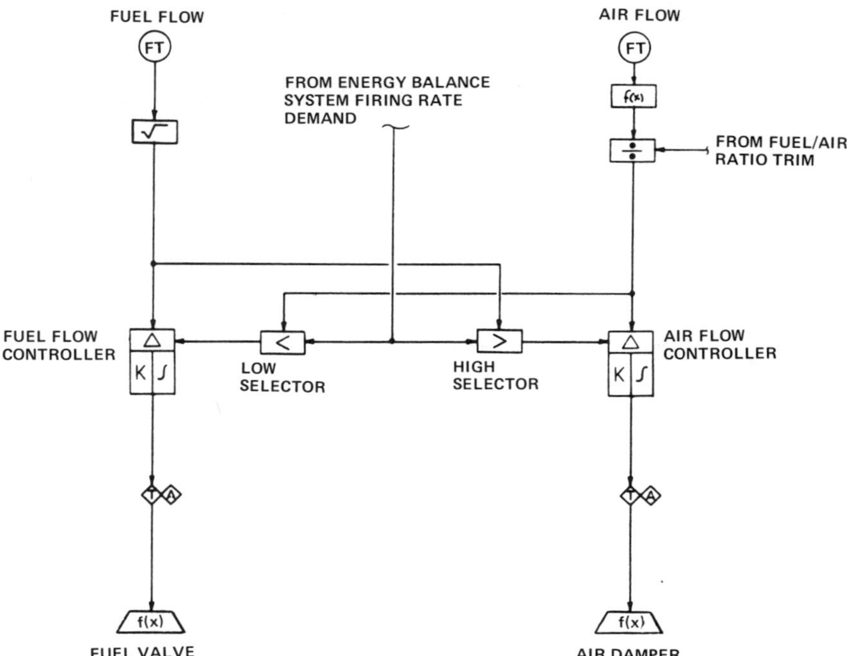

Figure 5-14. Basic parallel metering furnace control system—single measurable fuel flow interlock.

Figure 5-14 shows a parallel metering system incorporating both the fuel flow and airflow interlocks previously described. This system is known as either the *cross-limiting*, *flow-interlocked*, or *lead-lag parallel-series* metering system.

First, consider the steady-state condition of this system. The fuel and air controllers, which provide proportional-plus-integral action, act to continuously hold measurement equal to set point. Therefore, load demand, airflow set point, airflow measurement, fuel flow set point, and fuel flow measurement are all equal. The two flow loops are set in parallel.

In a load increase, the low selector will reject the load demand signal and accept the airflow measurement signal. Fuel flow demand (set point) becomes equal to airflow measurement. At the same time, the high selector rejects fuel flow measurement and accepts the increasing load demand signal. Airflow demand (set point) becomes equal to load demand. The system is acting as a series metering system with fuel following air.

Next, consider a load decrease. In this case, the low selector accepts the load demand signal and fuel demand (set point) becomes equal to load demand. Air demand becomes equal to fuel flow. Again, the system is acting as a series metering system, this time with air following fuel.

The parallel metering, flow interlock system provides the following features:

1. On a load increase, fuel demand cannot be increased until an actual increase in airflow measurement is sensed.
2. On a load decrease, air demand cannot be decreased until an actual decrease in fuel flow measurement is sensed.
3. On an inadvertent decrease in measured airflow, the fuel demand is immediately reduced an equal amount.

4. On an inadvertent increase in measured fuel flow, the air demand is immediately increased a like amount.

Multiple Fuel Firing. Where more than one fuel is to be fired in a unit, the most appropriate control system configuration depends on operating and economic factors. Whatever system configuration is used, however, any addition, subtraction, or calculations involving fuel flows must be done on a Btu basis.

Figure 5-15 represents the simplest and most economical approach. This system would be used where a second fuel is to be fired strictly on a manual basis. A manual loading station is used to set the position of the fuel valve for the base-loaded fuel (No. 2). The manually fired fuel flow is measured and totalized with the automatically fired fuel so that, as it is introduced, the total fuel flow controller will make an equivalent reduction in the flow rate of the automatically controlled fuel.

Since no provision is made for automatic cutback of the manually fired fuel, provision must be made external to the combustion control system to trip out the manual fuel on reaching some predetermined limit on the automatically fired fuel.

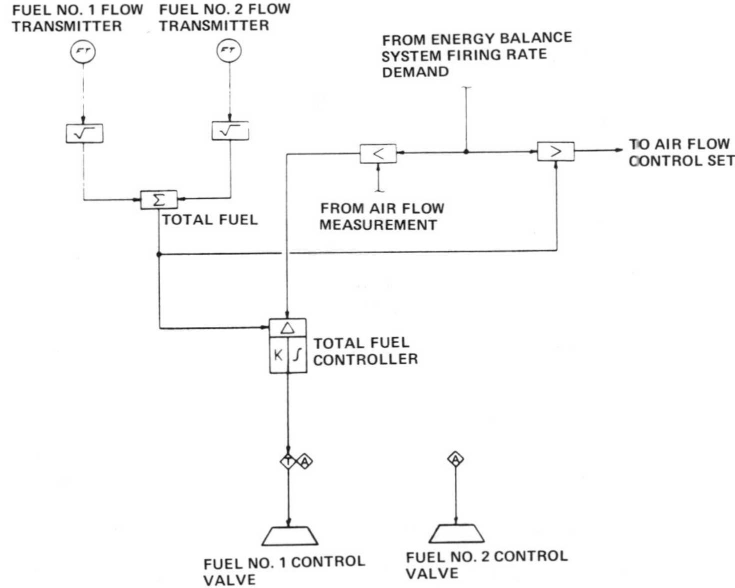

Figure 5-15. Dual-fuel firing—base-loaded fuels.

Figure 5-16 represents a system configuration generally used to fire two fuels optionally. The total fuel controller may manipulate either selected fuel valve with the alternate valve being base loaded manually. This system may also be used for firing both fuels automatically, but the required controller tuning adjustments for single-fuel versus dual-fuel firing may be considerably different for optimum response.

A system affording greater operating flexibility for dual-fuel firing is represented in Figure 5-17. In this system, the firing rate demand signal may be assigned all to one fuel or the other, or it may be apportioned between the two fuels in a relationship determined by the operator.

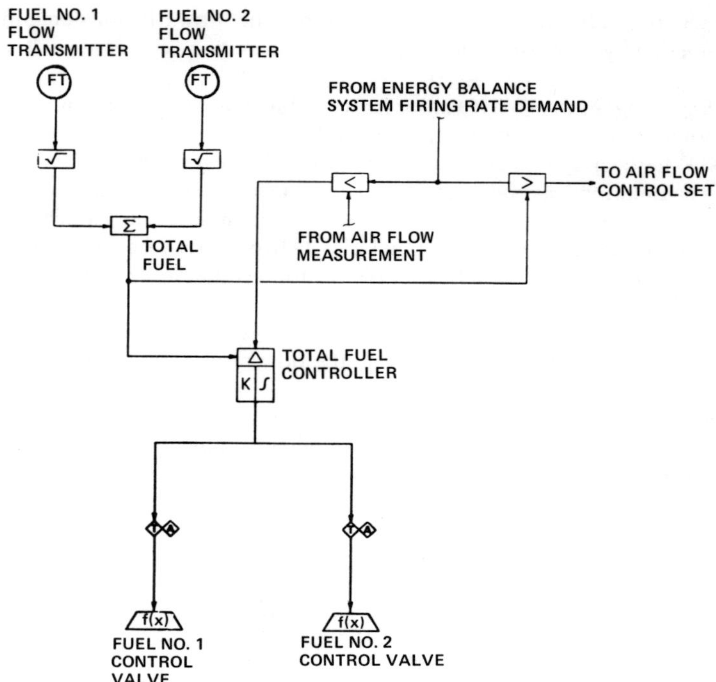

Figure 5-16. Dual-fuel firing—optional fuels.

Figure 5-17. Dual-fuel firing—ratioed fuels.

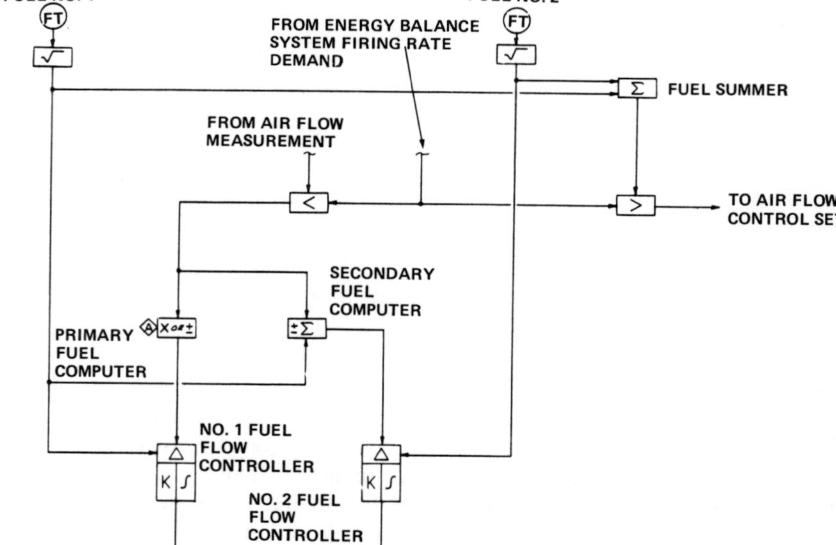

The firing rate determined from the low selector is first transmitted to the bias or ratio station for the primary fuel. A bias station here would provide for a constant firing rate of the No. 2 fuel (as set by the operator) at any load. A ratio station here would provide for a constant ratio of firing between No. 1 and No. 2 fuel at any load.

The No. 1 fuel measurement signal is subtracted from the firing rate demand by the secondary fuel computer. Any difference due to an operator-set negative bias or a ratio setting less than 1.0 appears as a demand for No. 2 fuel. Separate controllers are provided on each fuel loop to optimize response.

Assume that a ratio station is provided (the usual case). If the operator desires to fire all No. 1 fuel, he merely sets the ratio station at a ratio of 1.0. The firing rate demand signal is then transmitted unaltered through the primary fuel computer and the demand for No. 2 fuel, calculated by the secondary fuel computer, remains at zero.

If the operator desires to fire all No. 2 fuel, No. 1 fuel is shut down and the measurement is zero. The secondary fuel computer then transmits the firing rate demand signal unaltered since zero is subtracted from it.

If the operator desires to fire equal amounts of both fuels, he sets the primary fuel computer station at a ratio of 0.5.

The output from the station is then 50 percent of the input and the difference is calculated by the secondary fuel computer as the demand signal for No. 2 fuel.

Figure 5-18 represents a system for the firing of two fuels, one of which is to be fired preferentially up to some limit. This fuel system sequencing configuration is common where the prime or preferred fuel can be allowed to vary in response to load demand, but must not exceed a limit. An operator-set manual limit might be used, for instance, where a fuel purchase contract could require a premium price for demand exceeding a predetermined rate of consumption.

In other cases, fuel availability may vary arbitrarily and a cutback controller (such as header supply pressure) may be incorporated to limit prime fuel usage.

With this system, the firing rate demand is transmitted unaltered as the set point to the prime fuel controller until the availability limit is reached. As the firing rate demand further increases, the supplementary fuel computer calculates

Figure 5-18. Dual fuel—preferential firing.

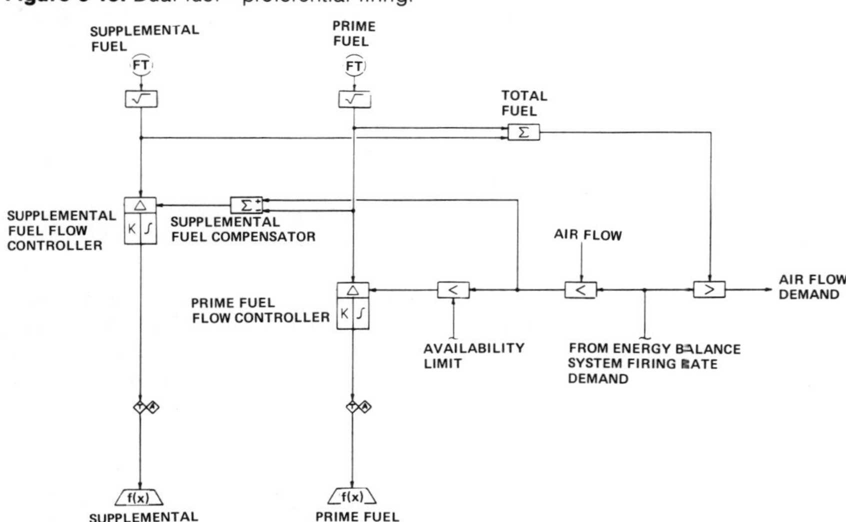

the difference between firing rate demand and prime fuel flow, and transmits this difference as the set point to the supplementary fuel controller.

Also with this system, 100 percent supplemental fuel may be fired by merely shutting down the prime fuel.

A system for dual-fuel firing which allows the operator to set fuel ratio and to set a limit on either fuel is shown in Figure 5-19. The firing rate demand signal is transmitted in parallel to the two fuel flow control loops via a multiplier with a low selector and limit override unit in each unit. The operator first selects the fuel ratio by adjusting the factors of the multipliers from the fuel ratio adjusting station. A 50 percent output provides a 0.5 factor to each multiplier, equally splitting the fuels.

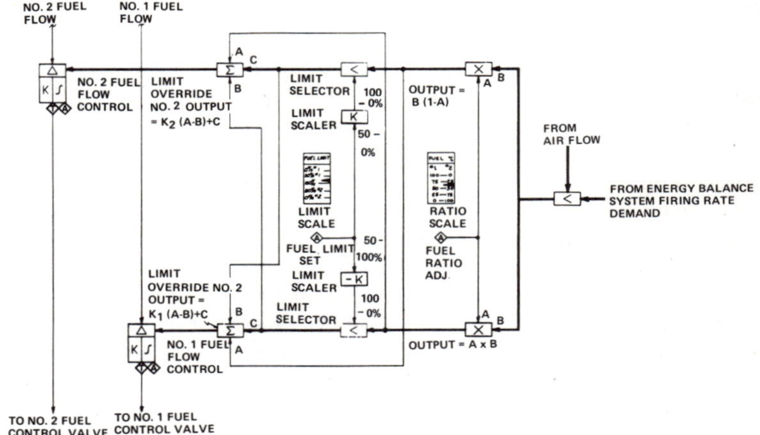

Figure 5-19. Dual-fuel firing with ratio and limit feature.

If the operator sets the fuel ratio adjust station at 100 percent output, the factor for fuel No. 1 is 1.0 and the factor for fuel No. 2 is zero; thus all No. 1 fuel will be fired. Conversely, if the operator sets the fuel ratio adjust station to zero output, the factor for No. 1 fuel will be zero and the factor for No. 2 fuel will be 1.0. Thus, all No. 2 fuel will be fired.

The operator sets a limit on either fuel via the fuel limit set station. If the output from this station is set at 50 percent, the outputs from both limit scalers are 100 percent and no limit is applied. If the output from the limit station is less than 50 percent, fuel No. 2 is high limited; if the output from the limit station is greater than 50 percent, fuel No. 1 is high limited. If either fuel is limited, an increase in firing rate demand above the limit must be satisfied by the nonlimited fuel. This is accomplished by the limit override unit in the following manner:

Assume that fuel No. 2 is limited and that the firing rate demand signal to the low selector exceeds the limit signal. Limit override No. 1 calculates the net difference between the firing rate demand and the limited demand from the limit selector and adds this net difference to the demand for No. 1 fuel. Thus, the actual total fuel demand is equal to the firing rate demand signal from the load control system.

A variety of configurations are available for multiple fuel firing. The systems described outline the philosophy to be followed in response to only one set of operating requirements.

Nonmeasurable and Variable Characteristic Fuels. Some fuels used by paper industry power boilers (such as pulverized coal) are either not measurable or have such widely varying characteristics that their flow measurements are unreliable indications of actual Btu release.

In a furnace control system, the important parameter is the heat release rate. Although the actual measurement is usually that of the fuel flow on a volumetric basis, the fact that the fuel heating value is relatively constant makes such a flow metering system applicable.

Where the fuel heating value is not constant, some form of compensation must be employed if a flow metering system is to be applied.

Where fuel flow is unmeasurable, a compensated inferential measurement will often yield satisfactory results. This is typically the case on pulverized coal applications.

Coal feeder speed, pulverizer primary airflow (air used for conveying the pulverized coal to the furnace), or pulverizer differential pressure are commonly used as inferential indications of fuel flow.

Except in unusual cases, these inferential measurements alone are generally not accurate enough to be used for fuel/air ratio control and, therefore, must be compensated. A common method of compensating is to use the boiler itself as a calorimeter and reason that the steam flow being generated bears a fixed relationship to the heat being released in the furnace. This is true and the concept can be utilized, subject to certain limitations:

First, this concept is an approximation. It will not yield precise control of the fuel/air ratio but, in many cases, will afford acceptable and reliable results.

Second, the approximation can be no better than the standard against which it is calibrated, in this case the boiler itself. Anything that affects the operation and performance of the boiler has a corresponding impact on the fuel/air ratio control calibration.

Third, the approximation is valid for steady-state conditions and cannot be expected to hold during transient conditions. Dynamic elements are included in the compensation system merely to minimize compensator action during transients.

Figure 5-20 illustrates the fuel side of a parallel metering system firing pulverized coal. Here, feeder speeds or primary airflows are summed to indicate total

Figure 5-20. Fuel measurement correction.

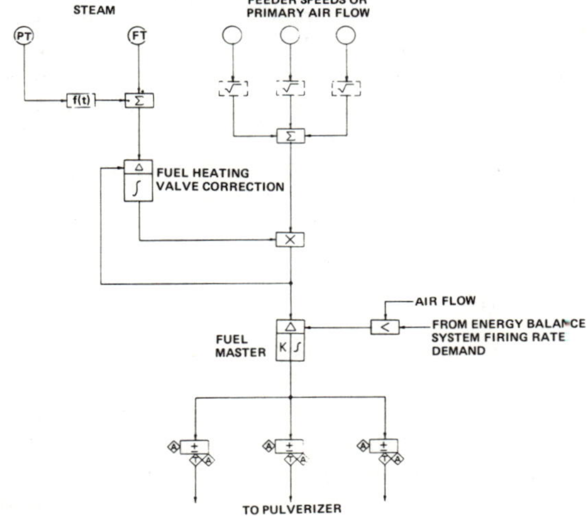

fuel flow. The fuel heating value correction loop operates as follows:

Boiler steam flow is summed with the rate of change of the steam pressure to obtain the heat release rate. (Compare this with Figure 5-5. Note that, on Figure 5-5, the rate of change of steam pressure was subtracted to obtain the firing rate demand.) This signal is now used as the set point to the controller for fuel heating value correction. The measurement signal to the controller is the corrected fuel flow measurement. An integral-only controller is used to minimize response to transients. In addition, the integration rate of this controller will be relatively long because its "process" includes the energy balance system for the boiler as a slave loop.

Fuel/Air Ratio Trim. Control of the fuel/air ratio by the parallel control of measured fuel and air loops will provide acceptable results in most cases, but will not yield the precise control sometimes required. This is particularly true if prime fuels are being fired or if a combination firing of fuels is being done.

The optimum solution to any control problem is obtained only when the variable under control is measured and monitored. In this case, measuring the flue gas oxygen content with an analyzing system is the answer.

Refer to Figure 5-14 again. Note that the divider in the airflow measurement signal affords "trimming" of the fuel/air ratio by an external signal. In a parallel metering, cross-interlocked system, fuel/air ratio trimming on the basis of oxygen analysis is introduced in this manner. (The correction is not placed in the set point signal line to the controller because this would make the fuel and air demand signals unequal and create a dead space in the action of the hi-lo selector safety interlock.)

Fuel/air ratio correction is done either manually or automatically. Manual correction is accomplished by providing the operator with a manual station which inputs to the divider. The operator makes correction on the basis of observed excess oxygen indication. A divider is used to preserve the operating sense of the system; for example, if the operator wishes to increase the airflow, he increases the output of the trimming station.

As was mentioned earlier, the fuel/air ratio (excess oxygen) is not held constant over the load range of the unit. Therefore, if automatic control is to be applied, the set point to the oxygen controller must be programmed either manually by the operator or automatically from the unit load or fuel flow. On measured fuel units, it is preferable to program from fuel flow because fuel flow and steam load may not correspond during transients such as a load ramp.

Oxygen analysis systems must be applied and maintained carefully. Probe locations, to obtain representative samples, are critical and sampling systems are complex because flue gas is hot, corrosive, and dirty. Frequent periodic maintenance is mandatory. For these reasons, it is sometimes preferable to apply manual trim based on operator observation of an analyzer recorder, rather than to apply automatic control. If automatic control is to be applied, limits to its action must be incorporated to prevent large deviations in the fuel/air ratio in the event of analyzer system malfunction.

Furnace Pressure Control

Before describing the techniques used for controlling furnace pressure, it would be well to review the behavior of this variable.

Figure 5-21 shows in schematic form the relationship between two variables that are said to be *coupled.* If separate control loops are applied to the flow and chamber pressure as is shown, an interaction develops between the controllers. If the flow controller, for example, makes a change in the position of the inlet valve to correct flow, a consequent effect is immediately felt in the chamber pressure. Similarly, if the pressure controller makes a change in the position of the outlet valve to correct pressure, a consequent effect is immediately felt in the system flow. The two variables have nearly identical time constants and are not independent. If the two controllers are tuned independently to match the response times of their respective variables, then one or the other may be placed in automatic. If both are placed in automatic, then the slightest disturbance will initiate a cycling between the two controllers which will persist. To regain stability, one controller will have to be *detuned* so that its response time is much slower than the other controller. Although the system is now stable, its recovery from an upset will be undesirably long. Thus, there has been a trade off of response for stability.

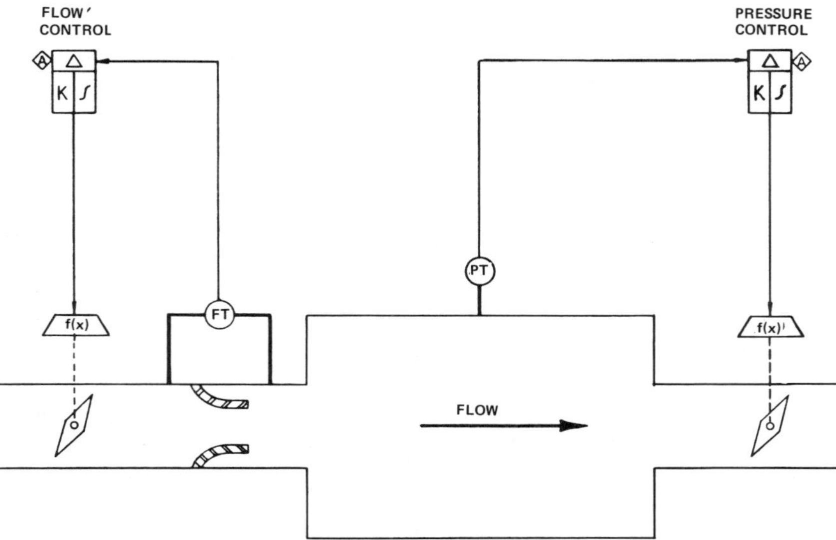

Figure 5-21. Furnace pressure control analogy.

Furnace pressure and boiler airflow behave exactly as described above. Many units are being controlled with single-element furnace pressure control systems, but in all cases response to upsets has been traded off for stability on either the airflow or furnace pressure loops, usually the latter. This is satisfactory provided load changes are infrequent and/or slow. To retain stability and response of both airflow and furnace pressure these two variables must be *decoupled.* The simplest way to accomplish this is shown in Figure 5-22.

In this case, the two dampers are operated in parallel by one controller, the flow controller. Now, when the flow controller makes a correction to flow, it also changes the position of the outlet (pressure) control damper in the direction of correcting chamber pressure. If the position versus signal relationship of both dampers is properly characterized to the process and to each other, little if any correction is required by the pressure controller. The pressure controller can now be detuned to provide the necessary stability since it merely serves as a trim on the outlet damper position.

There are other ways of decoupling dependent variables, such as controlling a third variable in relation to one of the dependent variables.

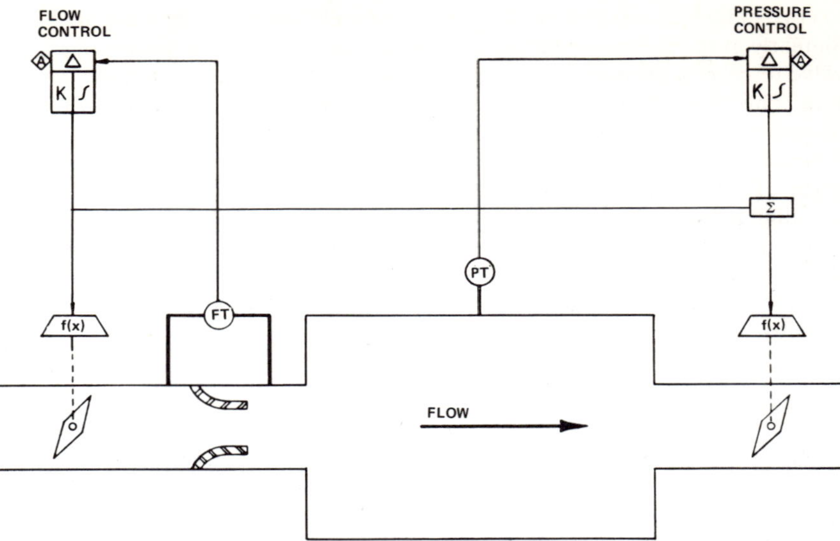

Figure 5-22. Decoupling dependent variables.

Complete Furnace Control Subloop

A typical complete furnace control subsystem is shown in Figure 5-23. This system controls both the furnace pressure and the boiler firing rate from a firing rate demand signal developed by any of the methods which have previously been described.

This system operates on a feedforward principle by transmitting the firing rate demand signal in parallel as the programmed set points to the induced-draft (I.D.) fan inlet pressure controller and also to the controllers for both the fuel flow and the airflow.

I.D. fan inlet pressure programmer CC-27 establishes the proper relationship between firing rate demand and I.D. fan inlet pressure to stabilize furnace pressure. Furnace pressure controller CC-13 trims this relationship through furnace pressure trimming unit CC-26 to maintain the furnace pressure at the manually set value for any load condition. The I.D. fan inlet pressure program compensates the system for single-fan versus dual-fan operation and also acts to decouple furnace pressure and airflow loops as previously described.

The fuel flow and airflow control loops are interlocked through hi-lo selectors CC-36 and CC-35 to ensure against a fuel-rich mixture development in the furnace. This flow interlocked system was previously described. Other features of this system include:

1. Automatic programming and control of excess oxygen in response to fuel flow (load) with appropriate limiting functions (CC-25, CC-31, and CC-32).
2. Individual excess oxygen set point programs for gas or oil firing.
3. Minimum boiler airflow limiting (CC-38).
4. Sequencing of induced-draft fan speed or outlet damper and inlet vane control (CC-39).
5. Individual automatic-manual and bias stations for all final operators where required.
6. Fan trip interlocks on all vanes.

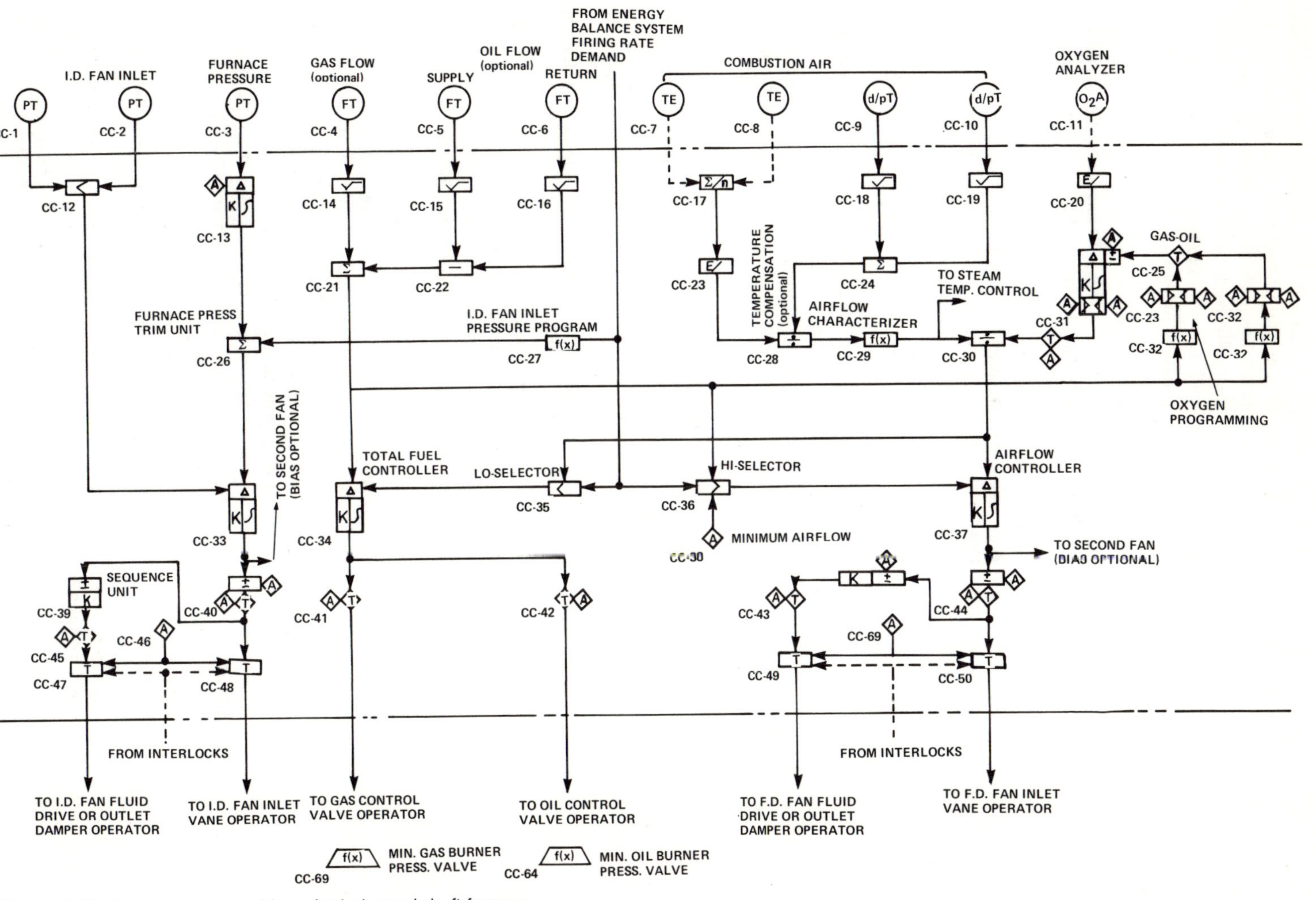

Figure 5-23. Furnace control subloop for balanced draft furnace.

Most of the applications encountered are single-element, proportional-plus-integral pressure controllers with a single- or two-fuel measurable fuel system. This simple configuration will do an adequate job in all but extreme cases, so there is no justification for unnecessarily complicating the system.

Burner Management and Safety Interlock Systems. There are important interrelationships between the burner management and safety interlock systems, and the combustion control system.

Before any boiler or furnace is lit off, it must be purged at a prespecified airflow rate for a prespecified period of time. This is to ensure that no combustibles have collected in the firebox due to fuel leakage, creating a possible explosion hazard.

After completion of the purge, certain prespecified conditions must be met before ignition is allowed. Among these conditions are requirements that: fuel valves and air dampers be in the proper *lite-off* position; there is water in the drum; and fuel pressure is at the correct value. These functions are provided by a *purge interlock* system which may require that solenoid valves and limit switches be included in the combustion control system for interfacing with the interlock system.

Usually, a *burner management* system is included. This system monitors for flame presence (by photoelectric or other sensors) and controls the startup and shutdown sequence of burners as well as tripping the unit in the event of a flame-out or other potentially dangerous condition. Usually required in the combustion control systems are:

1. Low fire start switches on fuel valve(s).
2. Limit switches on forced-draft fan and auxiliary register drives.
3. Solenoid valve(s) for lite-off-purge position.
4. Foolproof purge air switch.

There are three common types of safety systems used: manual, semiautomatic, and automatic.

Manual-Nonrecycle. The operator must manually position the forced-draft drive and fuel valves for purge and lite-off, manually opening the fuel shutoff cocks. Any limit trip requires complete manual restart. The combustion control system for the manual type includes only low fire start and purge damper position limit switches.

Semiautomatic—Nonrecycle. The operator initiates the purge program with a button. After the purge, he initiates the lite-off program and all valves and dampers are programmed through until the main flame is proven. On any limit trip, operator-initiation is required for restart. The combustion control system includes solenoids and airset loaders to tie into the flame safety system and program valves and drives. Limit switches are required.

Automatic—Recycle. The operator initiates the original cold start program for purge. After this initiation the sequence is automatic and all valves and drives are programmed to *main flame proving* and the control system is released to *demand firing*. This sequence requires the combustion control system to hold controllers in a steady state and to release on automatic. Also, when the load is reduced the system positions to *minimum fire* and remains there until steam pressure increases to *high-pressure trip*, then the unit is shut down. However, when pressure falls to a preset value the cycle will repeat, starting with purge, until back on automatic. All other interlock trip points require the operator to initiate for restart. The combustion control system must provide limit switches, external reset controllers, ramping functions, solenoids, and other safety devices.

General Comments

An almost infinite variety of other combustion control systems can be structured. Economic constraints, operating philosophy, safety codes, design of boiler and auxiliaries, type of fuels fired, behavior of load demand, and a variety of other factors influence the choice of hardware and system structure.

Bark and Hog Fuel Boilers

A common application found in pulp and paper mills are units burning bark and/or other wood waste, an abundant by-product fuel in this industry.

In some cases, an inferential measurement of wood waste "flow" is available. Some examples of this are: to measure the discharge pressure from the fan on a pneumatic conveying system and to appropriately scale and characterize this signal; to measure the speed and/or loading on a conveyor belt; or to measure the speeds of screws or other feeders. All of these schemes are, at best, rough approximations because of the type of wood, dirt content, moisture content, and other effects that upset the calibration. Nevertheless, these techniques have been successfully applied when the proper care is exercised. The most important requirement, and one that is not under the control of the instrumentation system designer, is the design and operation of the bark handling system. Bark is difficult to handle, and the storage, conveying, and distribution equipment involved is expensive. If an even, uninterrupted feed is not available, serious operating and control problems will result.

For these reasons, wood waste is seldom burned alone in a boiler; it is usually burned in combination with another prime fuel, generally gas or oil. If a reliable "bark flow" measurement is obtainable and the bark handling system operates smoothly, then a parallel metering, dual-fuel configuration, as previously described, can be used. The usual case, however, is not to make any attempt to measure bark flow. In this case, a combination metering-positioning system is employed.

Another approach is to treat the prime fuel as a completely separate firing process with measurements of fuel flow and a direct measurement of that portion of the air that actually is supplied to the prime fuel. This is not always possible because the configuration of the ductwork determines the airflow measurements that can be made.

The prime fuel is burned through burners located above a grate or stoker where the wood waste is burned. Economic considerations sometimes lead to a ductwork configuration that does not allow the best measurement of airflow.

Figures 5-24 and 5-25 illustrate two common methods of measuring airflow. The first method (Figure 5-24) obtains the prime fuel airflow measurement by subtracting the wood airflow from the total airflow because the duct to the prime fuel burners physically prohibits the installation of a measuring element. The problem with this arrangement is that when the wood burning is shut down and the damper to the stoker air is closed, leakage through this damper (there is never tight shutoff because some air leakage is required for grate cooling) causes an airflow differential just high enough to put this signal on the "hump" of the square root curve. Inaccuracies and possible instability of the prime fuel air measurement will result. Since the unit can be run on prime fuel only, the system operation deteriorates.

Figure 5-25 shows the preferred arrangement. Here, the prime fuel air measurement is completely independent, and if the unit is operated without wood waste firing a reliable airflow measurement is available. The fact that the total

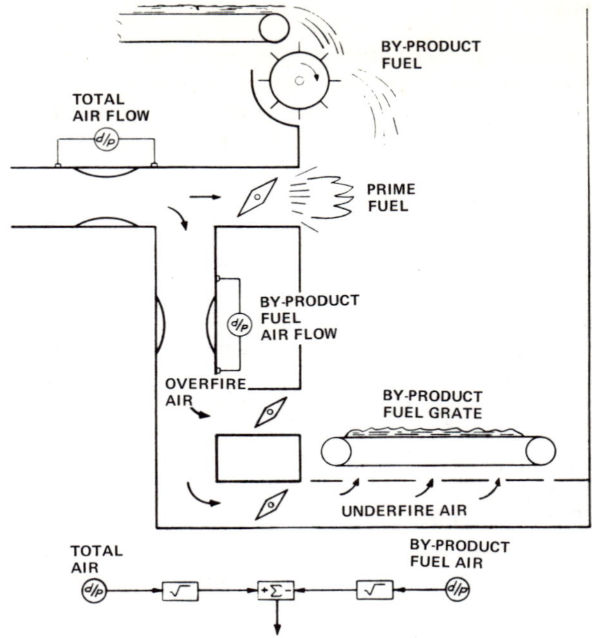

Figure 5-24. Airflow measurement—by-product fuel burning—method 1.

Figure 5-25. Airflow measurement—by-product fuel burning—method 2 (preferred).

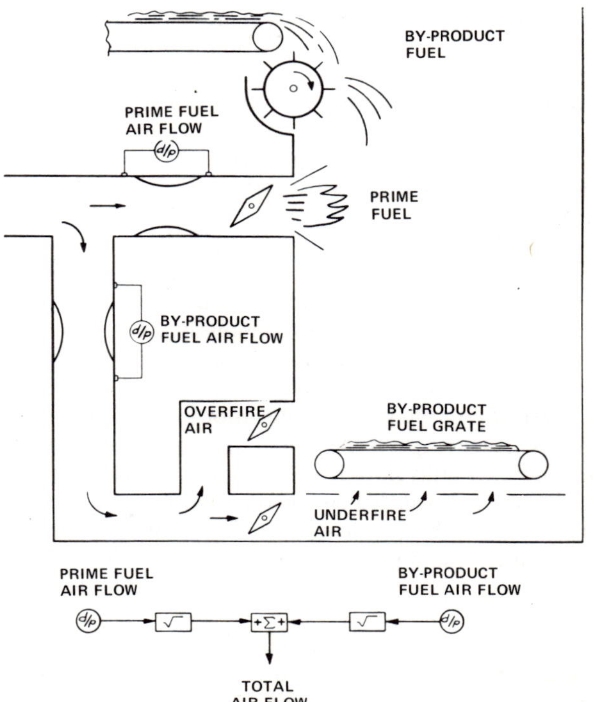

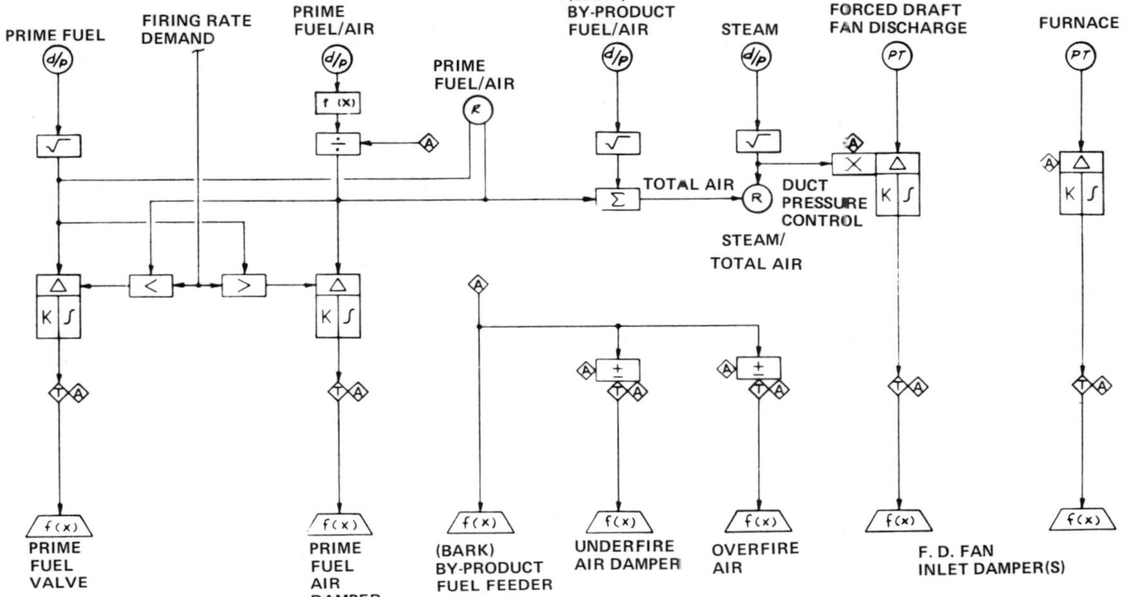

Figure 5-26. Furnace control (bark)—by-product fuel firing.

airflow measurement is deteriorated is of no consequence because this measurement is of no real value to the operator while firing prime fuel only.

A typical furnace control system for a wood waste firing unit is shown in Figure 5-26. Some modifications of this scheme may be found, but this general philosophy applies.

The following description of the system is based on the method of airflow measurement shown in Figure 5-25.

The prime fuel is treated as a completely separate firing system with comparison recording of prime fuel and prime fuel airflow. Wood waste feed is manually controlled with parallel control of underfire and overfire air through individual auto-manual-plus-bias stations. Steam flow is recorded and used as an inferential measurement of total fuel input. Total air is recorded with steam flow for operator guidance. The steam flow/total airflow relationship can be varied by the operator as a function of observed stoker conditions. Steam flow is also used to program the set point of the air duct pressure controller. A ratio station is provided on the set point for operator trimming of this relationship.

A single-element furnace pressure controller is usually applied on these installations. As mentioned earlier, this controller must be detuned somewhat to prevent interaction with the airflow and duct pressure controllers.

Recent cost increases and shortages of prime fuels have heightened interest in units burning wood waste only. If certain limitations are recognized and the proper precautions taken, satisfactory operation and control will result.

The first consideration is the unit's different response to steam pressure control when burning wood, prime fuel, or two fuels in combination. This is particularly important if the unit is the major contributor to steam header pressure control. Accordingly, the header pressure control system must be structured to account for the intended method of operation.

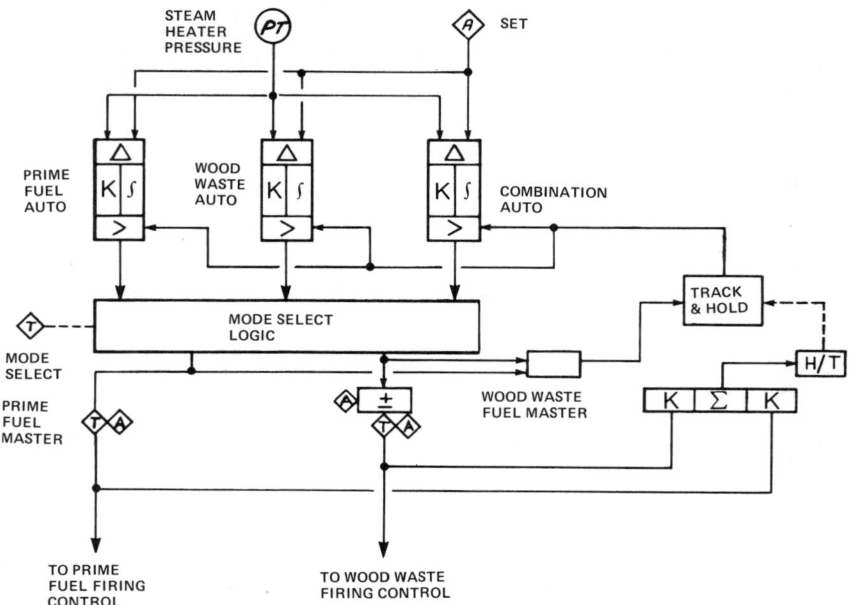

Figure 5-27. Steam pressure control by alternative modes.

Figure 5-27 depicts a system intended for a unit that is to be fired in any of four modes of steam pressure control. The alternatives are:

1. Prime fuel only. In this mode, the *prime fuel auto* pressure controller transmits a demand signal through the auto-manual station to the prime fuel firing control only. No demand signal is transmitted to the wood waste firing control.
2. Wood waste only. In this mode, the *wood waste auto* pressure controller transmits a demand signal through the auto-manual-plus-bias station to the wood waste firing control only. No demand signal is transmitted to the prime fuel firing control. The operator-set wood waste firing bias would be set to zero.
3. Base loading of one fuel. In this mode, either fuel may be base loaded from its master station, and the other fuel would be automatically controlled via one of the modes previously described.
4. Combination firing auto. In this mode, the *combination auto* pressure controller transmits a firing demand signal in parallel to both the prime fuel and wood waste fuel master stations. The proportions of fuels fired are established by the operator-set bias on the wood waste master station.

Features of this system are:

1. Each controller is tuned differently for the mode in which it is used.
2. Fuels can be fired in any mode, either automatically or base loaded, by operator selection. Steam pressure control is optimized by proper selection of the appropriate controller. Only one controller can be on *automatic* at one time.
3. Overfiring of the unit is prevented in any mode. The *track and hold* amplifier follows the output of whichever controller is in *automatic*. When the

sum of the firing rate demands to both prime fuel and wood reaches a predetermined value representing the maximum firing rate, the *track and hold* amplifier is tripped to *hold*. Any further increase in controller output is blocked by clamping its high limit. Control action in the direction of decreasing firing rate is not inhibited.

The system described is designed to provide the most versatile function, allowing any desired single or combination fuel firing with automaticallly compensating total fuel limit. Most applications do not require all of the described functions. Once the operating requirements are defined, usually appropriate simplifications can be made to the system.

If the unit is to be operated by firing wood waste alone with firing modulated in response to steam pressure demands, or with prime fuel plus wood waste modulation, certain limitations apply that must be recognized.

1. The same considerations apply as stated earlier for airflow measurement. Prime fuel air and wood waste air should be measured separately. A total airflow measurement is desirable and can be obtained by summing prime fuel air and wood waste air. Preferably, total airflow could be measured by pressure drop across the clean air side of the tubular air heater if the unit is so equipped.
2. Also, as stated earlier, the preferred system for firing any fuel that is to be continuously modulated is the parallel metering, cross-limited system described in the earlier part of this section. This configuration should be used, subject to the following precautions:

   Variations in the heating value of the wood waste can be compensated for by the circuit shown in Figure 5-20 if the wood waste moisture is controlled. If the wood waste moisture is not controlled, the action of this circuit would be the opposite of that desired on a change in moisture content.

   Accordingly, units designed to operate in this manner either have wood waste preparation systems that control the moisture content or are provided with manual correction for the fuel/air ratio based on operator observation of furnace conditions.
3. Single-element furnace pressure control systems are marginally acceptable. A feedforward influence to the induced-draft system is normally utilized. This signal is obtained by scaling and summing the fuel flow measurement signals.

Drum Level/Feedwater Control

The three main functions of a boiler drum are:

1. Storage of water feeding the evaporator section of the steam generator.
2. The last stage of feedwater heating where, by direct contact with the steam issuing from the evaporator section, the subcooled feedwater is heated to saturation.
3. Mechanical separation of vapor and liquid in order to deliver only dry, saturated steam to the superheater (or direct to the process).

Single-Element Drum Level Control. Single-element control is often used, particularly on small units. It will provide adequate control where load transients are not severe. Many older units are operating satisfactorily with so-called self-actuating proportional-only control. (See Figure 5-28.)

On single-element control, the measurement noise must be suppressed so that a very high gain (low proportional band) can be used in the controller. Filters with relatively long RC time constants (on the order of several seconds) can be applied very effectively since the time constant of the measurement itself is long, sometimes on the order of one-half minute or more.

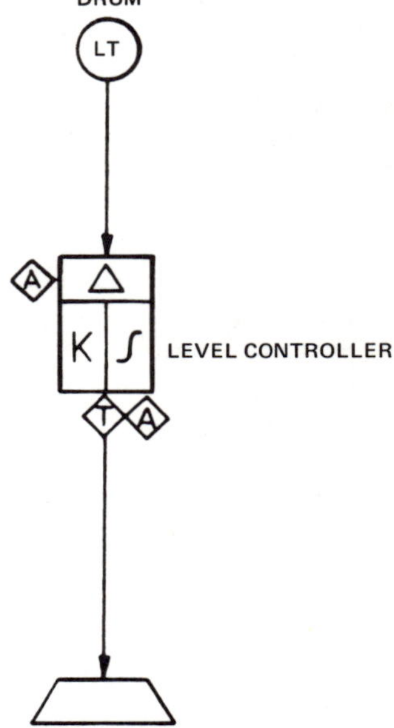

Figure 5-28. Single-element feedwater system.

Two-Element Drum Level Control. Where load changes on the unit are expected to be frequent and/or rapid, a feedforward element must be incorporated in either a two- or three-element configuration.

In the two-element configuration, steam flow and drum level are used. In the three-element configuration, steam flow, feedwater flow, and drum level are used.

From a performance standpoint, the choice between two- or three-element control is dependent on the final operator characteristic.

Figure 5-29 is a schematic representation of a typical two-element system. The steam flow signal is used as a feedforward influence to establish the demand signal to the feedwater regulation system. The drum level controller trims the feedforward demand to apply a vernier correction to maintain level at exactly the set point.

This configuration assumes that there is a linear and repeatable relationship between steam flow, feedwater regulation system demand signal, and hence, feedwater flow. If this relationship is true, then the addition of a third element (feedwater flow) is superfluous. (Note the similarity to the *positioning* combustion system.)

There are several types of feedwater regulation systems in common use. A feedwater regulating valve is often seen on smaller units. A method of regulating feed pump speed without the use of a valve is more common on larger units. Constant-speed motor-driven pumps with either hydraulic or magnetic variable-speed couplings are quite often used. Auxiliary steam turbine-driven feed pumps are also common, with speed regulation by control of the steam flow to the auxiliary turbine. These systems may or may not incorporate turbine governor speed control loops.

Systems utilizing pump speed control for feedwater regulation generally exhibit a very nonlinear relationship between feedwater flow and the speed demand signal. A two-element system would be inadequate in this case.

A two-element system is applicable where a feedwater regulating valve is utilized and its *installed* (not catalog) characteristic is reasonably linear. Since the available valve pressure drop usually decreases as boiler load increases, a linear (catalog characteristic) valve has a nonlinear installed characteristic. Often an equal percentage (catalog) valve shows a nearly linear installed characteristic.

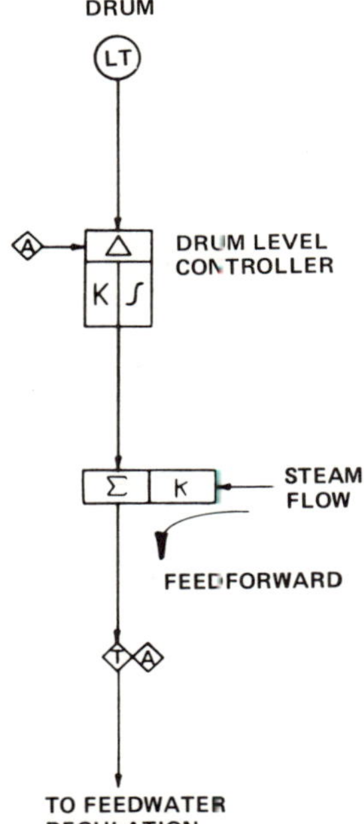

Figure 5-29. Two-element drum level control.

On some applications, characterized valves have been provided. This can be accomplished by the use of a characterizable valve positioner or by actual custom shaping of the valve plug to specified flow versus pressure drop data.

Next is the consideration of nonrepeatability of the position-versus-flow relationship of the final operator. The most common source of nonrepeatability problems occurs where multiple boilers are fed from a common header supplied by multiple pumps. If no header pressure control loop is applied, the header pressure is determined by the number of boilers and pumps in service and the plant's steam load. Taking a boiler or pump on or off the line will cause a drum level upset to all operating boilers.

To sum up, a two-element system will do a perfectly adequate control job provided proper consideration is given to the linearity and repeatability of the final operator as installed.

An alternative configuration for a two-element control system is shown in Figure 5-30. This system is functionally identical with the system shown in Figure 5-29.

The feedforward influence from steam flow is applied through a derivative unit installed in the signal line for drum level measurement. At steady state, the steam flow signal has no influence and the drum level is transmitted to the controller unmodified. When a change in steam flow occurs, a derivative component is subtracted from the drum level measurement. This apparent drum level error is immediately acted on by the proportional mode of the controller to move the valve to the new required position. The subsequent integration of the decaying derivative component over a period of time maintains the new valve position without requiring an actual deviation in drum level. Proper tuning of the proportional, integral, and peak height and decay time of the derivative unit to the boiler and to each other will essentially duplicate the function shown in Figure 5-29.

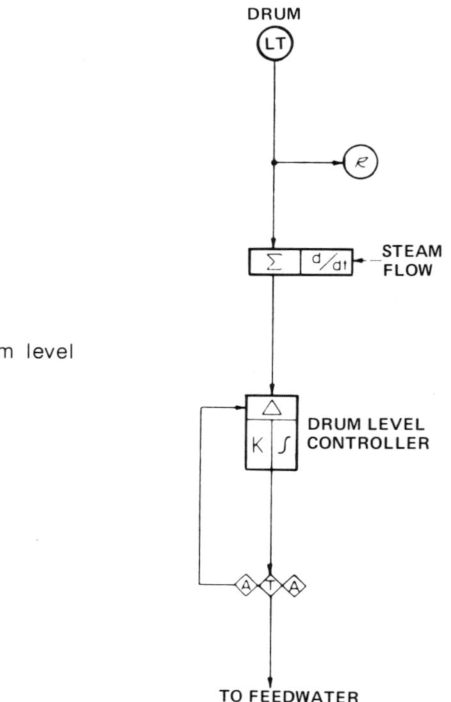

Figure 5-30. Two-element drum level control.

Three-Element Drum Level Control. The limitations inherent in the application of two-element drum level control systems are overcome by the three-element control system.

By adding the third element (feedwater flow) in a closed flow control loop, the usual problems of final operator nonrepeatability can be compensated for. Because of the relatively fast time constant of the feedwater flow measurement, a proportional-plus-integral controller can be applied to this loop.The proportional mode is tuned for the region of highest process gain to preserve stability, and the integral mode provides fast recovery in regions of low process gain.

There are several configurations for three-element control systems. The simplest configuration, the single-controller three-element system, is shown in Figure 5-31.

In this scheme, steam flow and feedwater flow are continuously compared and any difference between the two is added to the drum level signal. The resultant signal is transmitted to the controller. This system functions as a proportional level control loop. The gain of the flow control loop is equal to the product of the gain of the summing unit times the gain of the controller. The integral component is tuned to the time constant of the flow loop. The gain of the level control loop is equal to the gain of the controller. The summing unit gain is set very low; the controller gain is set very high.

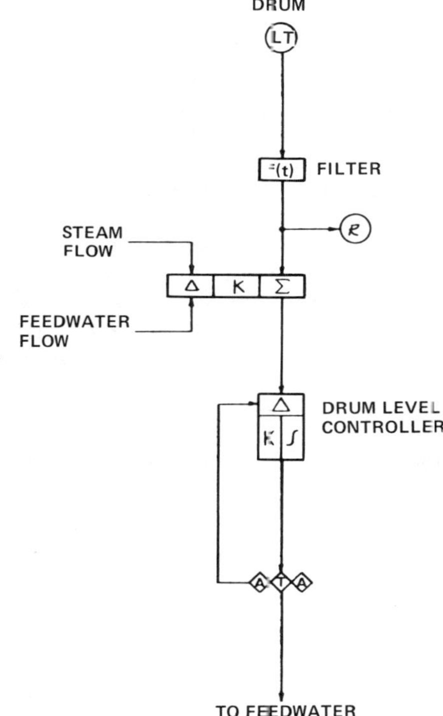

Figure 5-31. Single controller—three-element drum level control.

A change in steam flow appears as a deviation to the controller which acts to correct feedwater flow to eliminate the deviation. Control response is at a rate corresponding to the response time of the flow loop and is relatively fast.

A level deviation is acted on by the controller which creates a feedwater flow-steam flow deviation canceling the level deviation. This is in the proper direction to correct level. As the level responds at a rate significantly slower than the response of the flow loop, the controller gradually reduces the feedwater flow-steam flow deviation until the new balance point is reached.

Physically, it can be seen that for the drum level to remain constant, the feedwater flow (input) must equal the steam flow (output). Therefore, theoretically the net difference between these two signals is zero and the controller sees the unmodified level signal at steady state.

In actual practice, when the input and output of the drum are physically equal, the steam flow and the feedwater flow signals are seldom equal. There are factors causing this. They include:

1. Unmeasured flows in or out of the drum, such as blowdown or superheat spray flow.
2. Steam flow is a derived signal such as turbine stage pressure and, therefore, not accurate.
3. Temperatures and/or pressures at flow measuring elements differ from the design for different loads and are uncompensated.
4. Flow measuring equipment is out of calibration.

The effect of these factors is that the net difference in the flow signals is different for different loads and adds or subtracts from the drum level signal. Therefore, the balance point of the system may shift slightly with load. The operator must have an additional readout of the unmodified level signal and must trim the controller set point after a load change.

Figure 5-32 shows a three-element control system utilizing two controllers. In this system, a closed feedwater flow control loop is set by the feedforward index of the steam flow. A proportional-only controller on drum level trims the feedforward index as a level correction.

This system will provide very satisfactory control on units having relatively large drums (with respect to throughput), as many lower pressure units do. With proper noise suppression, as shown on the level measurement, very high gain can be used on the level controller. Because this high-gain proportional correction is applied as a trim to the feedforward index, proportional droop is almost entirely eliminated.

Figure 5-32. Dual controller, proportional level, three-element drum level control.

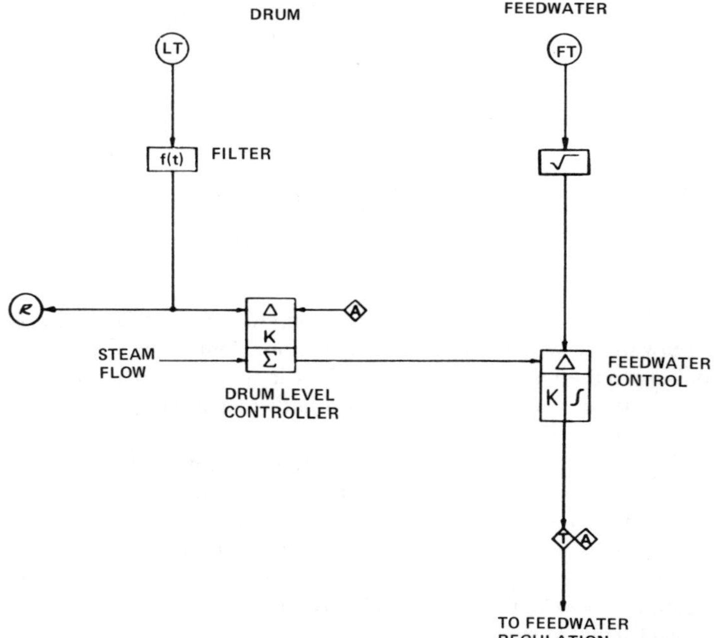

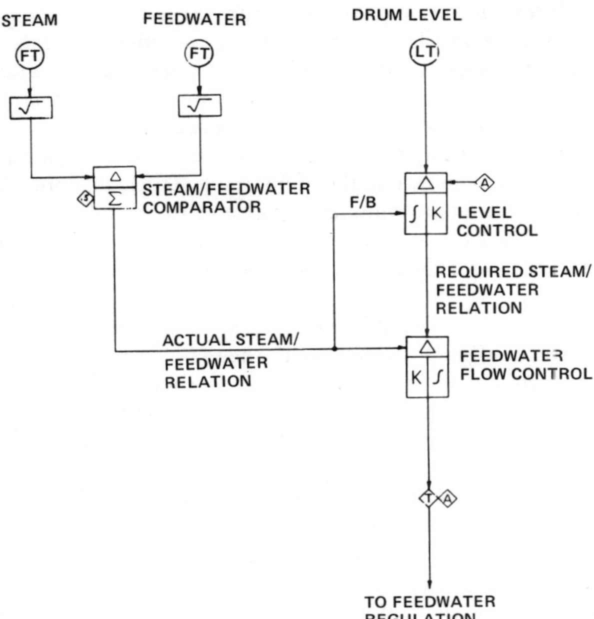

Figure 5-33. Dual-controller, three-element drum level control.

The features of this system are:

1. It is a relatively simple and inexpensive system.
2. It compensates for final operator nonlinearities that the two-element system cannot handle.
3. It provides two separate controllers, level and flow, and thus provides excellent dynamic response.
4. If steam and feedwater flow signals are not equal at steady state a small level offset will result. This offset will be proportional to the level controller gain.

The optimum configuration for the three-element drum level control is shown in Figure 5-33. This system continuously matches feedwater flow (supply) to steam flow (demand) to maintain the proper relationship of these variables. This relationship is trimmed by the drum level controller to maintain drum level at any desired manual set value over the load range of the unit.

Two separate proportional-plus-integral controllers are used. One, the feedwater flow controller, is matched to the relatively fast dynamic response of the feedwater flow loop. The second, the drum level controller, is matched to the slower dynamic response of the drum level loop. This provides for independent and more flexible tuning of the system for faster response and improved stability.

Steam flow is compared to feedwater flow in the steam-feedwater comparator. When steam flow and feedwater flow are equal the output of this comparator is at midscale.

Any change in steam flow is immediately sensed as a discrepancy in the actual steam-feedwater relationship by the feedwater controller. This controller, tuned to the response of the feedwater loop, will correct the feedwater flow to restore the steam-feedwater relationship.

The drum level controller receives a drum level signal as its measurement input and is manually set. The output of this controller, which is tuned to the dynamics of the level loop, sets the required steam-feedwater relationship as the demand to the feedwater controller. To maintain drum level, steam flow and feedwater flow must be essentially equal. When they are equal the output of the steam-feedwater relationship comparator is midscale and the output of the level controller will also remain essentially at midscale. The level controller affords a trimming action to maintain drum level. Since this controller has proportional-plus-integral action, drum level will be maintained exactly at the manually set value regardless of the steady-state relationship between the steam and feedwater flow signals.

The prevention of reset windup when in the manual mode is very simply accomplished in this configuration without the need for additional logic, antisaturation, or integral conditioning components. It is done by merely using the secondary measurement instead of the secondary set point for integral feedback to the primary controller. In the automatic mode this substitution has no effect because the ratio of the integral times of the controllers is large (greater than 10:1), and integral action of the secondary controller maintains its measurement equal to its set point.

In the manual mode, windup is prevented because the positive feedback loop (integral) is interrupted. The signal fed back to the primary integral circuit, instead of being the controller's output, is at a constant value determined by the steam-feedwater relationship established manually by the operator. Transfer to automatic closes both the level and flow loops simultaneously.

This system provides all the desirable features of a drum level control system:

1. It provides two separate controllers, level and flow, and thus provides optimum dynamic response.
2. It compensates for final operator nonlinearities.
3. At steady state, no level offset exists regardless of the relationship of the flow signals.
4. It provides the optimum control configuration without the complexity usually required to implement antiwindup functions.

Control Dynamics. Some comments are appropriate on dynamic disturbances to drum level measurement and control. Characteristically, an increase in load on a boiler is accompanied by a sudden increase in level, the opposite of what might be expected. This is commonly referred to as a *swell.* A decrease in load, conversely, is accompanied by a sudden decrease in level. Several variables have a significant effect on drum level which causes this *shrink* and *swell* phenomenon. The explanation lies in the fact that the evaporator section of the unit, which is part of the same closed fluid loop with the drum, contains a two-phase mixture of vapor and liquid, the percent composition—and therefore the average density—of which can vary. Anything that acts to change the density of this mixture will affect drum level. Some of the common effects are:

1. Effect on boiler load. At higher steaming rates, the percentage of steam in the evaporating mixture is higher; therefore the average density is less. If feedwater flow (input) is at all times held identically equal to the steam flow (output), the total mass of fluid stored in the boiler system will remain constant. Since the average density will decrease, the level will increase as the load increases. This is why a boiler must be underfired during a load increase.

2. Effect on heat release distribution. This is a function of the firing configuration and affects the distribution and percentage of steam in the evaporating mixture. Bringing burners or pulverizers in or out of service will affect fluid density, hence drum level.
3. Operation of soot blowers, burner tilts, manipulation of excess air or recirculation gas, changes in fuel characteristics, slag shedding, etc., all have an effect on the overall heat release distribution and, hence, on the drum level.
4. Unmeasured flows in or out of the drum system (soot blowing, steam blowdown, desuperheating water flow) will obviously affect drum level.
5. Boiler pressure affects saturation temperature and, therefore, density. An increase in pressure will cause water to flash. Any sudden change in boiler pressure will affect level.

The effects of all the variables listed above may contribute to steady-state and transient disturbances of the drum level. Any drum level system manipulates only one variable (feedwater flow) to control drum level. In evaluating the performance of a drum level control system, the effects of outside disturbances on the controlled variable must be properly understood. The system cannot correct for disturbances over which it has no control before they occur. All it can do is to correct a level offset after it happens.

Steam Temperature Control

On many smaller industrial boilers, steam temperature is not independently controlled. On boilers delivering saturated steam (no superheat), the steam temperature and pressure are not independent. Steam temperature will be constant, provided constant pressure is maintained.

Uncontrolled. Some boilers are equipped with uncontrolled or "wild" superheat. On these installations a relatively small superheater is installed to provide a slight degree of superheating. No attempt is made to control steam temperature; it is merely recorded. (See Figure 5-34 *A*.)

Figure 5-34. Steam temperature control modes, single-stage superheater.

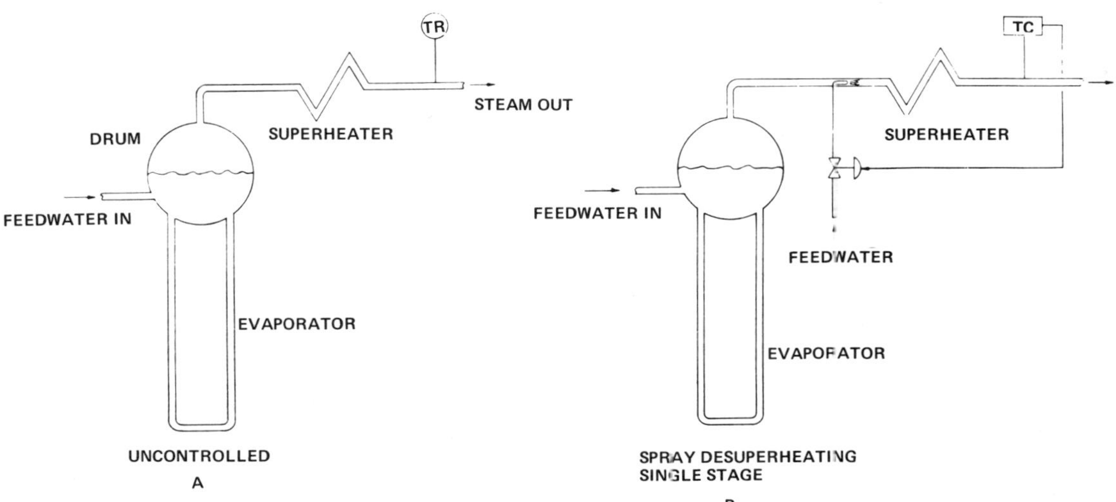

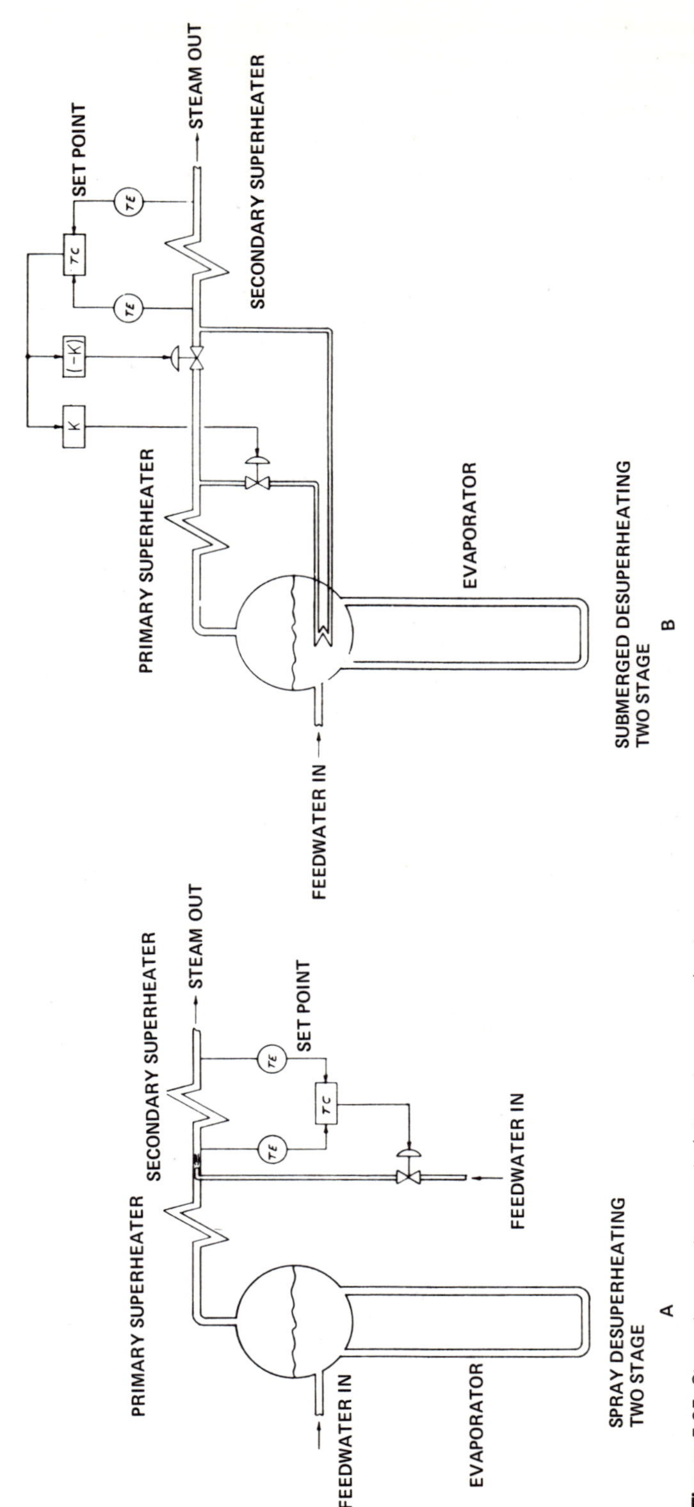

Figure 5-35. Steam temperature control, two-stage superheater.

Single-Element Control. There are two common methods of controlling superheat steam temperature:

1. A spray-type desuperheater introduces feedwater into the steam at the superheater entrance as shown in Figure 5-34 *B*, or between the primary and secondary stages on units having two superheat stages (see Figure 5-35).
2. Some units utilize a desuperheating heat exchanger submerged in the drum. This configuration is shown in Figure 5-35. Some of the steam leaving the primary superheater is diverted through the desuperheater heat exchanger and then reintroduced to the entrance of the secondary stage to control outlet temperature. Two sequenced control valves are utilized; one opens while the other closes. This type of process is very sluggish in response. Transport lags exist through the desuperheater heat exchanger and the secondary superheater. The time constants through both of these units are long. There is also a significant effect on transport lag from the flow rate (boiler load). At best, this process is very difficult to control; dead time may be several minutes. Single-element systems are often used on this application where economics prevail or where load changes are expected to be infrequent and/or slow.

Two-Element Control (see Figure 5-36). This is a conventional cascade system where inlet temperature to the final superheater stage is stabilized by a close slave loop controlling the desuperheater outlet temperature. The set point to this loop is established by the final or secondary superheater outlet temperature controller.

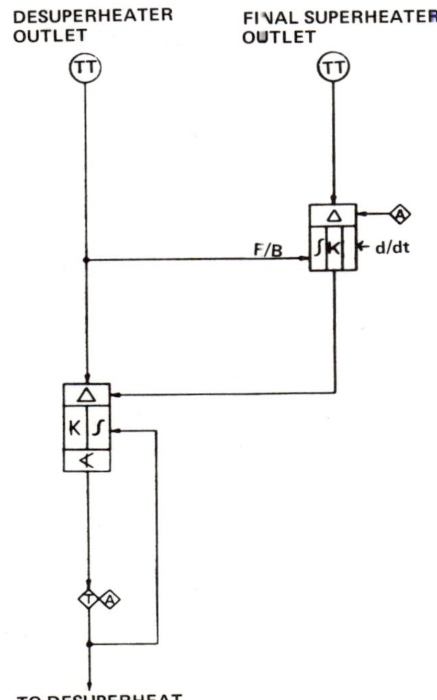

Figure Figure 5-36. Two-element steam temperature control.

At low boiler loads, it is generally impossible to achieve design outlet steam temperature because of limitations inherent in the superheater. This imposes a requirement for an antiwindup provision in the control system.

To accomplish this, the same integral feedback configuration is used as was described for the dual-controller, three-element drum level control system.

The temperature signal for the desuperheater outlet is used as feedback to the temperature controller for the final superheater outlet. When full load steam temperature cannot be met the superheater's control valve(s) will have reached limit(s), but the desuperheater's outlet temperature will fail to respond to the demand of the superheater's outlet temperature controller. Since the feedback to the latter controller is from the desuperheater's outlet temperature, integral action stops automatically and this controller will "wait" indefinitely without integral action until boiler load conditions are such that the designed steam temperatures can again be met.

Similarly, the low limiter and external feedback feature on the desuperheater's outlet temperature controller prevent this controller from integrating any further once the limit of valve travel is reached. Because of this antiwindup feature, this system may normally be left on automatic at all loads.

BLACK LIQUOR RECOVERY BOILERS

A type of boiler-furnace that is unique to the pulp and paper industry is the black liquor recovery furnace. The chemical recovery unit is used primarily to reclaim chemicals from the pulping process. Its secondary objective is the generation of steam to supplement the power boiler. The generation of steam from these units is a by-product since their primary purpose involves the chemical recovery cycle in the papermaking process. Therefore, unlike any other boiler application, steam flow or pressure is not controlled by manipulating fuel input. The black liquor flow to the furnace is a controlled constant and whatever by-product fuel is introduced generates a corresponding amount of steam.

For this reason, black liquor recovery boilers are always paired with power boilers on the same steam header. Steam header pressure is controlled by manipulating fuel input to the power boiler(s), and the recovery boiler(s) discharge an uncontrolled steam flow into the header.

The primary fuel of the recovery unit is black liquor from the process, which is a combination of lignin, spent chemicals, and water.

The strong black liquor must be concentrated by passing through multiple-effect evaporators and being further exposed to the hot stack gases in a direct-contact or cyclone-type evaporator.

There are major design and operating differences in the boilers and auxiliary liquor processing equipment used in the pulp and paper industry today.

Figures 5-37 through 5-41 illustrate the basic control schemes for the two most widely used black liquor boiler configurations as applied to the kraft pulping process. Although the control concepts are similar to those strategies used on the power boilers, the following general comments apply:

1. The magnetic flowmeter is the best device available for measuring liquor flow because, to make a measurement, it does not require any obstructions in the line that would have a tendency to build up with solids and distort the measurement.
2. Control schemes will vary from job to job as the major equipment varies.
3. The best feedwater control system is a three-element system.

4. The chemical recovery unit is essentially a base load device whose loading is established by the mill pulping rate. Steam is a by-product of chemical recovery. These units are never operated to follow changes in mill steam load and are usually piped into a common steam header with a conventional power boiler which follows the mill steam load swings.
5. Occasionally, the large units require a system with a simple superheater outlet steam temperature control.
6. The main airflows to be controlled are total airflow to the furnace, primary airflow, secondary airflow, and tertiary airflow when used. It is only necessary to measure any two of the three or four flows in order to design a control system. System design will vary slightly depending on which measurements are available.
7. Measurement of percent oxygen in the flue gas is always included for the purpose of monitoring the fuel/air ratio. This measurement is for recording purposes only, not automatic control. Percent combustibles may be measured for the same purpose.

Figure 5-37 shows one of the typical configurations used in an installation in which black liquor from the multiple-effect evaporator at about 45 to 50 percent solids is further evaporated by exposure to the hot gas stream from the recovery boiler in a cascade-type evaporator for further removal of water. The liquor serves as a wetting agent to remove dust from the furnace gas. Thick liquor feed passes through the cascade evaporators over rotating drums or discs where water is removed by contact with the hot flue gases. Liquor leaving the cascade evaporators at a concentration of 65 to 70 percent solids is combined with salt cake and sulfur to replace sodium salts and sulfur which are lost during the pulping and recovery cycle. After makeup, the fortified liquor is passed through the liquor heater which raises the temperature somewhere between 220° and 240°F. Liquor is introduced into the furnace through an oscillating spray nozzle which projects the liquor onto the furnace walls, where the remaining water content is removed. The dehydrated liquor falls to the smelt bed at the base of the furnace. An alternative firing system introduces the black liquor into the hot gas stream where it is dehydrated as it falls to the hearth. Air for combustion is supplied by a forced-draft fan and then passed through an air heater. It is then introduced into the furnace through primary air ports near the bottom of the furnace and through secondary air ports above the bed. Primary air is used in the combustion of carbon compounds in the black liquor, and it also maintains a reducing atmosphere for the reduction of the sodium sulfate to sodium sulfide in the smelt. The secondary air completes the combustion of volatile gases.

The hot gases, at about 1800° to 2000°F through the boiler tubes, enter the direct contact evaporator at 600° to 700°F. Approximately 95 to 98 percent of the gas-borne dust is removed by the cascade evaporator and precipitator.

The molten smelt, made up of soluble salts and some impurities, flows from the bed of the furnace through water-cooled smelt spouts into an agitated smelt dissolving tank. There it is mixed with weak green liquor from the recausticizing plant and pumped to the recausticizing plant as green liquor.

Instrumentation

Some of the more important instrumentation is shown in Figure 5-37.

Level controller LIC-1 maintains level in the back of the precipitator by regulating the recirculation flow of black liquor from the salt cake mixing tank. The level of the thick liquor storage tank is recorded on LR-2, and the black liquor

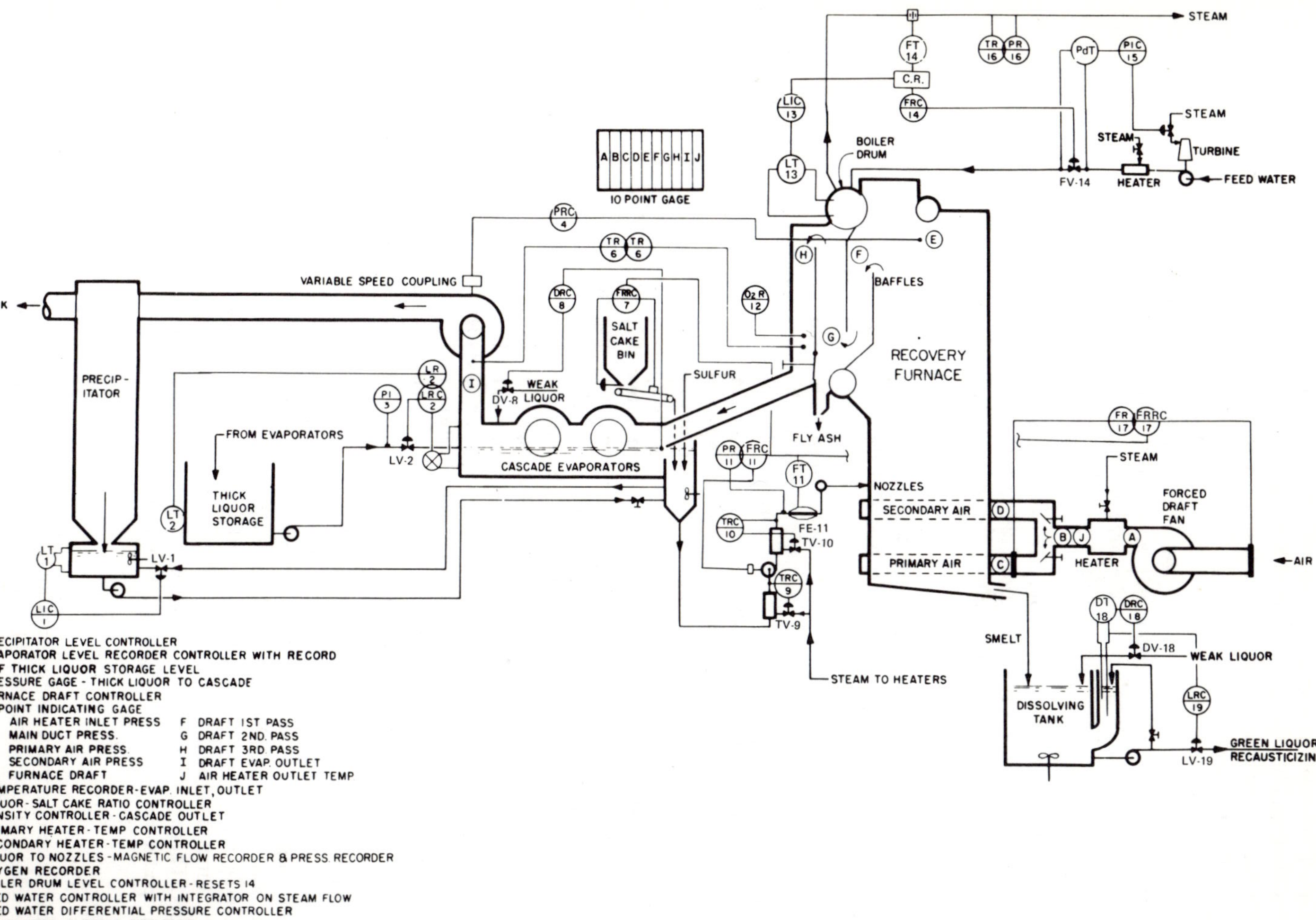

1 PRECIPITATOR LEVEL CONTROLLER
2 EVAPORATOR LEVEL RECORDER CONTROLLER WITH RECORD OF THICK LIQUOR STORAGE LEVEL
3 PRESSURE GAGE - THICK LIQUOR TO CASCADE
4 FURNACE DRAFT CONTROLLER
5 10 POINT INDICATING GAGE
 A AIR HEATER INLET PRESS
 B MAIN DUCT PRESS.
 C PRIMARY AIR PRESS.
 D SECONDARY AIR PRESS
 E FURNACE DRAFT
 F DRAFT 1ST PASS
 G DRAFT 2ND. PASS
 H DRAFT 3RD. PASS
 I DRAFT EVAP. OUTLET
 J AIR HEATER OUTLET TEMP
6 TEMPERATURE RECORDER - EVAP. INLET, OUTLET
7 LIQUOR - SALT CAKE RATIO CONTROLLER
8 DENSITY CONTROLLER - CASCADE OUTLET
9 PRIMARY HEATER - TEMP CONTROLLER
10 SECONDARY HEATER - TEMP CONTROLLER
11 LIQUOR TO NOZZLES - MAGNETIC FLOW RECORDER & PRESS. RECORDER
12 OXYGEN RECORDER
13 BOILER DRUM LEVEL CONTROLLER - RESETS 14
14 FEED WATER CONTROLLER WITH INTEGRATOR ON STEAM FLOW
15 FEED WATER DIFFERENTIAL PRESSURE CONTROLLER
16 STEAM TEMP. & PRESS. RECORDER
17 FORCED DRAFT & PRIMARY AIR FLOW RECORDER
18 DISSOLVING TANK DENSITY CONTROLLER
19 DISSOLVING TANK LEVEL CONTROLLER

Figure 5-37. Complete typical system and control schematic—configuration 1.

pumping pressure to the cascade evaporator is monitored by pressure indicator PI-3. Level controller LRC-2 maintains uniform pickup by the drums or discs of the evaporator, while density recording controller DRC-8 maintains a constant liquor density going to the burners by sensing the load that is on the drum drive motors.

Salt cake makeup to the liquor leaving the cascade is added by flow ratio recording controller FRRC-7 in accordance with the rate of flow of black liquor to the furnaces measured by an electromagnetic flow element, FE-11. Reduction of the salt cake in the furnace yields sodium sulfide which replaces the sulfide lost in the cooking and washing operations. The liquor is further heated in the primary and secondary heaters, whose temperatures are controlled by temperature recording controllers TRC-9 and TRC-10. The liquor flow rate is measured by an electromagnetic flow element, FE-11, and is controlled by FRC-11 which regulates a variable-speed pump. The liquor is then sprayed through burner nozzles into the furnace where it ignites and burns.

To ensure proper liquor-air mixture, the total airflow is ratioed to liquor flow and controlled by FRRC-17.

The boiler drum level measurement is made with a differential-type level transmitter, LT-13, and is fed to a level indicating controller, LIC-13, whose output biases the boiler steam flow signal from transmitter FT-14 through a combining relay. The output of the combining relay regulates the set point of the feedwater flow controller, FRC-14, to maintain feedwater flow based on boiler outlet steam flow and trimmed by the drum level controller.

Oxygen recorder O_2R-12 in the rear pass of the boiler records combustion efficiency so that proper fuel/air ratio settings can be maintained. Draft controller PRC-4 regulates the induced-draft fan speed to ensure proper negative pressure in the furnace. Various temperatures, pressures, and drafts are recorded or indicated, depending on the type of furnace. This gives the operator all the necessary information for complete furnace control.

The smelt passes from the furnace floor into a dissolving tank where it becomes green liquor. The tank level is maintained by level recording controller LRC-19. Density recording controller DRC-18 maintains constant density of the green liquor in order to ensure the uniform gravity necessary for efficient operation of the next process, liquor causticizing. One way of measuring density is by using bubble tubes in two water chambers continuously supplied with a small water purge to maintain a water leg in each chamber equivalent to the hydrostatic level of liquor. (See Figure 5-38.) This arrangement eliminates all contact of liquor with the bubble tubes and prevents crushing and plugging. This method also permits the full utilization of the instrument range of the density controller for a relatively small range of density.

Liquor level changes in the dissolving tank do not affect the density measurement, although the optional controller, LRC-19, may be desired for plant operation.

Referring to Figure 5-38, specifications for a differential pressure transmitter to cover a specified range of specific gravity are easily calculated. The following example will illustrate the procedure.

Given:

H = tap spacing
d_{min} = minimum specific gravity
d_{max} = maximum specific gravity
h = height of liquor in tank above bottom tap

Bubble-tube ends, always submerged, both at same height.

In example:

$$H = 25 \text{ inches}$$
$$d_{min} = 1.098\ (13°\text{Bé})$$
$$d_{max} = 1.208\ (25°\text{Bé})$$

Differential pressure at the instrument, in inches of water, equals tap spacing times the difference between the specific gravity of the liquor and that of the water which fills the standpipes, i.e.,

$$\text{d/p} = H \times (d - 1)$$

At 13° Bé:

$$\text{d/p} = 25 \times (1.098 - 1)$$
$$= 2.45 \text{ inches of water}$$

At 25° Bé:

$$\text{d/p} = 25 \times (1.208 - 1)$$
$$= 5.20 \text{ inches of water}$$

Span of instrument for 13° to 25° Bé

$$= 5.20 - 2.45$$
$$= 2.75 \text{ inches of water}$$

Elevation of instrument to read 0 at 13° Bé

$$= 2.45 \text{ inches of water}$$

In designing the system, certain requirements must be kept in mind.

The vertical distance between taps (shown as *H* in Figure 5-38) must be great

Figure 5-38. Dissolving tank density control.

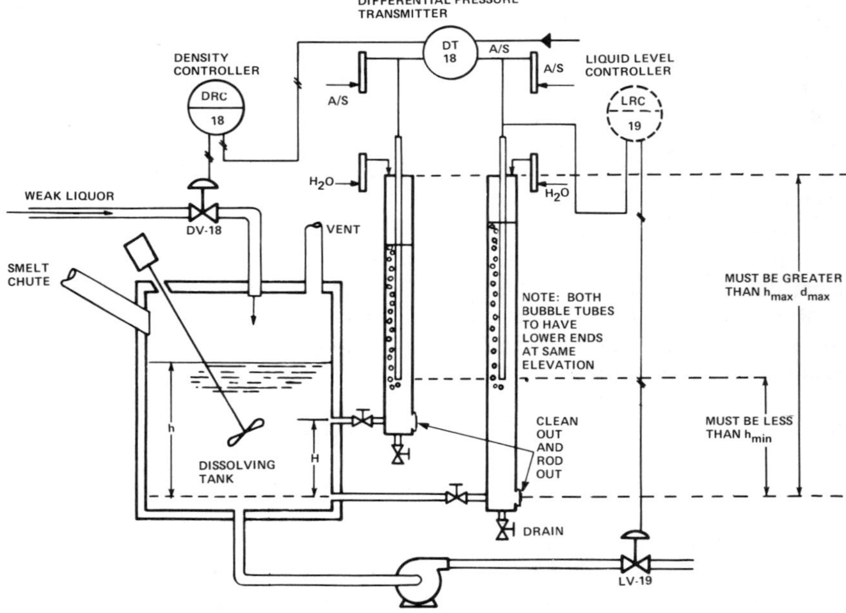

enough to produce a change in differential pressure within the range of the selected instrument.

The standpipes must be high enough not to overflow with the highest level and density of the liquor expected in the tank. Figure 5-38 states that the height of the standpipes must be greater than $h_{max} \times d_{max}$.

The bubble tubes must always be immersed in water. Figure 5-38 states that they must go below the minimum level expected in the tank, i.e., that their height above the bottom tap must be less than h_{min}.

Level of the liquor must never go below the upper tap. i.e., h_{min} must always be greater than H.

Dissolving tank density can also be made with a nuclear radiation gamma gauge.

Figure 5-39 is a schematic diagram showing the instruments and basic control schemes used in a typical combustion control system on this type of recovery boiler configuration.

In this case, the black liquor flow to the furnace is controlled. The black liquor flow set point and the total airflow controller set point are adjusted by the same manual station. Primary air control is ratioed to the total airflow controller/black liquor flow controller set point. Secondary air is manually controlled and no tertiary air is used in this type of recovery boiler configuration. Furnace pressure is being controlled by a single-element control loop and, as in the previous system, must be detuned to prevent interaction with the total flow controller.

Figure 5-40 shows the configuration, instruments, and basic control scheme of another typical configuration widely used on sulfate recovery boiler systems in the paper industry. The objectives of the installation are the same as, and the operations are similar to, the first system described.

In this arrangement, a cyclone-type evaporator is used to concentrate the liquor from the multiple-effect evaporators. When a cyclone-type evaporator is used, the level control of the cyclone base section regulates admission of 50 percent liquor to the venturi spray rings. Concentrated liquor is pumped through rotating screens. A portion of this liquor recirculates to provide cyclone wall wash. Recirculation through the second set of venturi sprays increases the discharge density of the black liquor. The concentrated liquor is pumped to the salt cake mix tank.

In this example, the black liquor flow to the furnace is controlled, but there are also installations of this type of recovery boiler where the black liquor flow to the furnace is recorded only and the total air is controlled. This method is shown in a basic control scheme used in a combustion control system of this type (Figure 5-41).

An electromagnetic flowmeter is used to measure the flow of liquor to the boiler. The black liquor to the burners is pressure controlled and the black liquor feed spray nozzles are designed to give a constant feed under constant pressure. Total air is controlled and primary airflow is ratio controlled to the total airflow. The secondary and tertiary airflows are manually set, with the tertiary being measured and recorded.

Furnace pressure is a single-element type which operates a damper to the stack through a variable coupling operator. This loop must be detuned to prevent interaction with the total flow controller.

General Comments

Some installations have a density controller which is not shown in the typical system illustrations. This density controller operates in sequence with a bypass

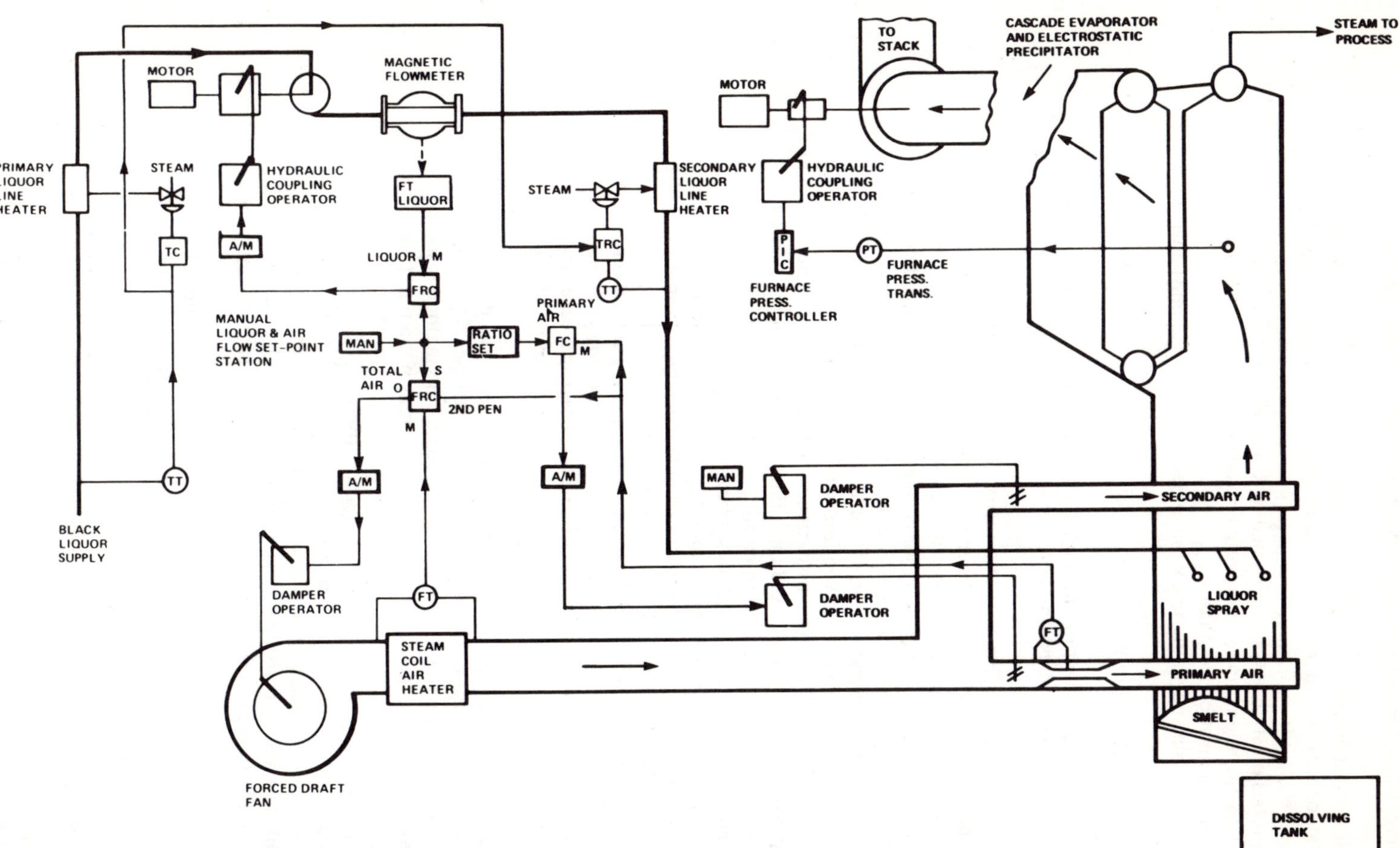

Figure 5-39. Black liquor recovery boiler combustion control schematic—configuration 1.

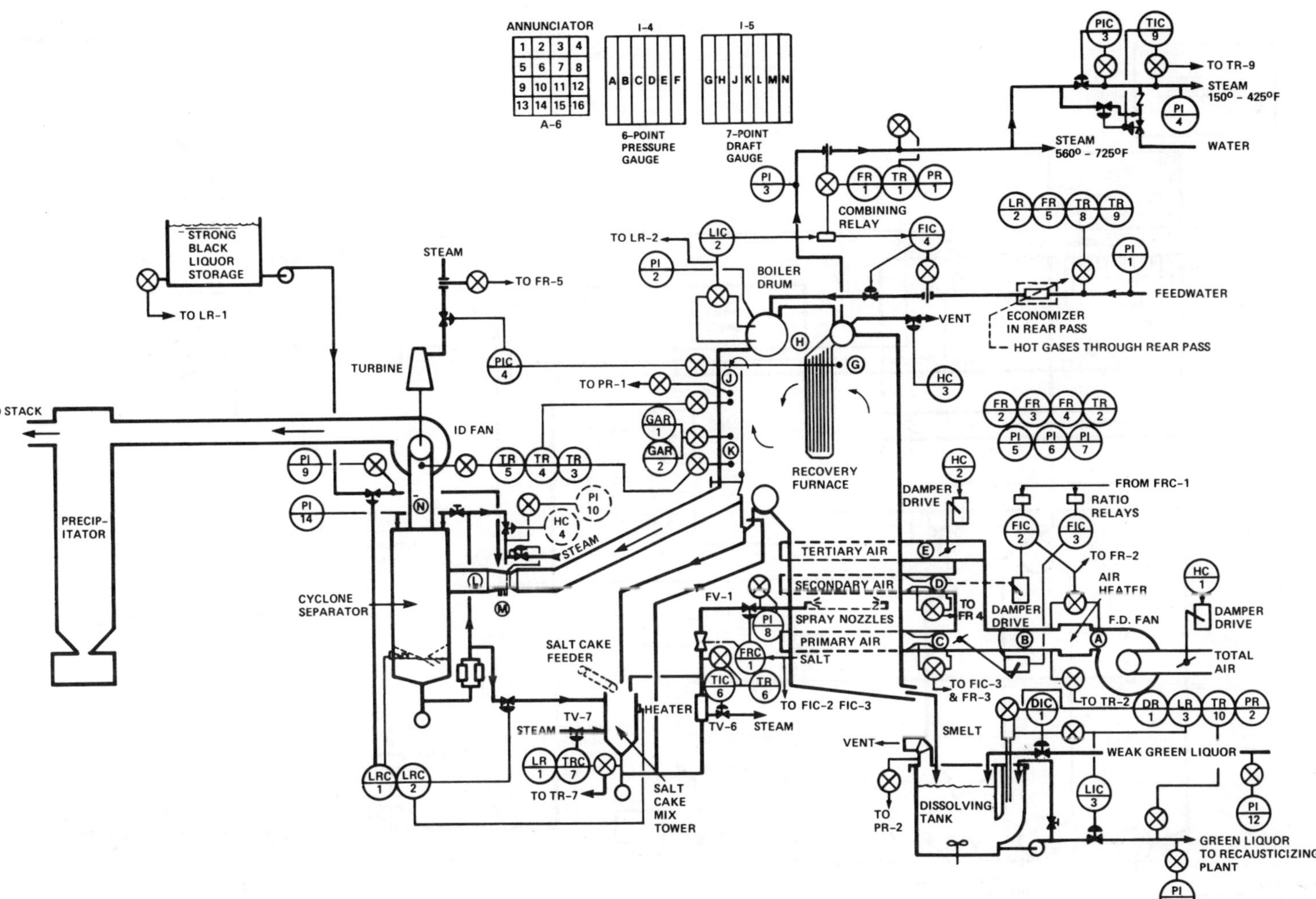

Figure 5-40. Complete typical system and control schematic—configuration 2.

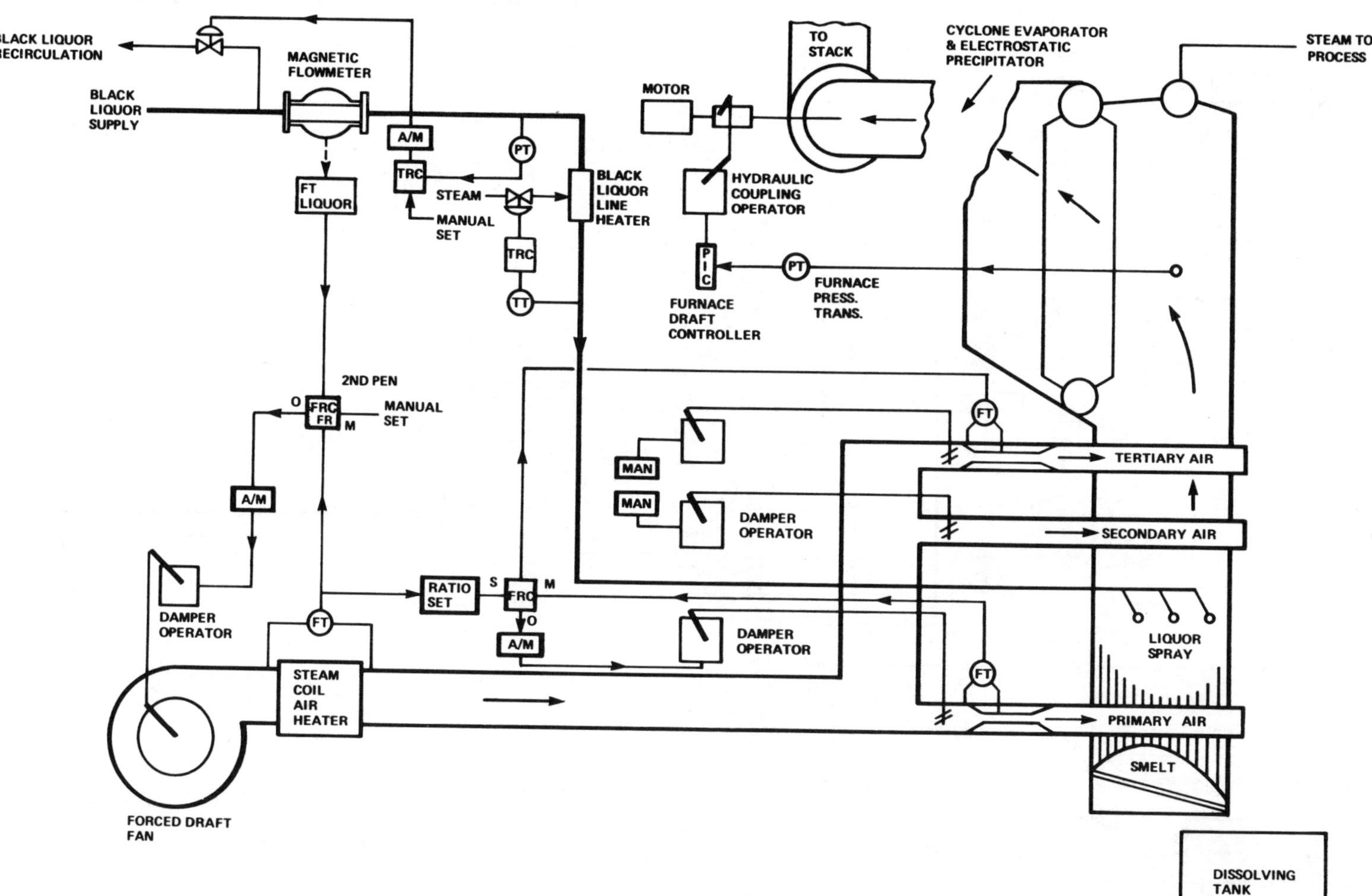

Figure 5-41. Black liquor recovery boiler combustion control schematic—configuration 2.

damper to divert part of the hot gases around the evaporator and a weak black liquor dilution valve which weakens the concentrated black liquor if its density is too high. The concentrated liquor then passes through a flow box to the salt cake mix tank.

Many plants are now using either a nuclear gauge or a refractometer to monitor the concentrated black liquor to the boiler. These devices sound an alarm whenever density changes adversely affect the firing of the black liquor and, even more significantly, when they disable the operation of the recovery furnace unit.

In the general operation of the recovery furnace:

1. The remaining water is evaporated from the black liquor in the upper section of the furnace. This section, called the oxidation zone, is the stage where complete combustion occurs.
2. The lower portion of the furnace is a reduction zone where incomplete combustion of the char (organics) provides carbon and carbon monoxide which reduce the sodium sulfate to sodium sulfide (smelt) and impurities. This reduction occurs in the presence of a limited amount of primary air that is injected around the smelt bed. At temperatures of 1500° to 1800°F, the smelt is drawn from the bottom of the furnace into the smelt dissolving tank. A density controller regulates the admittance of the weak green liquor solution to provide the correct density of strong green liquor outflow. Some plants use the opposite approach whereby level control admits the weak green liquor solution to the dissolving tank and a density controller regulates the outflow of strong green liquor.
3. The generated heat maintains a chemical reaction of the smelt bed and produces steam.
4. Inorganic ash is transported by the flue gas out of the boiler and is reclaimed as it passes through a dust hopper, direct-contact evaporator, and the precipitators.

Auxiliary fuels are generally used during startup until the smelt bed is high enough to support combustion, during blackouts, and when the boiler must generate supplementary steam to meet plant requirements.

Because the recovery unit does not ordinarily take load swings, the combustion control system is simpler than that of the power boiler. In a typical base-loaded combustion control system, the black liquor and total airflow are set manually. The ratio set point of the airflow controller modifies the relationship of air to black liquor. The presence of secondary or tertiary levels of combustion air determines how the remaining air is controlled. Generally, secondary air is ratioed from total air so that the percentage required for the reduction zone is established and the remainder of the air is utilized in the secondary and tertiary levels for complete combustion.

The same type of drum level control system is used on the chemical recovery unit as on a power boiler because control requirements are just as critical. Although this unit does not take the steam flow load changes, it is subjected to steam pressure changes in the drum which cause the drum level measurement to shrink and swell.

As in most industries, the pulp and paper industry has moved toward centralization of all operator functions. This has inspired a trend to electronic instrumentation because of its shorter systems' response time and the hope of integrating many of the mill (primary process area) control functions into a digital system.

The papermaking process requires large amounts of steam. Steam is typically provided by recovery units, bark burners, and power boilers burning prime fuel only. These will all discharge steam into a header distribution system which may be operated at several different pressure levels depending on process requirements. Pressure "letdown" stations may be used to control the various lower pressure legs.

It is also not uncommon to encounter applications where steam is generated at a high pressure to operate a noncondensing turbine-generator set. The turbine exhaust is then supplied as process steam. This type of turbine is sometimes called a *topping* turbine.

In recent years there has been increasing emphasis on the use of recovery equipment for the sulfite pulping processes. While the general equipment and control problems for kraft and sulfite units are similar, there are also variations in configurations of chemical recovery boilers used in the sulfite processes.

Causticizing 6

Raw green liquor, consisting mostly of sodium carbonate and sodium sulfide formed by dissolving the residual ash (smelt) from the recovery furnace in weak wash liquor, is pumped to the causticizing area of the pulp mill. Here the sodium carbonate contained in the green liquor is converted to an inactive chemical (sodium hydroxide) to produce white liquor.

The resultant white liquor is separated from the sludge and pumped to the digester house to be reused as cooking liquor in the sulfate pulping process. The sludge, which is primarily calcium carbonate, is pumped to the lime recovery area where it is converted to lime which is used to recausticize more green liquor (see Figure 6-1).

The liquor causticizing process can be divided into a number of separate functions:

1. Green liquor clarification where foreign material (dregs) is removed.
2. Washing of green liquor to remove entrained sodium salts.
3. Mixing lime and green liquor in the slaker.
4. Allowing the lime and green liquor to react in agitated tanks that are called causticizers.
5. Settling out and separation of sludge (lime mud) from the white liquor in a clarifier.
6. Thickening of lime mud.
7. Removing excess moisture from lime mud before burning in the lime kiln.

CONTINUOUS CAUSTICIZING

As shown in Chapter 5, the green liquor from the dissolving tank at the recovery furnace is regulated to a uniform concentration by controlling the amount of dilution weak wash liquor added.

Weak wash liquor from the causticizing process is used instead of fresh water to minimize dilution of the cooking liquor. The green liquor is clarified to remove the suspended solids, such as unburned carbon, plastic chrome iron, and iron compounds, in a continuous causticizing system (see typical system, Figure 6-1). The solids removed by settling are called dregs. They are pumped to a washer where they are washed with water to recover any entrained sodium carbonate and sulfide carried over. The wash water is decanted and pumped to the weak liquor storage to be used as dilution at the smelt dissolving tank. The washed green liquor dregs are discharged to the sewer.

The mixing of green liquor with lime to form a slurry is done in the slaker. Reburned lime from the lime kiln is conveyed to the slaker to suit the amount of sodium carbonate in the slurry. Fresh lime is added to the reburned lime from the new lime storage bin to replace losses. The slaked lime is classified to remove sand, stone, and improperly burned lime. The classified slurry is then pumped to the causticizers. The formation of sodium hydroxide begins on contact of the

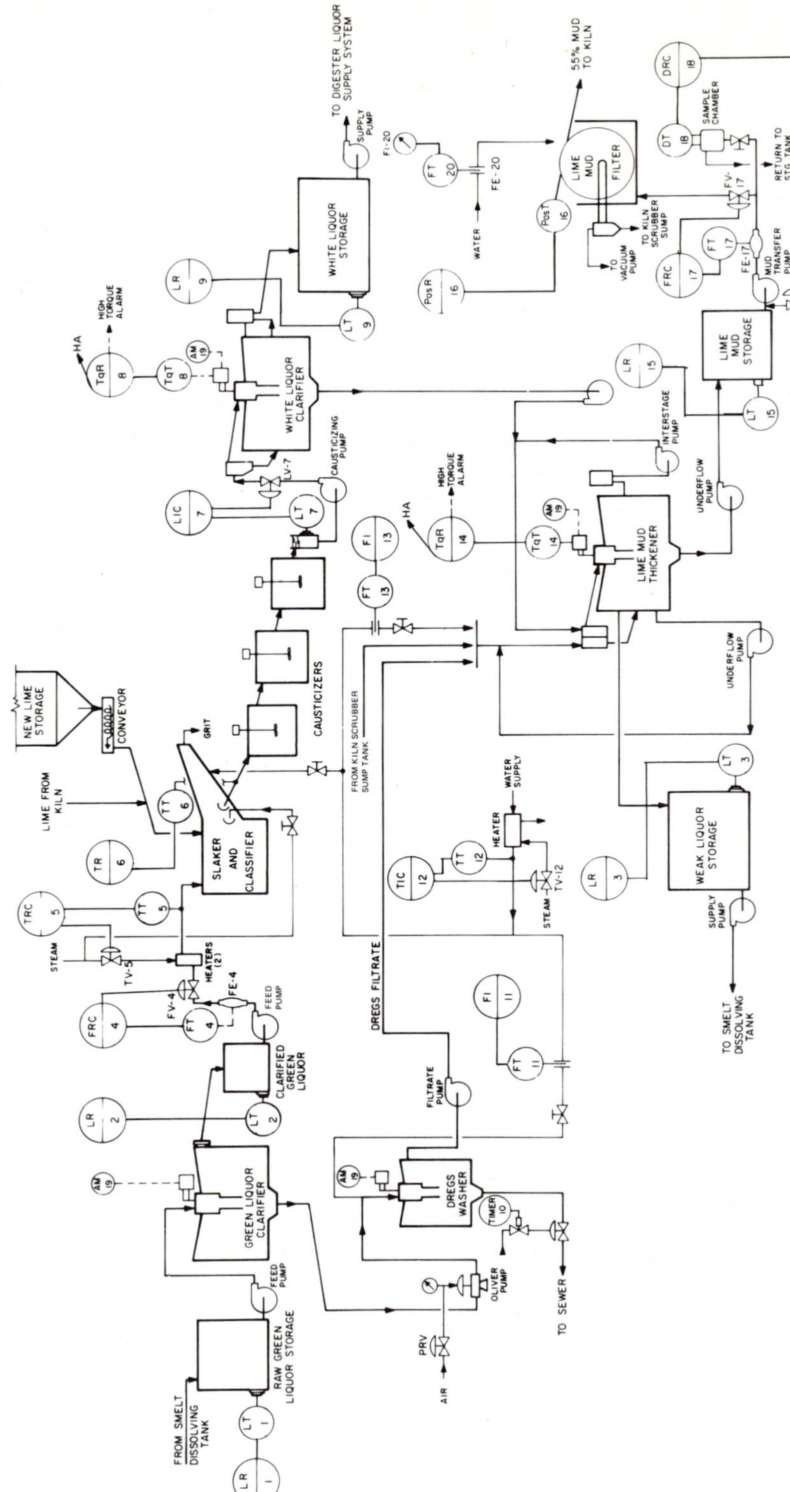

Figure 6-1. Instrumentation for controlling liquor causticizing.

green liquor with lime. Insoluble calcium carbonate precipitate is also formed.

The first stage of the causticizing process is made up primarily of mixing and classifying operations. The equipment consists of a cylindrical mixing section with an agitator and an inclined trough where reciprocating rakes slowly move the undissolved solids up the incline through wash sprays to waste discharge.

In order to achieve maximum conversion of the causticizing reaction, additional agitation and retention in three causticizing units arranged in series are employed to minimize short circuiting and improve efficiency.

The white liquor mud slurry from the causticizers is then sent to the white liquor clarifier for separation of the clear white liquor (which contains sodium hydroxide and sulfide suitable for the digesters) from the lime mud slurry (which consists primarily of calcium carbonate). This is accomplished in multicompartmented, gravity-settling type clarifiers. As with the green liquor clarification, this stage consists primarily of settling the solid calcium carbonate or decanting off clear liquor. The clarified white liquor is pumped to storage for reuse in the pulping of wood in the digesters.

The lime mud from the white liquor clarifier, containing between 35 and 40 percent solids, is pumped to the lime mud thickener which is also a washing operation designed to extract any remaining soda chemical for reuse in the process. Liquid used for mud washing is shown as being made up of filtrate from the green liquor dregs washer, water from the lime kiln scrubber sump tank, and fresh heated water. Other sources such as evaporator condensate are also used. The overflow from the lime mud thickener is directed to the weak liquor storage tank and, subsequently, to the smelt dissolver tank at the recovery furnace.

The underflow from the lime mud thickener is pumped to a lime mud storage tank from which it is pumped to a rotary-drum type vacuum filter. There it is dewatered and produces a cake containing about 40 to 45 percent moisture. Additional washing is accomplished by displacement with the use of water showers directed on the drum after the cake is formed.

Some mills have gone to the use of continuous centrifugal filters for this purpose. In this system, mud from the mud washer is introduced into the large end of a conical drum, which has a conical screw inside to move the mud along, as the entire drum rotates at high speed. This "throws" the water to the sides and out the cone end of the drum. The mud that is discharged from the small end of the conical drum contains approximately 60 to 65 percent solids.

The lime mud, which is essentially calcium carbonate, is now ready to be sent to the lime kiln for conversion to lime (calcium oxide) which will be reused for slaking, and causticizing green liquor.

Instrumentation

Typical instrumentation required for control of the liquor causticizing process for alkaline pulp cooking liquor is also shown in Figure 6-1. The raw green liquor flow to the slaker and classifier is regulated by recording flow controller FRC-4. The primary measuring system, FE-4 and FT-4, is an electromagnetic flow transmitter; its lined straight-through metering tube makes it suitable for handling the suspended solids and corrosive liquor. Flow control of the green liquor supply, of uniform density, stabilizes the system input to secure optimum clarification and slaking efficiency.

Lime slaking is best done within fairly close temperature limits. At the optimum temperature, the reaction proceeds more completely, unreacted lime settles more rapidly in the white liquor clarifier, and there is less of it to be recirculated to the kiln. Recording temperature controller TRC-5 controls the system input to the green liquor heater to maintain the optimum temperature of the green liquor entering the slaker, regardless of the load variation. Temperature recorder TR-6 shows the temperature of the slurry leaving the slaker. The record will show if the final desired temperature has been reached, indicating completion of the reaction. A drop in temperature will reveal any interruptions in the lime feed.

Wash water temperature is important. Optimum temperature in the wash accelerates the setting of the lime mud and the dregs, ultimately resulting in clearer white liquor. It reduces soda losses occurring through incompletely washed lime mud or green liquor dregs. Indicating temperature controller TIC-12 maintains the desired wash water temperature, regardless of load fluctuations, by controlling the steam input to the heater.

At times, the lime mud may build up in the white liquor clarifier or lime mud thickener, overloading the rakes and resulting in excessive maintenance costs and production losses. Torque recorders TqR-8 and TqR-14, with high load alarms, automatically warn the operator of damaging overloads on these rakes. The lime mud filter operates best if supplied with constant density slurry. Density recording controller DRC-18 maintains this desired density, thus stabilizing filter operation and improving calcination in the lime kiln.

PosT-16 detects cake thickness variations on the lime mud filter and transmits the signal to PosR-16 where a record is made for an operational guide. Timer-10 operates the dump cycle of the dregs washer.

A scheme for making the density measurement using a differential pressure transmitter is shown in Figure 6-2.

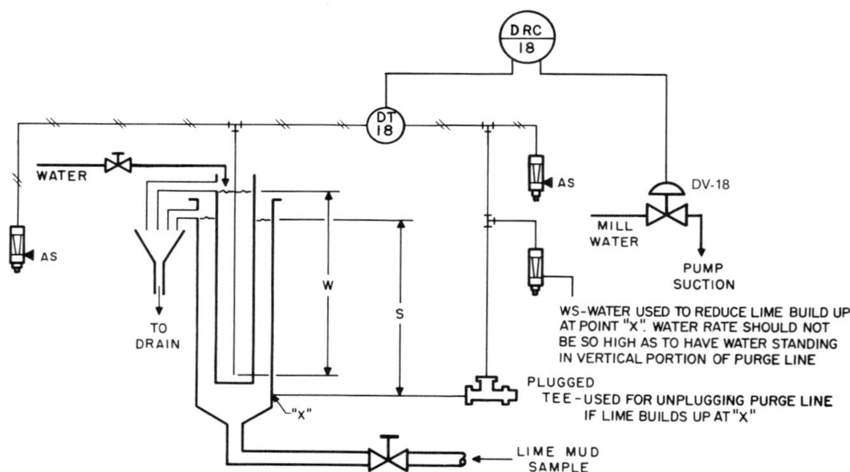

Figure 6-2. Lime mud density control.

A sample of the lime mud slurry from the pipeline to the lime mud filter is fed into the bottom of a chamber with constant level overflow. A second, smaller chamber, also with constant level overflow and filled with water, is immersed in the slurry chamber. The water is always at the same temperature as the slurry, obviating density correction for variation in slurry temperature.

The location of the slurry chamber air purge connection and the depth of the bubble tube immersion in the water chamber are chosen so that the hydrostatic head on each will be the same—zero differential—when the slurry is at minimum density. This permits the selected range to be spread across the full scale of the density controller.

In operation, if the slurry density tends to vary from the desired value, the density controller throttles a valve in the water line to admit more or less water into the slurry being pumped to the filter.

There are certain values which must be known—either determined or assigned—before an installation can be made. They are:

d_{min} = minimum slurry density (sp. gr.)
d_{max} = maximum slurry density (sp. gr.)
S = depth of slurry column
W = depth of water column

The differential range of the d/p Cell transmitter

$$= S \times d_{max} - S \times d_{min}$$
$$= S\,(d_{max} - d_{min})$$

Density range of recorder controller

$$= d_{min} \text{ to } d_{max}$$

Example:
The installation is based on premise:

$$W \times 1 = S \times d_{min}$$

Assume: $d_{min} = 1.25$
$d_{max} = 1.35$
$S = 50$ inches
then, $W = 50 \times 1.25$
$= 62.5$ inches
diff. range $= 50\,(1.35 - 1.25)$
$= 50\,(0.1)$
$= 5$ inches (use d/p Cell transmitter E13DL or 15A)
density range $= 1.25$ to 1.35 sp. gr.

Lime mud density measurement can also be made with a nuclear radiation gamma gauge.

The lime mud filters also require a stabilized vat level for best operation; otherwise, a filter cake of varying moisture will be produced and will constantly upset kiln operation. (Refer to Figure 6-1.) Flow recording controller FRC-17 maintains the slurry feed at a constant flow rate to the filter by use of an electromagnetic flow transmission system, FE-17 and FT-17, thereby regulating the level in the vat. Filter cake thickness recorder PosR-16 records the variations in cake thickness as transmitted by position transmitter PosT-16. The operator, guided by this record, can maintain the desired thickness.

Showers are used to give the lime slurry a final washing on the mud filter drum. If enough hot water is used, the soda chemical will largely be recovered before the sludge enters the kiln, with reduction in formation of the undesirable "mud rings" in the kiln. On the other hand, excessive water usage may produce a "sloppy" filter cake which will not be readily calcined. This water flow is indicated on FI-20 with orifice plate FE-20 and differential pressure transmitter FT-20 used as the measurement system.

Chemicals used in the causticizing operation are expensive and any loss from the overflow of storage tanks should be avoided. For this reason, recording level instruments LR-1, LR-2, LR-3, LR-9, and LR-15 are used for the various storage

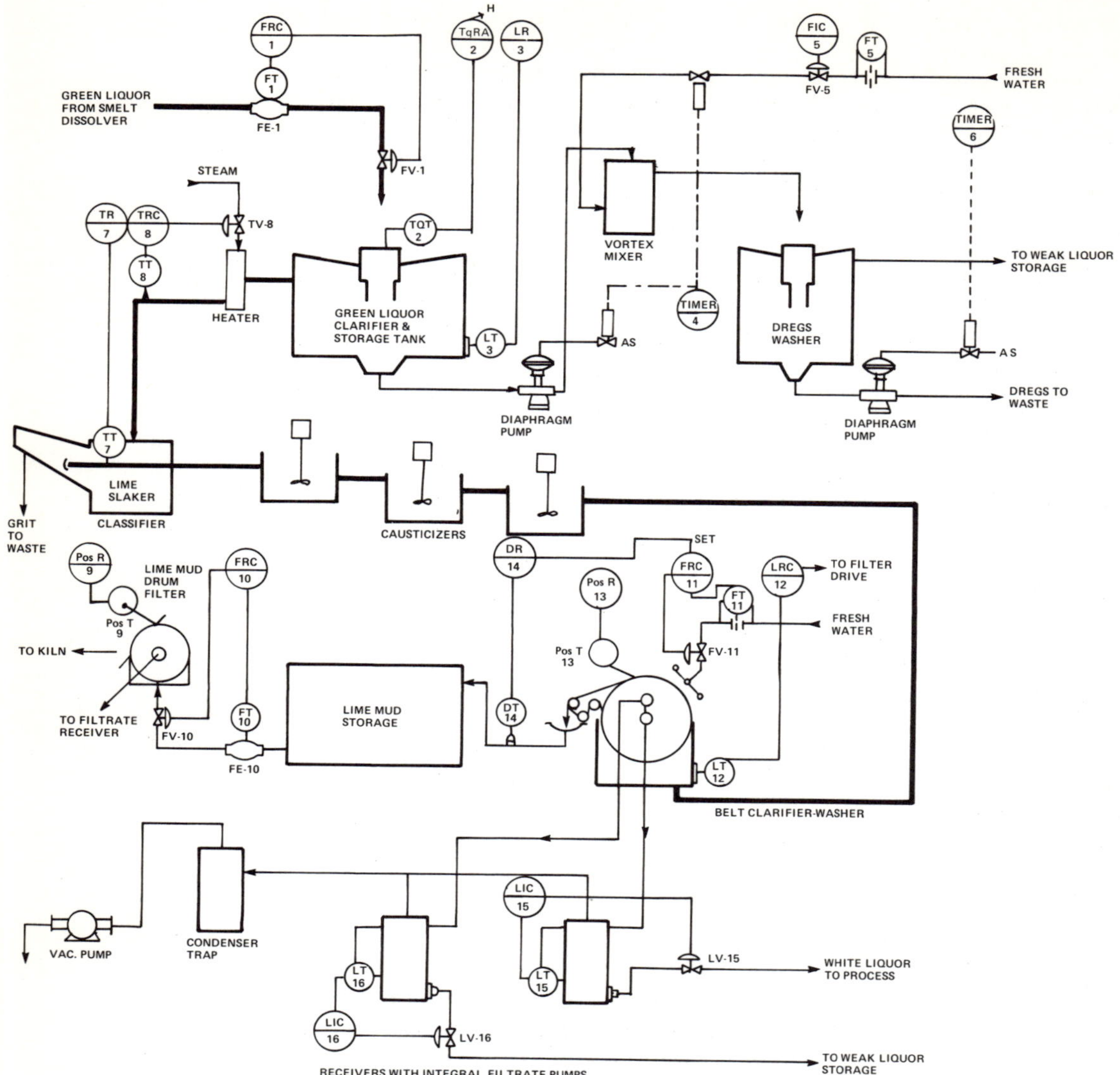

Figure 6-3. Liquor causticizing system with belt clarifier-washer.

tanks in the process. The flange-mounted level transmitter with flush-diaphragm element is normally used for the primary measurement in these cases: LT-1, LT-2, LT-3, LT-9, and LT-15. The level recording instruments can be equipped with appropriate high and low alarms, or independent alarm systems can also be used with level sensing electrodes or level switches installed on the tanks.

CAUSTICIZING WITH BELT FILTER

A modification of the typical continuous causticizing system utilizes a belt filter for both white liquor clarification and lime mud washing (Figure 6-3).

By performing both the white liquor clarification and the lime mud washing in one step:

1. There is a saving in floor space requirements.
2. Upsets caused by peak or surge flows do not affect clarity of the white cooking liquor.
3. Less wash water is needed for displacement-type washing on the filter.
4. The causticizing process is simplified.

The instrumentation used is similar to that used with the typical continuous system with raw green liquor flow to the green liquor clarifier regulated by recording flow controller FRC-1. The primary measuring system, FE-1 and FT-1, is an electromagnetic flowmeter. Torque recorder TqRA-2 can be used as a guide to inflow green liquor density variation and to automatically alarm on high loads on the rakes. LT-3 and LR-3 are used to record the level in the clarifier. Recording temperature controller TRC-8 controls the system input temperature for optimum slaking and causticizing. Temperature in the lime slaker is measured and recorded by TT-7 and TR-7.

Level in the vat of the belt clarifier-washer is controlled by LRC-12 which adjusts the drive speed of the filter as the feed varies. If the density of the lime mud is measured, it can be transmitted by DT-14 to be recorded on DR-14, and also to remotely set the control point on the fresh water flow controller to the filter showers, FRC-11, to maintain a constant density to the lime mud storage tank.

The flow from the lime mud storage tank to the lime mud drum filter is measured by electromagnetic-type flowmeter FE-10 and is controlled by FRC-10. Cake thickness variations on the belt clarifier-washer and lime mud drum filter are detected by transmitters PosT-13 and PosT-9, respectively, and recorded on PosR-13 and PosR-9.

The fresh water flow to the vortex mixer is regulated by indicating controller FIC-5. Timers 4 and 6 determine and automatically program the dump cycle of the green liquor clarifier and dregs washer.

LT-15, LIC-15 and LT-16, LIC-16 are level control loops on white liquor and weak liquor receivers from the belt clarifier-washer and are sometimes locally mounted units in the field.

7 Lime Recovery

The last operation in the cycle of chemical recovery is the reburning of the lime mud (calcium carbonate) cake from the filter to produce lime (calcium oxide). This is returned to the causticizing plant for use in recausticizing additional green liquor from the sulfate chemical recovery boilers. This not only solves the lime mud disposal problem but, more importantly, as with the recovery of other chemicals, the economics of the sulfate pulping process demands the recovery of the lime. The use of only fresh lime for recausticizing purposes would increase the cost of operation.

Reburning, or calcining, is usually accomplished in a conventional rotary kiln but the fluid bed principle has also been used for this purpose.

LIME KILN

Lime sludge kilns are used in pulp mills to recover CaO from precipitated $CaCO_3$. This precipitated $CaCO_3$, containing about 70 percent water, is pumped from lime mud storage tanks to either centrifuges or filters, where it is reduced to about 35 percent water. The resulting filter cake, or sludge, from the centrifuges is usually fed into the kiln by means of a screw conveyor. The lime being discharged from the kiln is conveyed to a storage bin by means of drag conveyors and bucket elevators.

The gases coming out of the kiln are passed through a dust collecting system, usually a venturi-type scrubber. Part of the scrubbing liquid is recycled.

A typical lime burning system is shown in Figure 7-1. The basic rotary lime kiln is a refractory-lined, open-end steel cylinder, inclined from the feed end to the discharge end. It is mounted on rollers and rotated about its longitudinal axis by a variable-speed electric motor drive through a girth gear assembly around the kiln. Lime mud from the filter in the causticizing plant, with solids content of about 65 percent, is conveyed into the upper feed end of the rotary lime kiln. As the lime mud works down the kiln flowing countercurrent to the heat, moisture is evaporated. The carbon dioxide is driven off, converting the mud which is primarily calcium carbonate, into burnt lime, which is calcium oxide.

The fuels that have been used to produce the necessary heat in rotary lime kilns are pulverized coal, natural or manufactured gas, and No. 6 fuel oil. Today, the most common fuels used in the burners at the lower end of the kiln are natural gas and fuel oil. The diagram shows an installation set up to use either of these two fuels. Some heat is also generated by the burning of the carbonate itself in the lime mud.

The lime mud goes through several definite phases as it travels through three kiln zones while being converted to burned lime. The first zone consists of the initial drying stage where moisture is eliminated. The second zone is where the temperature is increased, during which time the dry lime passes through a plastic state. The third and final zone is the stage in which the actual calcining occurs. At the elevated temperature that exists in this zone, carbon dioxide is evolved from

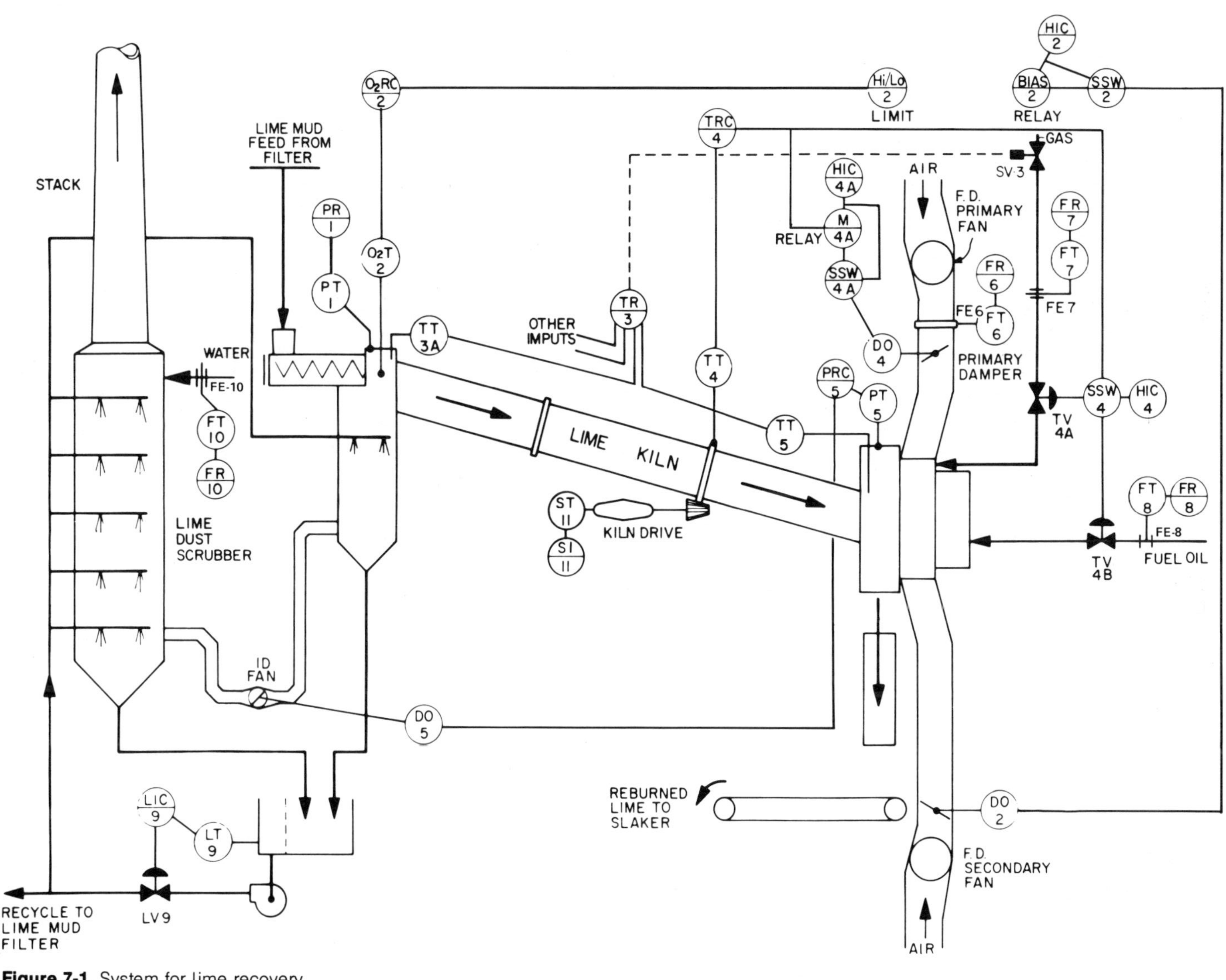

Figure 7-1. System for lime recovery.

the calcium carbonate to yield calcium oxide or burned lime.

Many lime kiln installations are equipped with chains attached to the shell; these chains hang in the hot gases of the first zone, thus providing an extended surface for the transfer of heat from the gases to the lime mud. The mud is taken in thin layers or coatings on the chains as the kiln rotates.

Kiln sizes are designated by inside diameter and outside length of the steel shell as measured when the kiln is cold. Normally, sizes vary from 6 to 13 feet in diameter, and 100 to 350 feet in length. The downward slope of a kiln can be from 5/8 to 3/4 of an inch per foot, depending on diameters and lengths. The rate of product travel is determined by the slope of the kiln and its speed of rotation. Many kilns are equipped with either two-speed, four-speed, or slip-ring motors in order to vary the rotational speed of the kiln.

Instrumentation

In order to produce good-quality lime at minimum fuel consumption, the instrumentation shown in Figure 7-1 holds the burning temperature as constant as possible by controlling the air necessary to support combustion in order to completely consume the fuel but, at the same time, avoiding any excess air. The control system also maintains the moisture content of the entering fuel as constant as possible and maintains constant optimum retention time in the kiln.

Temperature recorder TR-3 keeps a record of the temperature of the entire system from measurements made by thermocouple sensing elements located in strategic locations, such as the feed end of the kiln, the firing end of the kiln, gas stream, and other critical areas. These temperatures can be used in an alarm and safety system to automatically shut off the fuel supply when undesirable or dangerous temperature conditions develop. A continuous oxygen recording controller, O_2RC-2, is used at the feed end of the kiln to establish and maintain optimum oxygen levels by adjustment of secondary air damper DO-2 at the firing end of the kiln. The burning zone temperature is controlled by a recording temperature controller, TRC-4, which adjusts a flow control valve on the gas or fuel oil supply as selected by selector switch SSW-4. Although a thermocouple sensing element is shown, this measurement can also be made with a radiation-type pyrometer "looking" into the firing end of the kiln.

The draft at the firing end of the kiln is recorded and controlled by PRC-5 through the regulation of the induced-draft fan speed by the adjustable operator, DO-5. The draft on the feed end is recorded on PR-1.

The kiln speed is monitored by an indicator, SI-11, which receives its signal from a magnetogenerator-type speed transmitter, ST-11, mounted on the shaft of the kiln drive. Flows of gas and fuel oil are recorded and totalized by instruments FR-7 and FR-8. The primary flow element for gas, FE-7, is usually an orifice plate, while for fuel oil, FE-8, it is usually a target-type or positive displacement-type flowmeter. The flow of water to the lime dust scrubber is recorded on FR-10 with an orifice plate flow element, FE-10, and with a differential type of transmitter, FT-10.

Although not shown on the basic diagram, some kilns are equipped with continuous carbon dioxide and carbon monoxide analyzers. TV cameras mounted behind the firing end of the kiln have been used to "see" the fire and load the kiln. The picture on a remotely located TV screen actually gives the operator a better view than if he were personally looking into the kiln. This picture aids him in making operator adjustments. Other TV cameras focused on the mud filter and a remotely located TV screen also allow the operator to inspect this operation without leaving his normal station.

To meet insurance requirements when natural gas is used as fuel, it is necessary to incorporate safety features within the entire kiln system, which will shut off the flow of fuel in the event of induced-draft fan failure, secondary-air fan failure, subnormal gas pressure, or flame failure. Such a safety system must incorporate an automatic purge cycle so designed that the induced-draft fan will clear the kiln of any residual gas prior to the operator's relighting the fire.

Feed Control

In the lime recovery process, constant throughput of the lime kiln is maintained by measuring mass flow and controlling the delivery of the solids content of the lime mud feed to the kiln. This is shown in Figure 7-2.

The system provides constant loading of the lime kiln. This assures a uniform product and permits fuel economies through optimum kiln firing conditions.

Additional economies are realized through reduced water evaporation in the kiln because the maximum amount of water is extracted in the filter. This is in contrast to operation at an arbitrary high water content; it is necessary when the kiln feed is controlled from volumetric flow measurement alone and constant consistency is required.

The system also eliminates the need for dilution of the mud from storage, sometimes done in order to regulate the solids content of the filter output.

Figure 7-2 illustrates a typical system for control of mass flow lime mud. Lime mud of varying solids content is pumped "as is" from storage. Volumetric flow is measured with a magnetic flowmeter, FE-1. Density is measured with a gamma density gauge, DE-2, calibrated in degrees Baumé. The two measurements with suitable conversion factors are multiplied in an analog computing relay, X-3. The output of the relay is used as the measurement signal to an indicating controller,

Figure 7-2. Instrumentation for lime kiln feed.

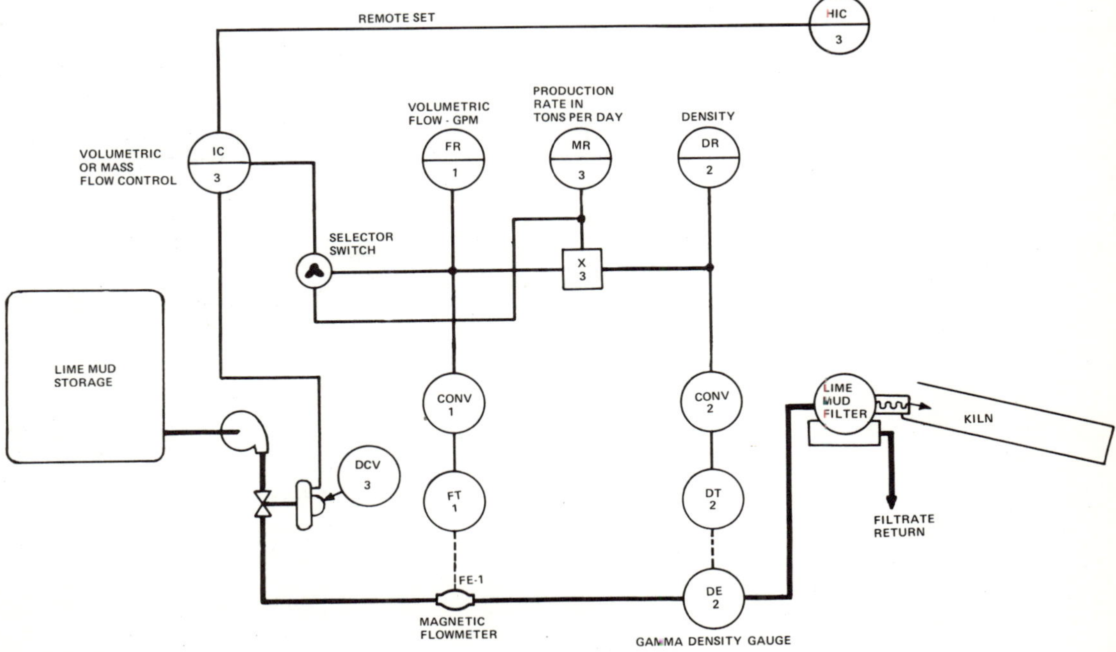

IC-3, which controls the feed in terms of tons of dry calcined lime per day to allow for lime losses in the recovery system.

A remote set point manual station, HIC-3, is provided so that the production rate can be set at the operator's station.

The flow control valve is regulated by an indicating controller, IC-3, on the basis of either volumetric flow or mass flow.

FLUID BED CALCINER

The use of the fluid bed principle to convert lime mud into reburned lime has been a more recent method of recalcining in the lime recovery process. A generalized flow diagram is shown in Figure 7-3.

Fluid bed calciners are vertical units in contrast to rotary kilns which are essentially horizontal with a slight incline. The unit operates on the basis that, when solids of a uniform size are suspended in a gas stream, their flow properties are similar to liquids. This dynamic mixture of solids and gases is known as the fluid bed. The term *fluidization*, as used here, is the unit operation in which a mass of finely divided solids is maintained during treatment in a turbulent, dense state by being dispersed in an upward moving gas stream. This imparts to the mass a turbulence resembling that of a boiling liquid.

In fluid bed calcination, the fine calcium carbonate particles pelletize because of the action of a small amount of alkali present in the feed or otherwise added in

Figure 7-3. System for reburning fluid bed lime mud.

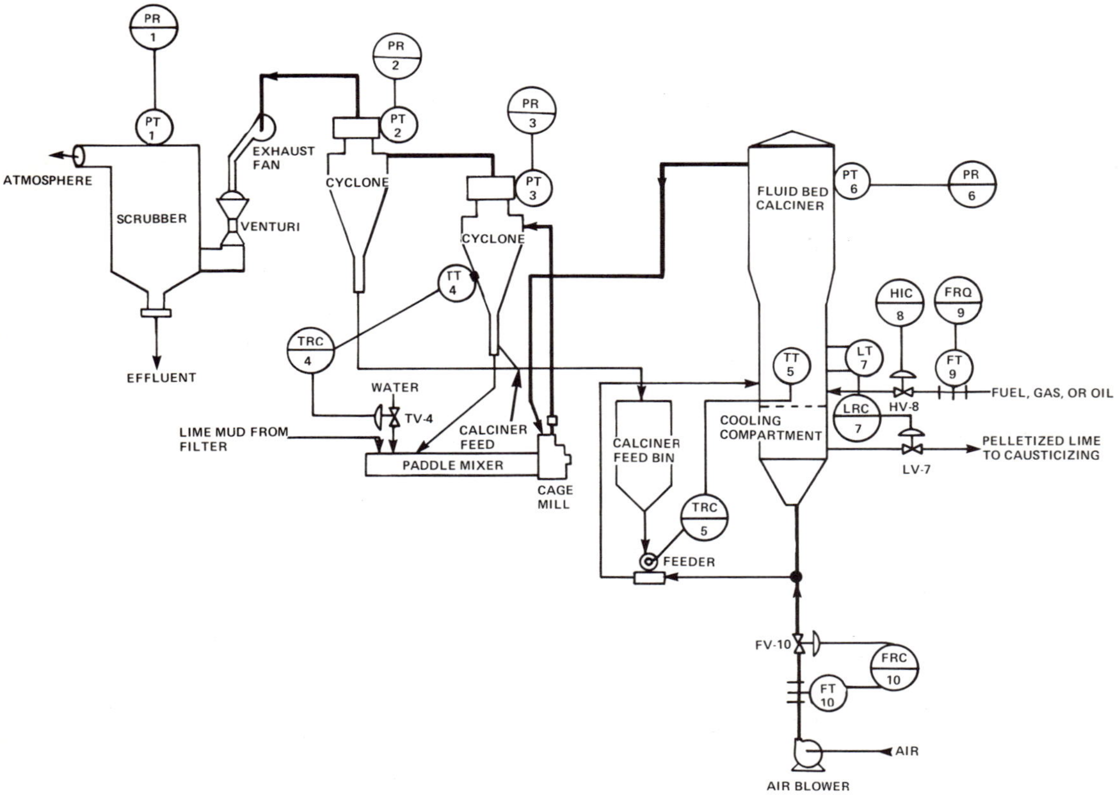

the form of soda ash. The lime burning system is divided into two basic parts. The first is the drying system in which the wet lime mud filter cake is dried to form dry, powdery feed for the fluid bed calciner. The second part is the calcination system which produces uniform, dense pellets of calcined product.

In the drying system, the lime mud from the causticizing system filter is conveyed to a paddle mixer where it is mixed with dry, recycled calcium carbonate along with a small amount of water as required for control of the temperature of the drying system. The lime mud from the paddle mixer is fed to the cage mill along with the calciner gases. There the moist solids are dried and agglomerates are dispersed. The cyclone stack gases pass through an exhaust fan into a wet scrubber where they are cleaned before they are discharged to the atmosphere.

The calcination system consists of a two-compartment vessel. The top bed is for calcining the carbonate and pelletizing the calcium oxide; the bottom bed is for cooling and heat recovery. The dry feed is metered by a special rotary valve to an air-swept pipeline and blown into the bottom of the calcination bed through a series of feed guns located around the periphery of the calciner. Heat for the calciner is obtained by direct burning of a fuel which is distributed through fuel guns, also located around the periphery of the calciner. The calciner can be equipped to burn either heavy No. 6 fuel oil or natural gas. Simultaneous calcination and pelletization of the carbonate feed occurs in the calcining bed. The flow of calcined lime pellets from the calcining section to the cooling compartment is through an internal underflow pipe and is regulated at a rate to maintain predetermined bed depth.

The hot pellets are cooled in the cooling compartment. The discharge rate of cooled pellets from this chamber is controlled to maintain the cooling bed depth constant. The product pellets, conveyed to a product bin, are then ready for use in the causticizing system.

Instrumentation

For proper and efficient operation, calcination temperature, drying system temperature, and fluid bed levels are controlled. The calcination temperature is maintained by temperature recording controller TRC-5. The airflow is controlled at a given capacity by recording flow controller FRC-10. The fuel is manually adjusted by HIC-8 to optimum fuel/air ratio for efficient fuel combustion. The fuel consumption is recorded and totalized on FRQ-9. Temperature recording contoller TRC-5 automatically adjusts the carbonate feed by a special rotary valve to maintain calcination temperature at the desired preset level.

The drying system temperature is controlled by temperature recording controller TRC-4, automatically adjusting control valve TV-4 on the water added to the paddle mixer. The fluid beds are maintained at desired levels by bed level recording controller LRC-7. Pressure taps sense the fluid bed depth by measuring a differential pressure across the fluid bed by LT-7. This pressure signal is transmitted to the level controller and automatically adjusts the transfer valve, LV-7, controlling pellet flow from the bed, thereby maintaining the fluid bed at the desired level of depth.

8 Bleach Chemicals Preparation and Pulp Bleaching

After cooking, screening, and washing, the next and last step in the process of manufacturing most pulps, depending on their intended use and the type of paper and paperboard finally produced, is bleaching. Pulp bleaching is essentially the removal of coloring impurities in the fibers by: (1) destruction and dissolving of coloring matter, and (2) removal of residual lignin material. Thus, in this sense pulp bleaching can be regarded as a continuation of the stepwise isolation and purification of wood fibers begun in the wood chip cooking process. The lignin materials and colored compounds are converted into a water-soluble form, while an attack on the desired carbohydrate filter material is held at a minimum. This is accomplished by a bleaching agent which is usually an oxidizing chemical, but which may also be a reducing compound. Some of the chemicals used for bleaching can be shipped to the local mill from chemical preparation plants, but the majority of the bleaching chemicals are prepared on site.

BLEACHING CHEMICAL PREPARATION

Chlorine and its compounds, such as sodium hypochlorite, calcium hypochlorite, and chlorine dioxide containing *available chlorine*, are commonly used for bleaching wood fibers. Other chlorine-containing compounds that have been used but not as frequently are hypochlorous acid and sodium chlorite. The available chlorine content of these types of bleaching agents is a measure of the oxidizing capacity to react with the residual lignin and coloring materials in the unbleached fiber.

Other oxidizing chemicals used primarily on color improvement of pulp to a lesser degree are sodium peroxide and hydrogen peroxide. Reducing chemical agents such as zinc hydrosulfite and sodium hydrosulfite are also used on pulps requiring partial bleaching.

Sodium hydroxide and sulfur dioxide are other chemicals used not only in bleaching but also in the preparation of other bleaching compounds.

Chlorine

Chlorine (Cl_2) is usually purchased in tank cars or cylinders as a liquefied gas under pressure, although a few pulp mills produce chlorine by the electrolysis of saturated sodium chloride brine solution. One of the by-products of this electrolysis is sodium hydroxide, which is also used in the manufacture of sodium hypochlorite and in the pulp bleaching process. Due to the difficulty of metering or controlling the rate of flow of liquid chlorine in a continuous flow operation, most bleaching operations utilize chlorine in gaseous form. Normally, chlorine in liquid form is transferred from tank cars under pressure to an evaporator station. Gaseous chlorine can then be metered and controlled readily and accurately.

Figure 8-1 is a simplified instrument diagram of a typical chlorine vaporizing operation when liquid chlorine is supplied to the mill in tank cars. It is unloaded

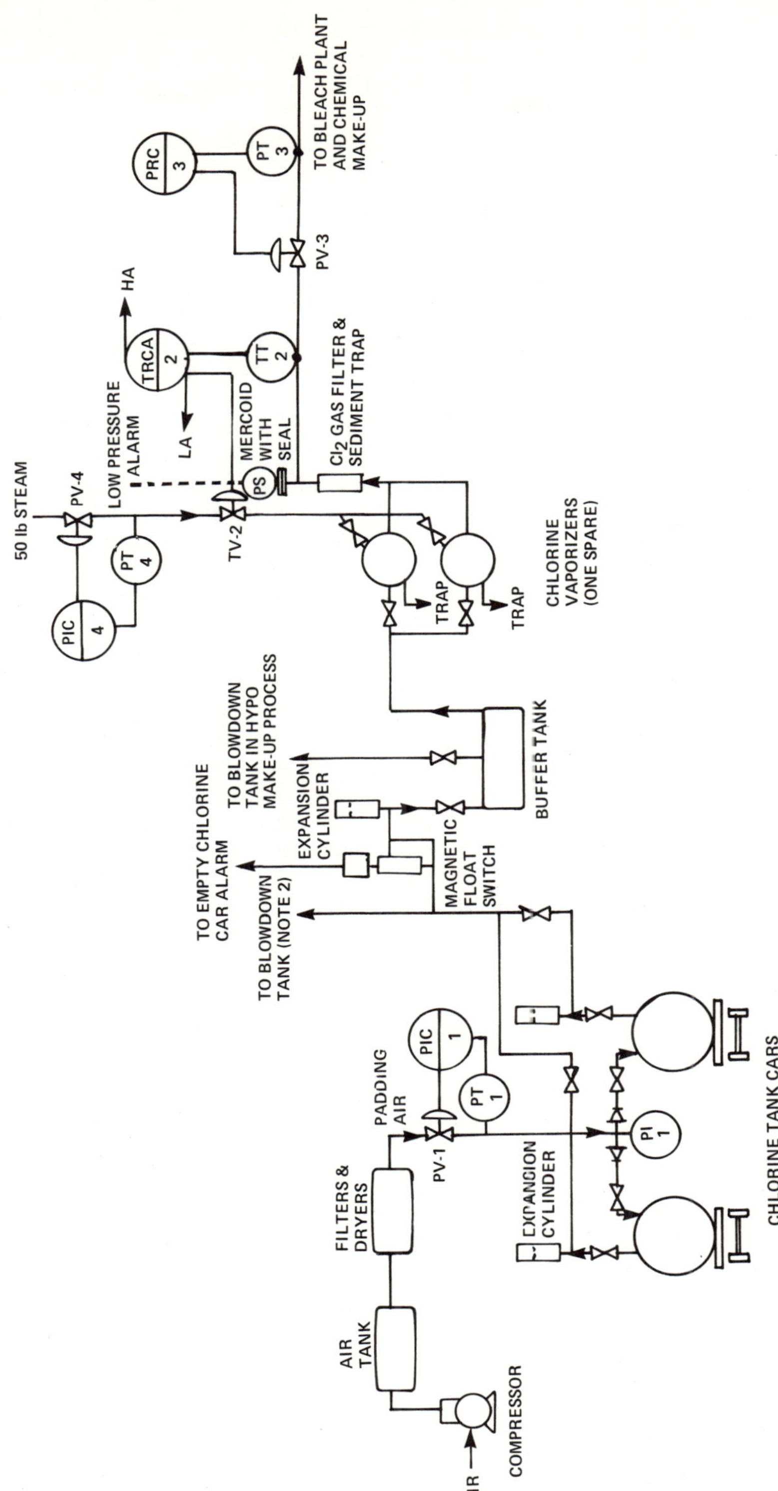

Figure 8-1. A typical chlorine vaporizing system.

from the tank car by admitting dry, oil-free air from a compressor filter and dryer system through a pressure regulating valve, PV-1, operated by a local indicating controller, PIC-1, into the top of the tank cars. PI-1 is an indicating gauge showing pressure at the tank car location.

When the liquid chlorine is withdrawn from a tank car to an evaporator, it is necessary to know when the tank car becomes empty so that a new one can be brought into service without interrupting the chlorine flow. Various methods have been used. The method shown in the diagram uses a magnetic-type float control switch, installed in parallel, with a vertical riser in the liquid chlorine line between the tank cars and the buffer tank. When liquid chlorine stops flowing through the line the float chamber will drop, causing a corresponding drop in the float chamber level, energizing the switch, and actuating the empty-car alarm.

Other methods have included the use of a capacitance probe mounted in the liquid chlorine line ahead of the buffer tank. This device detects changes in capacitance between liquid and gaseous chlorine and activates the empty-car alarm system.

Another method that has been used involves the mounting of a pressure capsule so that the buffer tank and associated piping rest directly on it. The combined weight is zeroed out when the buffer tank is empty. An indicating gauge attached to the capsule contains high and low alarm switches which can be set to correspond to any desired weight of liquid chlorine sufficient to enable tank car switching before complete exhaustion.

Depending on the size of the particular mill, various types of evaporators have been used to vaporize liquid chlorine into the gaseous state. The instruments shown in the diagram are based on a steam-jacketed type. The temperature of the chlorine is controlled by sensing the temperature in the gas outlet line from the vaporizer with the sensing element of transmitter TT-2, and activating valve TV-2 in the steam line to the steam jacket of the vaporizer by controller TRCA-2. The indicating pressure controller maintains constant steam pressure to the vaporizer to produce a constant rate of evaporation and prevent excessive buildup in the vaporizer caused by response lag in the temperature controller. Pressure recording controller PRC-3 regulates the pressure of the chlorine vapor leaving the vaporizer to assure accurate flow of gas either to the bleaching stage or to facilitate the ratio of chlorine gas in preparation of bleach liquors.

Sodium Hydroxide

Sodium hydroxide or caustic soda is normally received at the mill in 50 or 73 percent liquid form. When utilized for the caustic extraction stage in the bleach plant it must be diluted to about 5 percent. When utilized for making hypochlorite bleaching chemicals it is diluted to 3 to 6 percent depending on the desired strength of the bleaching chemical. In dissolving pulp mills, for instance, the caustic soda is used at a strength of 12 to 15 percent or higher.

A number of methods of accomplishing this dilution have been used quite successfully in the past and are still being used. Some of these methods include specific gravity control utilizing differential elements activated by air-purged bubble tubes mounted on a dilution tank, ratioing caustic and water volumes with flowmeters or volume pumps, and differential temperature measurement of the reaction heat of dilution.

Caustic Dilution by Specific Gravity. Figure 8-2 shows how specific gravity is used as a basis for caustic dilution control. The strong caustic from storage and water is introduced in a mixing tee, and the mixture then flows through a simple pipe coil to assure a completely uniform solution. A portion of the dilute caustic

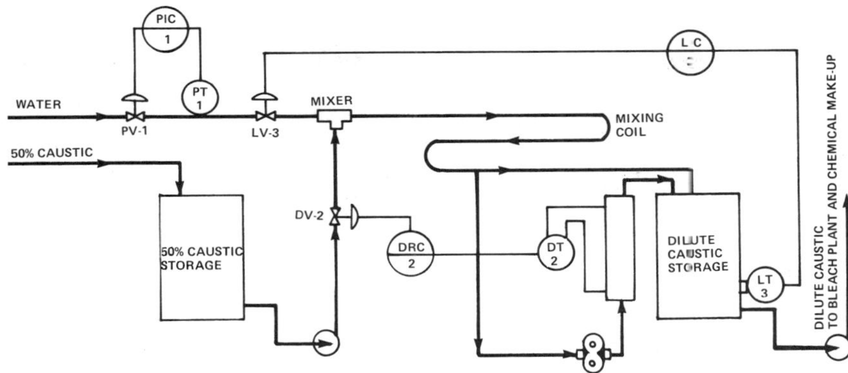

Figure 8-2. Caustic dilution by specific gravity.

is diverted to a bypass pipeline with an expanded section on its way to the dilute caustic storage tank. The specific gravity measuring probes are mounted in this section. There are a number of types of probes used, the most common being bubble pipes, with specific gravity or density transmitter DT-2 being of the differential pressure type. The specific gravity is controlled by recording controller DRC-2 at a value which is related to percent sodium hydroxide in accordance with the graph in Figure 8-3. The level in the dilute caustic storage tank is con-

Figure 8-3. Specific gravity of caustic soda solutions at various temperatures.

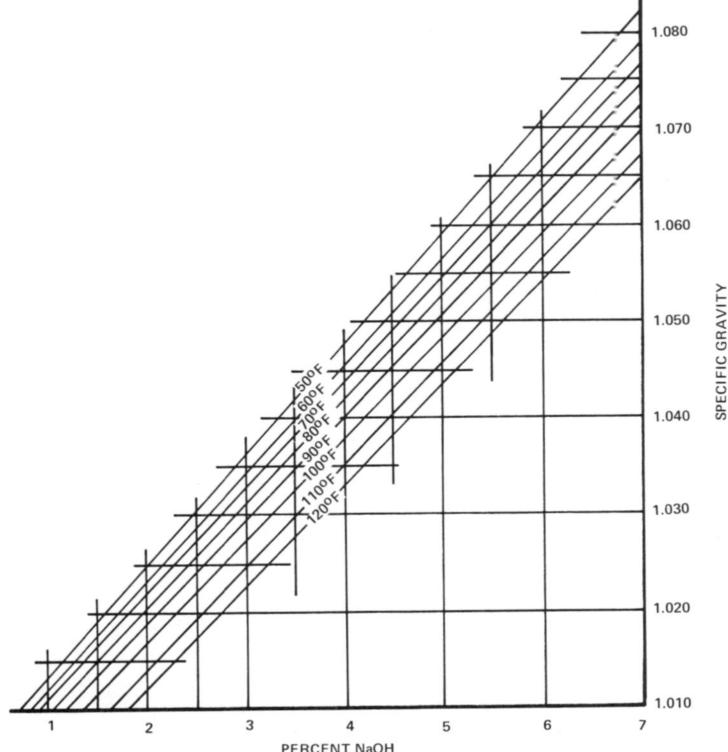

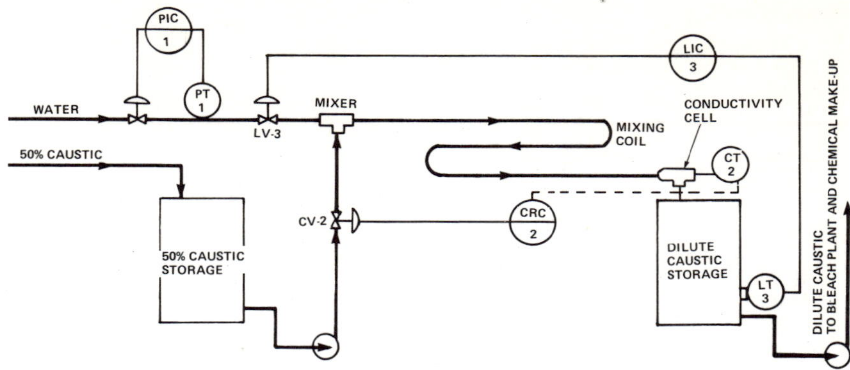

Figure 8-4. Caustic dilution by conductivity.

trolled by indicating controller LIC-3 throttling the dilution water to the mixer. The pressure in the water dilution line is controlled by PIC-1.

Caustic Dilution by Conductivity. The use of a conductivity measurement to dilute caustic to a specified percentage is practicable up to a concentration of about 14 percent This configuration is shown in Figure 8-4 and is similar to that used when the specific gravity measurement is used, except that the dilute caustic flowing from the mixing tee to the dilute caustic storage passes through a flow cell. There the conductivity is continuously measured. The conductivity controller, CRC-2, adjusts the addition of strong caustic to the mixing tee to maintain the solution concentration in accordance with the graph in Figure 8-5. Level controller LIC-3 on the dilute caustic storage tank regulates output of the system by adjusting the water flow to the mixing tee. Dilution water pressure to the system is controlled by PIC-1.

Figure 8-5. Conductivity-specific gravity relationship in the control of dilution by conductivity.

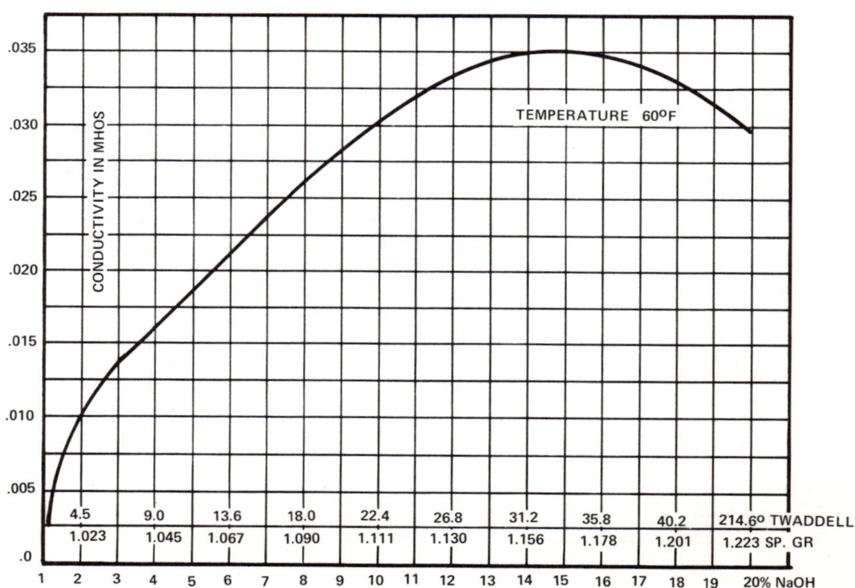

Hypochlorite

Many of the chemicals used in pulp bleaching are manufactured in-plant for both economic and technical reasons. Chlorine, one of the more important of the bleaching chemicals, is used both in the elemental form as chlorine gas in the chlorination stage, and as calcium or sodium hypochlorite in the hypochlorite stage.

Calcium Hypochlorite. At one time only calcium hypochlorite was available. It was delivered in drums as a basic salt in powder form, containing excess free lime that brought the available chlorine content to approximately 35 percent. Later, when in-plant calcium hypochlorite manufacturing began, it was first done on a batch basis. In batch preparation a milk of lime slurry at a specified concentration and temperature is put into a large vat or tank equipped with a suitable agitator. The chlorine is introduced near the bottom and, although the chemical reaction is rapid, agitation is provided to prevent escape of the gas in case that short circuiting occurs.

The introduction of continuous systems to make calcium hypochlorite resulted from the application of two facts: that the chlorine-lime reaction is exothermic, and that two unlike electrodes immersed in a hypochlorite solution generate a small electrical potential which can be used as a measure of the available chlorine and excess lime, known as oxidation-reduction potential (ORP).

At the present time the ORP principle is most widely used in the manufacture of calcium hypochlorite bleaching chemical on a continuous basis, as typically shown in Figure 8-6.

The process begins with lime from the lime storage bin being fed to a slaker for continuous slaking with water at about a 6:1 water/lime ratio. The ratio of lime and water is automatically controlled by a ratio controller measuring the feed rate of lime to the slaker with FT-1 and controlling the flow of water to the slaker in relation to it. The feed rate to the slaker is controlled by level controller LRC-2 to maintain the desired level in the slaker. Slaker temperature is controlled by TRC-3 which regulates steam flow during slaking operations. The slaked lime is then pumped to a classifier and classified at a solids content of about 10 percent. Dilution water flow to the classifier is controlled by density recording controller DRC-4 and the 10 percent slurry entering the 10 percent slurry storage tank. The bubble-pipe, specific-gravity type measurement element is used in Figure 8-6 for this measurement, but other measurement methods such as radioactive nuclear devices have also been used.

The storage tank level is maintained by regulation of the lime slurry input with level control loop LT-5, LRC-5, and LV-5. The 10 percent slurry overflowing from the classifier is passed through a shell and heat tube exchanger where the slurry is cooled to 90°F.

The 10 percent slurry is then diluted continuously to 4.2 percent. Dilution water flow is controlled by density recording controller DRC-6, the same as on the 10 percent lime slurry. Mixing of the water and 10 percent slurry is obtained in the centrifugal pump which delivers diluted slurry to another classifier for further classification. The overflow from the classifier goes to a storage tank whose flow rate is controlled by level recording controller LRC-7, maintaining the level in the tank by recycling a portion of it, thus supplying a more or less steady flow to the classifier. The rejects from the classifier containing unslaked lime, calcium carbonate, silica, and other inert grits are pumped to the kraft mill lime recovery system where complete utilization of the unslaked lime and calcium carbonate is obtained. The 4.2 percent lime slurry is then pumped to a recycle tank whose level is controlled by a locally mounted controller, LC-8.

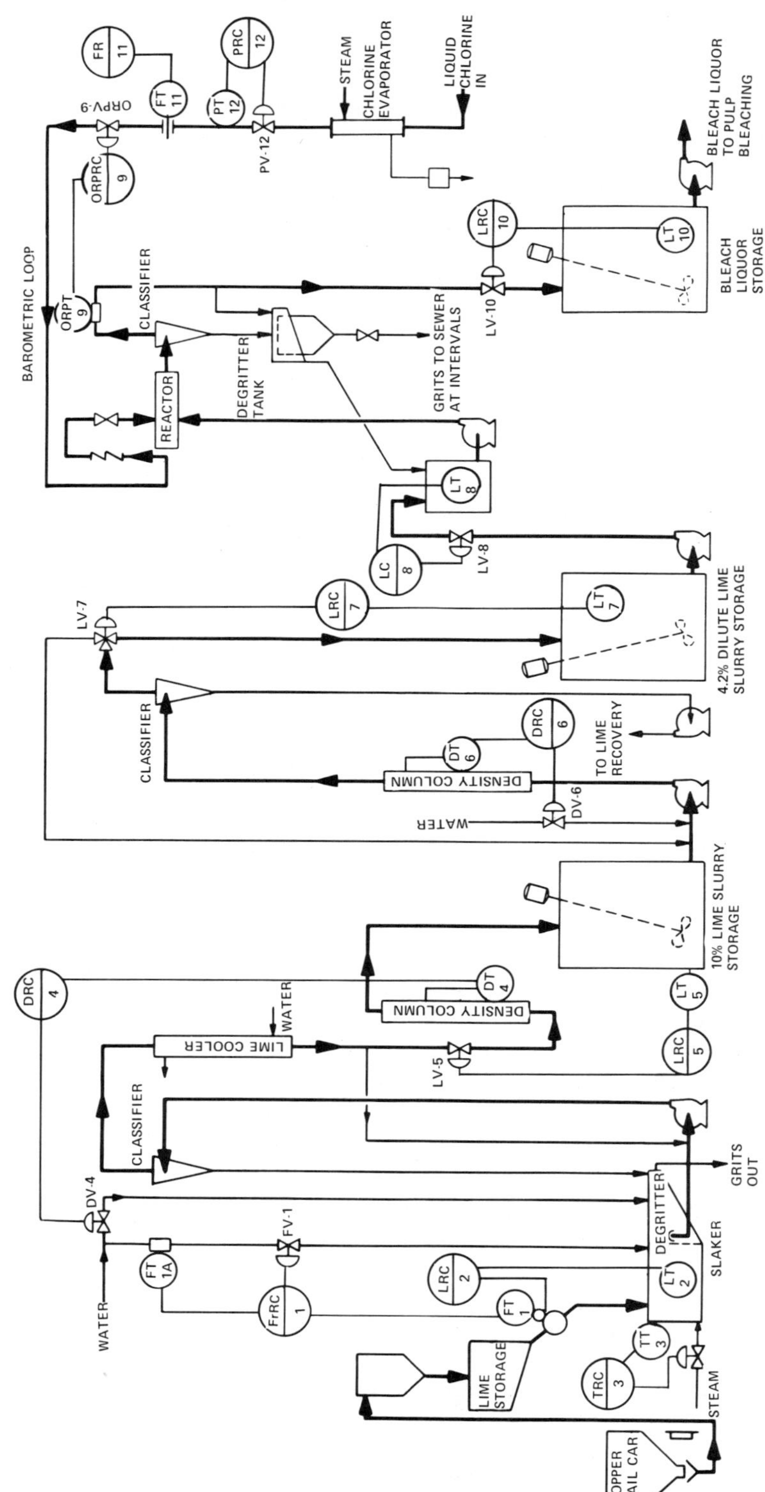

Figure 8-6. Manufacturing of calcium hypochlorite bleach liquor.

The recycle tank supplies partially chlorinated slurry to a reactor where it is intimately mixed with gaseous chlorine, the chlorine vaporizer system forming calcium hypochlorite bleach liquor which flows from the reactor to another classifier. There the residual lime solids are continuously removed. The semiclarified liquor passes through an electrode flow cell where the oxidation-reduction potential is measured by ORPT-9 with a silver-silver chloride and platinum electrode measuring element. A millivolt signal is produced and transmitted to controller ORPRC-9 which activates the automatic valve, ORPV-9, controlling chlorine gas flow to the reactor. The purpose is to maintain the desired ratio of excess lime to available chlorine concentration or strength of the bleach liquor in accordance with the properties shown on the graph in Figure 8-7. Some bleach liquor is recycled to the reactor and classifier for maximum grit separation with constant mixing in the reactor and constant time lag in the control circuit at varying production rates. Bleach liquor flow is controlled by level recording controller LRC-10 in the bleach liquor storage tank.

Sodium Hypochlorite. The manufacture of sodium hypochlorite, or soda bleach as it is sometimes called, follows the same chemical principle as of manufacture of lime bleach, but it is somewhat simpler because of better solubilities resulting in the lessening of the solids removal problem.

Figure 8-7. Oxidation potentials for various calcium hypochlorite solutions.

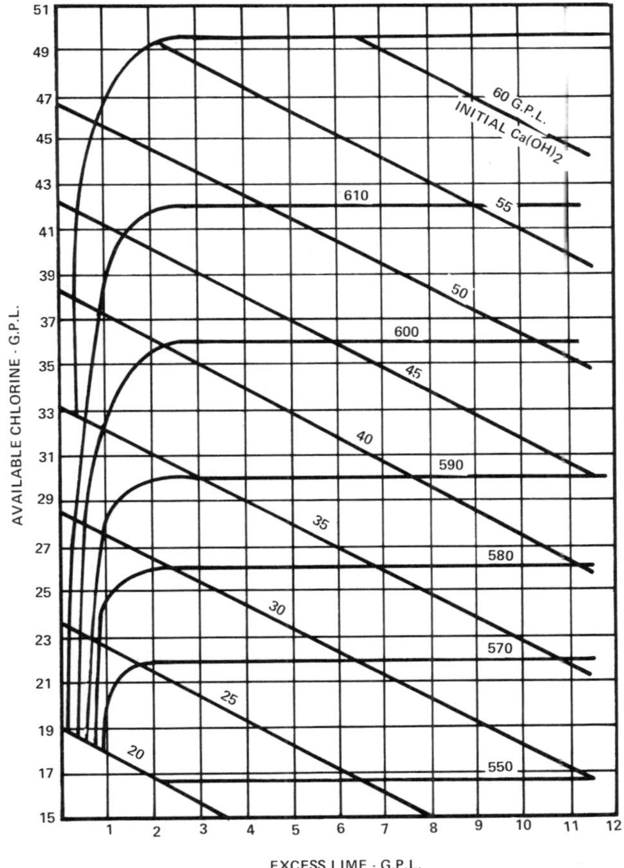

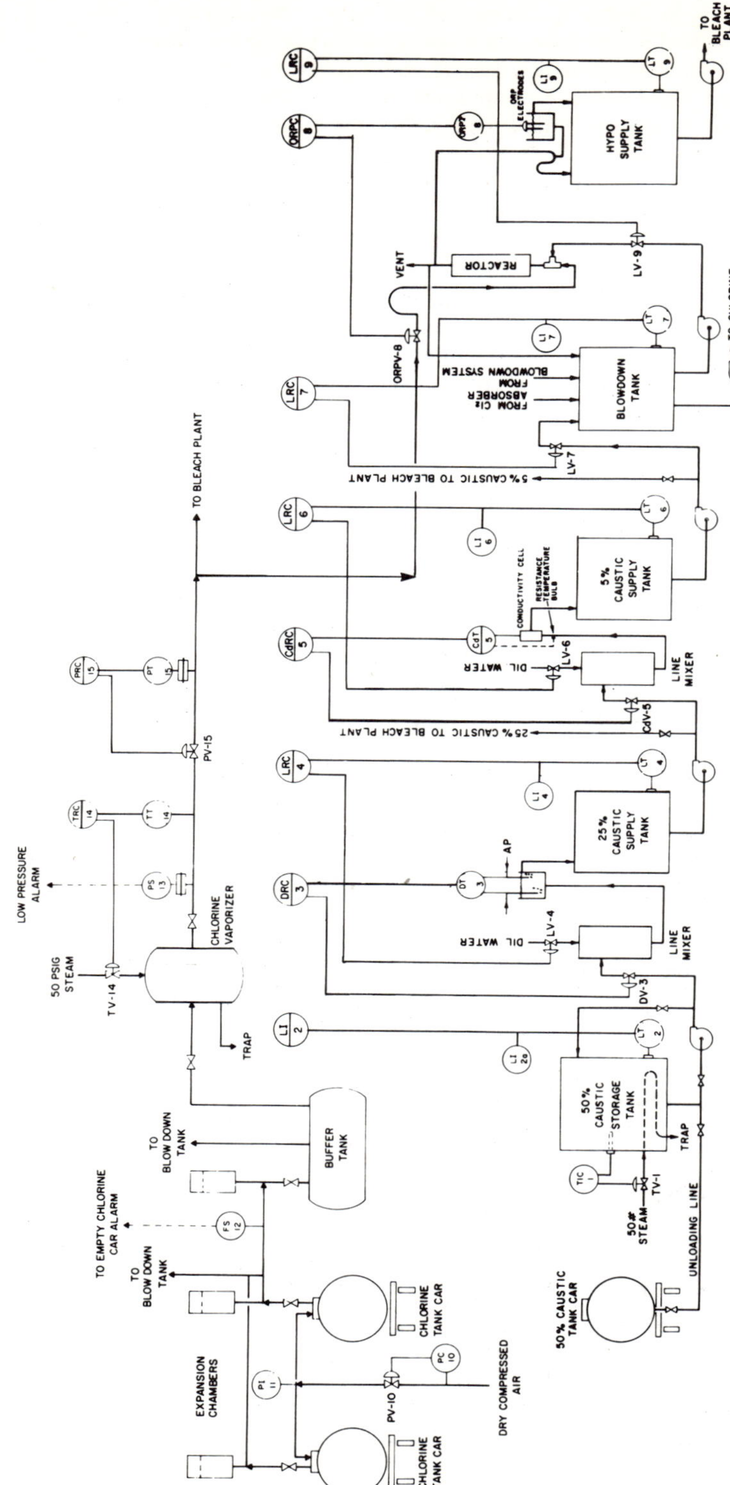

Figure 8-8. Preparation of sodium hypochlorite bleach liquor.

A typical flow diagram of a continuous sodium hypochlorite manufacturing process is shown in Figure 8-8.

Chlorine gas from the vaporizing system is piped directly to the bleach plant for use in pulp bleaching and for sodium hypochlorite bleach liquor makeup.

Caustic soda solution arrives at the pulp mill in tank car lots at a concentration of 50 percent. The desired concentration of 5 percent dilute caustic solution is continuously produced by mixing with proper amounts of water in two steps. The first step dilutes from 50 percent to 25 percent and the second from 25 percent to 5 percent. The dilute caustic solution and gaseous chlorine are reacted in proper proportions to produce the finished sodium hypochlorite bleach liquor for use in the plant. Some of the 25 percent and 5 percent solutions are also used directly in the bleaching process.

The primary control requirements for the instrumentation in the manufacture of sodium hypochlorite bleach liquor are to maintain proper:

1. Pressure of air padding to the chlorine cars.
2. Operation of the chlorine vaporizer by controlling temperatures of chlorine gas discharge.
3. Pressure of the chlorine gas to the bleach plant and hypochlorite bleach liquor makeup.
4. Dilution of the caustic solutions.
5. Mixing of the chlorine gas with caustic solution to produce hypochlorite bleaching chemical of the desired strength.

Pressure of the padding air to the chlorine tank cars is controlled by in-line adjustable-type regulator PC-10, PV-10, and is indicated locally by a direct-connected indicating gauge, PI-11. Flow switch FS-12 actuates an alarm on low flow, signaling an empty chlorine car, and pressure switch PS-13 actuates an alarm on low pressure in the chlorine gas line to the bleach plant and hypochlorite makeup. Control loop TT-14, TRC-14 controls the temperature of the chlorine gas by regulating the steam flow valve, TV-14, to the chlorine vaporizer. Control loop PT-15, PRC-15, PV-15 controls the chlorine gas pressure. All pressure measurements must be protected from the corrosive effect of the chlorine by chemical seals constructed of proper materials.

The caustic is received at the mill in tank cars containing a 50 percent concentration solution and pumped to a storage supply tank. To prevent the 50 percent caustic solution from crystallizing in the storage tank it is kept warm by steam controlled by a locally mounted temperature indicating controller and valve, TIC-1, TV-1. The level is measured and indicated both locally and remotely by LT-2, LI-2, LI-2a. It is then diluted to 25 percent concentration by mixing with water in a line mixer. A low differential pressure type of transmitter, DT-3, measures the line mixer discharge density. This signal is used by density recording controller DRC-3 to regulate the flow of 50 percent caustic to the line mixer to maintain the proper density level equivalent to 25 percent caustic concentration. The flow of dilution water to the line mixer is regulated by level loop LT-4, LRC-4, LV-4, maintaining the desired level in the 25 percent caustic supply tank. The same method is used to dilute the 25 percent caustic solution to 5 percent concentration except that a conductivity measurement, CdT-5, is made after mixing with water. A conductivity recording controller maintains a conductivity value equivalent to 5 percent caustic concentration by throttling the 25 percent caustic flow to the line mixer. The dilution water to the line mixer is regulated on the basis of the level in the 5 percent caustic supply tank by instruments LT-6, LRC-6, LV-6.

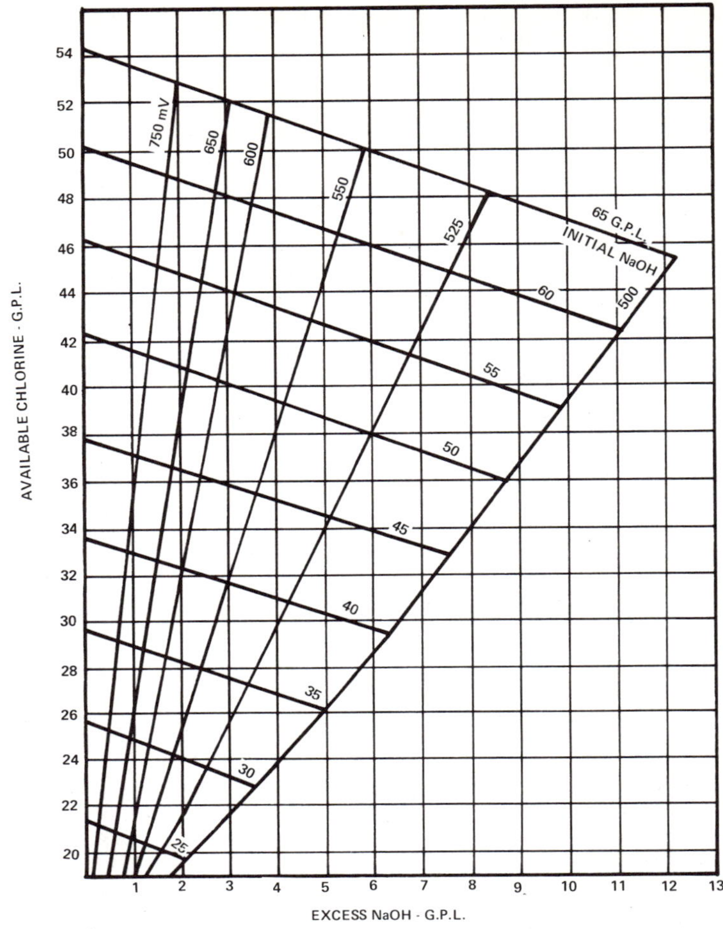

Figure 8-9. Redox potentials for typical NaOCl solutions.

Due to the type of chemical reaction that occurs when chlorine is mixed with a caustic solution, it being similar to the reaction with a lime solution, a relationship exists between the oxidation-reduction potential millivolts generated and the excess NaOH (caustic) to available chlorine that exists, which determines the final strength of the hypochlorite produced. This measurement is made by ORPT-8 and used by controller ORPC-8 whose control point is set in accordance with the graph in Figure 8-9 to regulate the amount of chlorine being reacted with the 5 percent caustic solution. This is accomplished by adjusting ORPV-8 to maintain the corresponding hypochlorite strength being made. Level in the hypochlorite supply tank is maintained by loop LT-9, LRC-9, LV-9, regulating the flow of 5 percent caustic to the reactor. This level is also indicated locally by level indicator LI-9.

Chlorine Dioxide

Due to its unique ability to oxidize residual lignins and extraneous coloring matters to water soluble, colorless materials with minimum detrimental effect on the desirable cellulose fiber component of pulp, the use of chlorine dioxide gas has spread very rapidly and is today an essential element of sulfate pulp bleaching. It is a gas of such explosiveness that it cannot presently be commercially prepared safely in undiluted form and is so unstable that it cannot be liquefied for transport and storage. Therefore, bleach plants that use this bleaching agent must manufacture it on the spot. It is imperative to hold its partial pressure in the gaseous phase below 100 millimeters of water. All chlorine dioxide produced so far has been ultimately derived from chlorates by reduction. A number of processes have been developed for its production. Although they differ in design and method, most of them are similar in that the process involves the basic chemical reaction of reducing sodium chlorate in the presence of an acid and absorbing the chlorine dioxide gas in water to form a solution. Sulfuric acid is the most common acid used, while the reducing agents that have been used are methyl alcohol, sulfur dioxide, sodium chloride, hydrochloric acid, and chromic sulfate.

Methyl Alcohol Process. The flow diagram in Figure 8-10 shows instrumentation involved when the two-reaction-vessel method of manufacturing chlorine dioxide with methyl alcohol is used. The methyl alcohol, sodium chlorate, and sulfuric acid for one day's operation are shipped to the plant site and stored in tanks. The tank levels are indicated by LI-1, LI-2, and LI-3. Both reactors are filled with reagent to a certain height, leaving a gas chamber above the level. In the reactor the methyl alcohol reduces the sodium chlorate to produce chlorine dioxide gas. The flows of methyl alcohol to the reactors are controlled by flow control loops FT-5, FRC-5, FV-5, and FT-6, FRC-6, FV-6. The flow measuring elements used are of the electromagnetic type. The flows of sodium chlorate and sulfuric acid solutions are controlled in the same manner by flow recording controllers FRC-7 and FRC-8. The sulfuric acid is diluted on its way to the reactors in a cooled acid dilution tower with water addition being controlled by flow recording controller FRC-4.

The reagents enter the first reactor through nozzles from the bottom and are mixed thoroughly with compressed air which is metered to each reactor by flow control loops FT-9, FRC-9, FV-9, and FT-10, FRC-10, FV-10. The partially exhausted reaction liquor then flows into the second reactor where further amounts of methanol and sulfuric acid are added. Exterior heating jackets permit necessary heating of the reaction mixture.

Sulfur Dioxide Process. A typical process in which chlorine dioxide is generated by the reduction of sodium chlorate with sulfur dioxide in the presence of sulfuric acid is illustrated in Figure 8-11.

The sodium chlorate and sulfuric acid feeds from the storage tank to the chlorine dioxide reactor are measured with electromagnetic-type flow elements FE-1 and FE-2, and regulated by flow recording controllers FRC-1 and FRC-2. Sulfur dioxide is shown as being shipped to the plant site in tank cars. Many mills manufacture their own sulfur dioxide by burning elemental sulfur in various manners similar to processes used to produce sulfur dioxide gas for making up sulfite pulping liquor, described in Chapter 2 on pulping. Pressure in the sulfur dioxide gas line to the reactor is controlled by recording controller PRC-13. The sulfur dioxide is ratioed to the flow of air to the generator by ratio flow recording controller RFRC-3. If the temperature of the generator gas rises too high, temperature recording controller TRC-10 opens valve TV-10 on the purge air line to the top of the generator. The purge air line pressure is controlled by venting it to atmosphere by controller PRC-14. The temperature in the generator is maintained constant by cooling water flowing into a jacket around the generator body and is under the control of recording temperature controller TRC-11. The chlorine dioxide gas from the generator is absorbed in water in an absorption

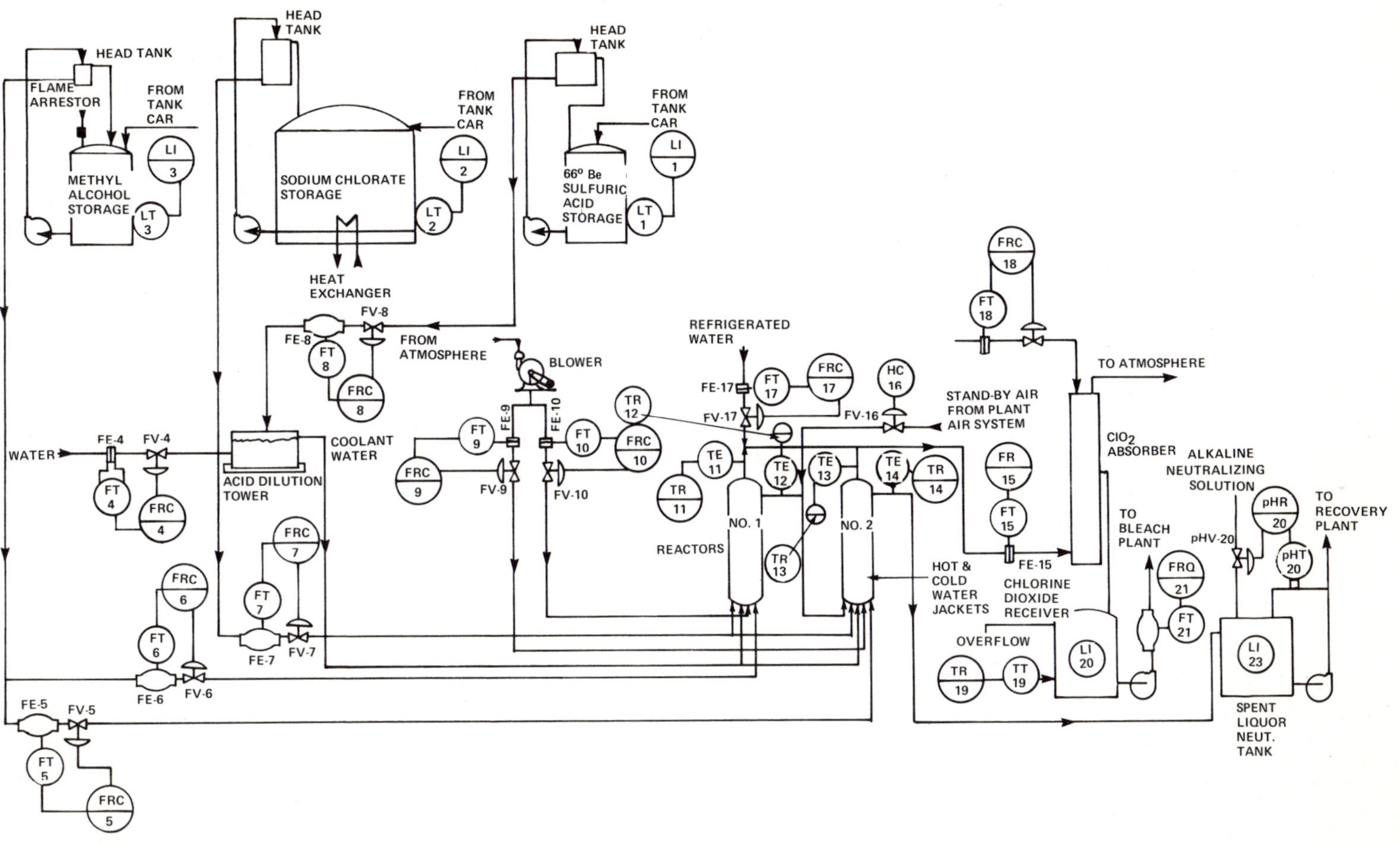

Figure 8-10. Chlorine dioxide manufacturing using methyl alcohol.

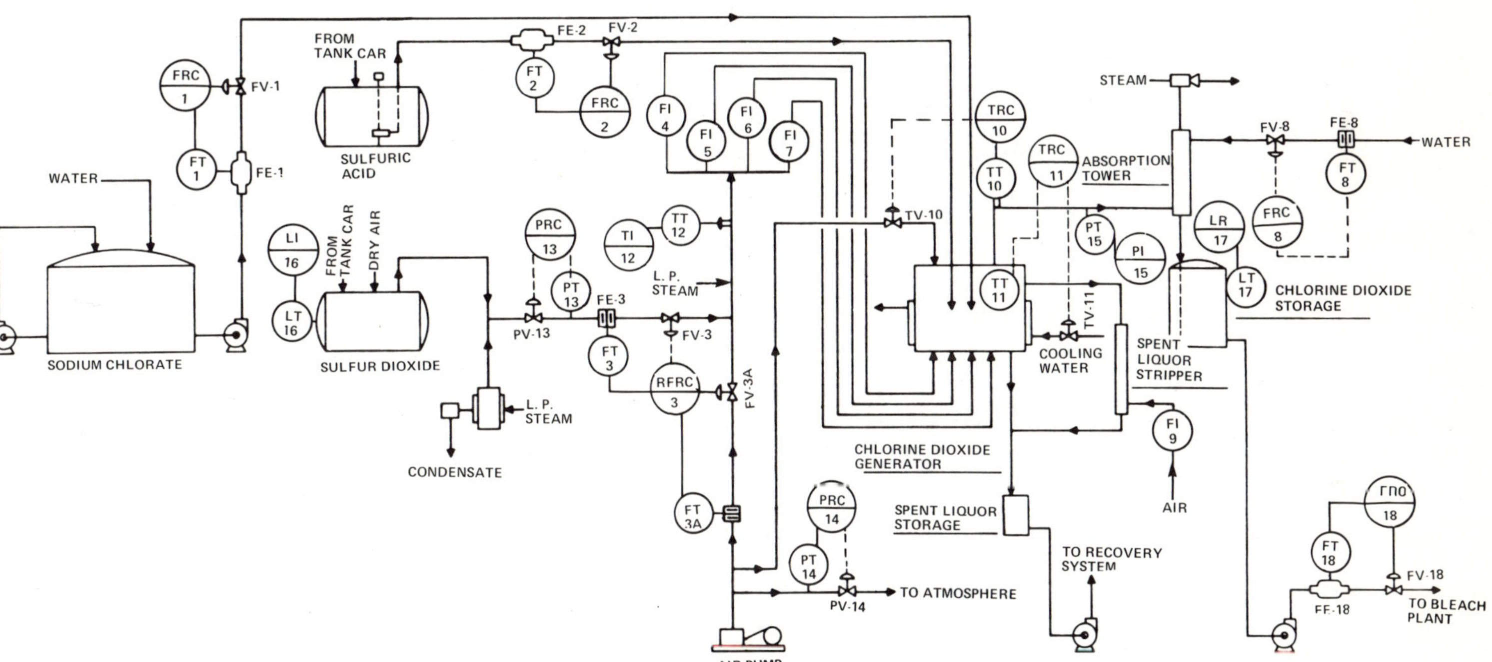

Figure 8-11. Chlorine dioxide manufacturing using sulfur dioxide.

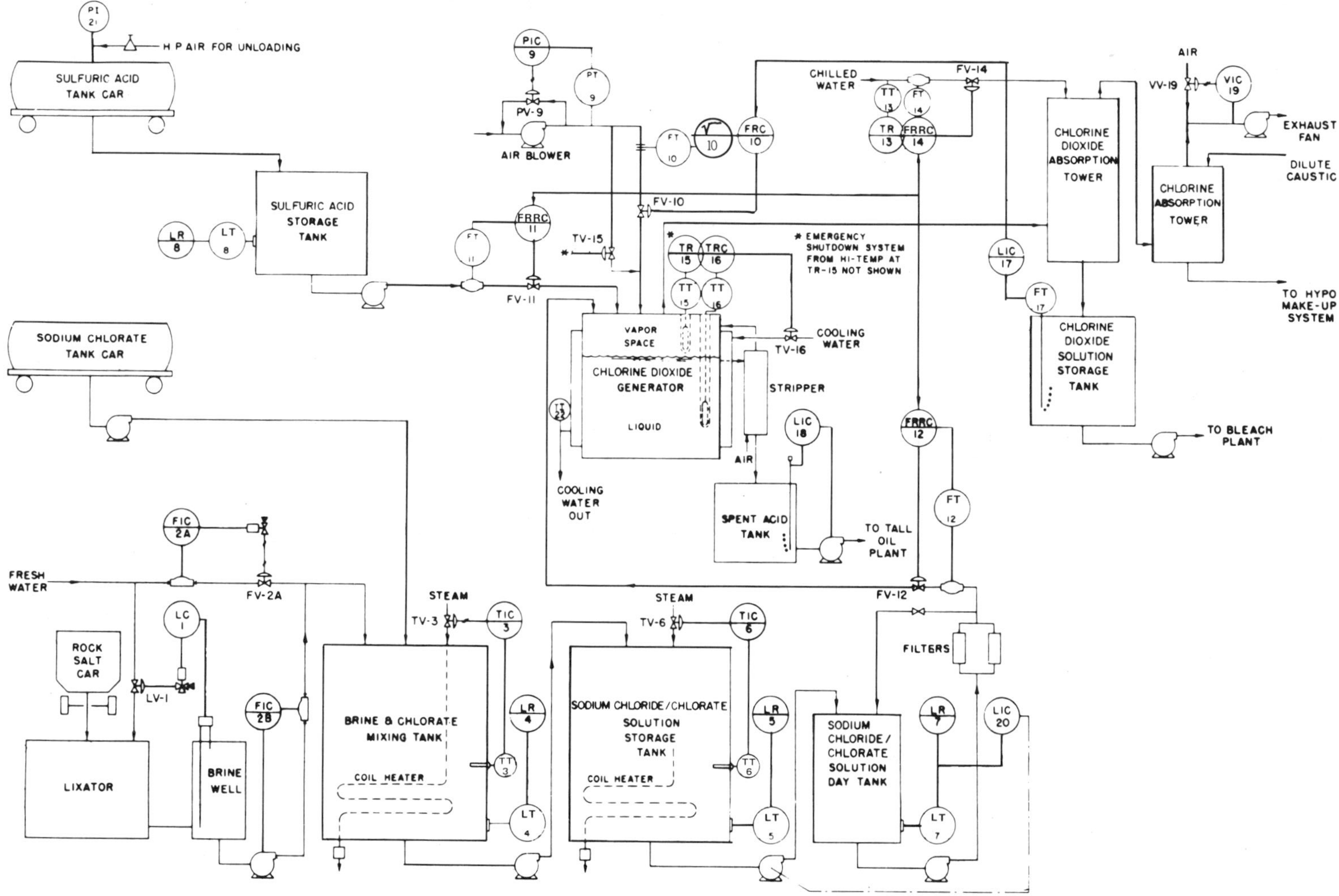

Figure 8-12. Chlorine dioxide manufacturing using sodium chloride-chlorate solution.

tower. The resulting solution is sent to a short storage tank and pumped as soon as possible to the bleaching operation for immediate use. Spent liquor from the generator is stripped and pumped to the recovery system for burning in the recovery boiler. Pressure and level indicators are located strategically around the process as guides to the operation of the equipment.

Sodium Chloride Process. Figure 8-12 illustrates a typical chlorine dioxide bleach liquor manufacturing process based on the reduction of sodium chlorate by sodium chloride. Sulfuric acid and sodium chlorate solutions are delivered to the mill site by tank car and pumped into storage tanks. The sodium chloride (salt), brought by truck or rail car, is dissolved in a lixator. The sodium chloride solution (brine) is mixed with the sodium chlorate solution to form a sodium chloride-chlorate solution containing approximately equimolar concentrations of sodium chloride and sodium chlorate. This solution and sulfuric acid are fed to the generator where, under acid conditions, the sodium chloride reacts with the sodium chlorate reducing it to form chlorine dioxide and chlorine in about a 2:1 ratio. The gases that are formed are stripped out with air and diluted to a safe concentration.

The gas mixture first passes through an absorption tower where the chlorine dioxide and about 25 percent of the chlorine are absorbed in chilled water. The gas leaving the absorber, which still contains valuable chlorine, is then passed through a smaller packed tower irrigated with dilute caustic soda, or through a scrubber-eductor which is fed with dilute caustic solution to recover the chlorine as hypochlorite.

Control requirements of the instrumentation system include:

1. The control of acid flow to maintain proper acid conditions in the chlorine dioxide generator.
2. The maintaining of necessary flow of brine and levels in sodium chloride-chlorate storage tanks to assure proper ratio of sodium chloride to sodium chlorate.
3. The regulation of flows of sodium chloride-chlorate solutions and air to maintain a balanced stripping of chlorine dioxide-chlorine gas produced.
4. The monitoring and control of temperatures in order to initiate emergency safety measures on the development of hazardous conditions causing sudden temperature increase in the process.

The level device on the brine well is a single one-point capacity probe on-off controller, LC-1, which admits fresh water to the lixator to maintain the level below a maximum in the well. Indicating flow controller FIC-2A receives a flow signal from a positive displacement-type measuring element in the fresh water brine and operates an on-off valve, FV-2A, to batch fixed quantities into the mixing tank. Controller FIC-2B performs the same function by starting and stopping a pump on the brine line. Temperatures of the mixing and storage tanks are controlled by instrument loops TT-3, TIC-3, TV-3 and TT-6, TIC-6, TV-6 by regulating steam to the coil heaters located in the tanks. Their levels are recorded by loops LT-4, LR-4 and LT-5, LR-5. LT-7 and LR-7 are used to measure and record the level in the day tank of the sodium chloride-chlorate solution. It is controlled by indicating controller LIC-20, starting and stopping the pump between the storage and the day tank.

Pressure gauge PR-21 is used to locally monitor the pressure in the sulfuric acid car as a guide to when it is empty. LT-8, LR-8 are used to measure and record the level in the sulfuric acid storage tank. The stripping air pressure to the generator is controlled by throttling a bypass valve, PV-9, around the air blower with an indicating pressure controller, PIC-9, from a pressure measurement being made with transmitter PT-9. The airflow is measured by an orifice plate and flow transmitter FT-10, whose output signal is used by flow recording controller FRC-10 to operate valve FV-10 to control the airflow to the generator. This same signal is transmitted as the primary signal to flow ratio controllers FrRC-11, FrRC-12, and FrRC-14. A controlled amount of sulfuric acid and sodium chloride-chlorate solution passes to the generator, and chilled water goes to the chlorine dioxide absorption tower in relation to the stripping airflow. With this arrangement, all flows would automatically shut down if there were a failure of the airflow to the generator, preventing the development of a dangerous accumulation of unpurged gases in the generator. Any malfunction causing an excessively hazardous temperature in the generator is detected by temperature recorder TR-15, which actuates an emergency shutdown system stopping all flows to the generator and flooding it with water. Temperature recording controller TRC-16 operates control valve TV-16 throttling the cooling water to the generator jacket to maintain temperature at optimum operating and safety values. Level indicating controller LIC-17 on the storage tank for the chlorine dioxide solution remotely sets the control point of flow controller FRC-10 on the air line to the generator. This adjusts the flow of air and all other ratioed flows to the generator and absorption tower, as the level tends to vary in the storage tank (caused by the amount of chlorine dioxide bleach solution being used by the bleach plant). The outlet temperature of the generator cooling water is measured by a locally mounted temperature indicator, TI-22, while a locally mounted bubble-type measurement level indicating controller, LIC-18, maintains a constant level in the spent acid tank. Another local controller, VIC-19, regulates the vacuum of the evacuating system on the chlorine absorption tower.

Electromagnetic flowmeters are used to make the chemical and water flow measurements to the chlorine dioxide generator.

Single Vessel. It is noted that, in the manufacture of chlorine dioxide, there is a spent liquor effluent from the reactors that is primarily sulfuric acid. Some mills use this effluent to split tall oil from the black liquor evaporators or send it to the recovery area for adding to the black liquor being burned. Many mills do not have tall oil plants and it is preferable to have the effluent sent to the recovery area in the form of sodium sulfate (salt cake). Any acid in the excess spent liquor ending up in the plant effluent must be neutralized before allowing it to flow out of the mill so that it will not become a source of pollution.

A chlorine dioxide manufacturing system which produces no sulfuric acid in the effluent is shown in Figure 8-13. The heart of the system is a single vessel which acts as a generator for chlorine dioxide gas, an evaporator for the water formed, and a crystallizer of the anhydrous sodium sulfate. These functions are performed at an elevated temperature and a reduced pressure.

For simplification, only the major components and instrumentation of a typical system are shown in the figure. In operation, a stream of sodium chlorate and sodium chloride solution and a separate stream of sulfuric acid are fed to the generator-evaporator-crystallizer reactor under flow control of loops FT-1, FRC-1, FV-1 and FT-2, FRC-2, FV-2. The reactor liquid is circulated through a heat exchanger which supplies the heat needed to sustain chemical reaction and evaporate the water in the system. Temperature recording controller TRC-3 is used to control this temperature by regulating the steam flow to the heat exchanger with control valve TV-3. The gaseous chlorine dioxide and chlorine formed together with the steam are drawn out of the reactor vessel and into a condenser where the water vapor is removed. Condenser temperature is controlled by temperature indicating controller TIC-4. The gaseous chlorine dioxide and chlorine are then

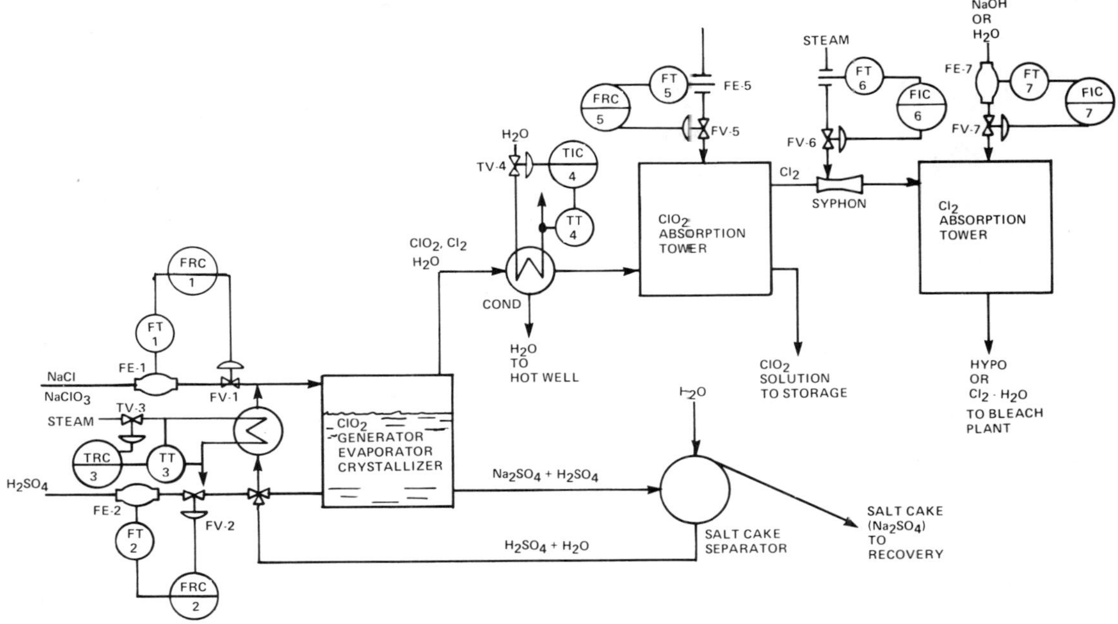

Figure 8-13. Chlorine dioxide manufacturing using a single vessel.

drawn into a chlorine dioxide absorber where chilled water selectively absorbs the chlorine dioxide, producing the bleaching solution which is sent to storage for use by the bleach plant.

The remaining gaseous chlorine is drawn into a chlorine absorber with water or sodium hydroxide to produce chlorine water or sodium hypochlorite solution which is used in the pulp bleaching process.

A continuous stream of salt cake slurry, made up primarily of sodium sulfate crystals, and sulfuric acid is drawn from the generator reactor to a separator where the crystals are washed and filtered. The washings and mother liquor containing the effluent sulfuric acid are returned to the generator. The separated sodium sulfate (salt cake) crystals are sent to the recovery area where they are dissolved in the black liquor for chemical makeup before burning.

Peroxide Solution

Peroxide bleaching solutions are prepared by either batch or continuous processes. The processes consist of mixing sodium silicate, caustic, and hydrogen peroxide solutions shipped to the plant site by tank cars. As shown in Figure 8-14, the process is relatively quite simple with individual flows of each chemical being measured with magnetic flow elements FE-1, FE-2, and FE-3. They are recorded, controlled, and totalized by integrating flow recording controllers FRCQ-1, FRCQ-2, and FRCQ-3. Orifice plate FE-4 is used to measure the flow of water and it is recorded, controlled, and totalized by FRCQ-4. The mixed bleaching solution typically goes directly to the pulp mixer with no intermediate storage.

Sometimes epsom salts are dissolved and added as a stabilizer, depending on mill operating conditions. Peroxide solutions are used extensively for bleaching mechanical- and chemimechanical-type pulps.

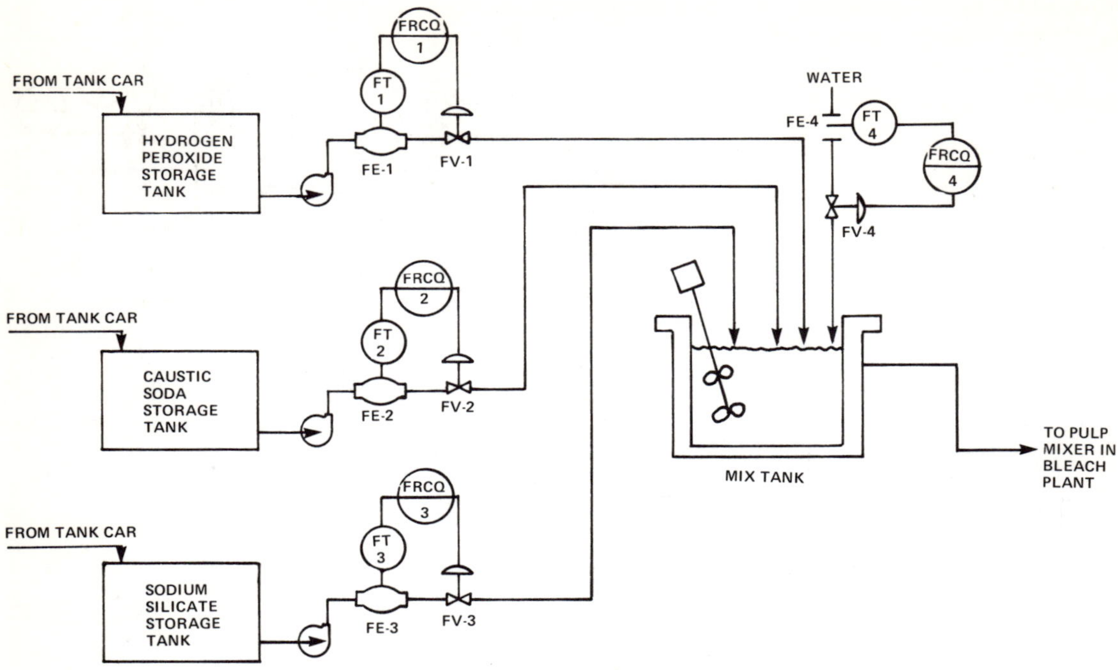

Figure 8-14. Makeup system for peroxide bleaching solution.

PULP BLEACHING METHODS

Recent history in the pulp and paper industry has experienced a veritable revolution in the wood pulp bleaching processes during which the simple batch single- and two-stage hypochlorite processes have been largely replaced by continuous multistage processes in which chlorination, caustic extraction, chlorine dioxide, and peroxides have been notable additions to the simple oxidation with hypochlorites. Even more recently, oxidation with the use of elemental oxygen has found an important place in some pulp bleaching procedures. The continuous stepwise purification in the bleaching of pulp has now become common practice. The continuous multistage process has not only been used on the bleaching of sulfite-type pulps but, more importantly, it has made practically feasible the successful bleaching of other types of pulp, primarily sulfate and semichemical.

It has now become practice to refer to the various stages in the bleaching process by letters representing the primary chemical used in that stage, such as:

C-chlorination stage
E-caustic or alkali extraction stage
H-hypochlorite stage
D-chlorine dioxide stage
P-peroxide stage

There is considerable variation in the technique and chemicals used between plants in the bleaching operation depending on the fibrous raw material from which the pulp has been produced, such as wood species, agriculture residues, and grasses; the pulping process used (sulfite, sulfate, semichemical, etc); and the final use of the bleached pulp, i.e., fine papers, coarse papers, heavy weights, light weights, dissolving pulps, etc.

Sulfite Pulp Bleaching

At one time, the bleaching of sulfite pulp was predominant. It was originally done in a batch-type operation, but practically all of these operations have been replaced by the continuous multistage process.

The first stage in bleaching sulfite and sulfate pulps is, in most cases, chlorination. The instruments for controlling a chlorine stage are shown in Figure 8-15. In this particular installation, the washed stock storage tank level is measured and recorded on LR-1 so that the operator can see how his bleaching operation rate is keeping up to the washing and screening rate ahead of him. High and low alarms on the recorder warn him when he is approaching the upper and lower limits of his storage tank. Consistency of the stock from the storage chest is measured by transmitter CT-4 and controlled by recording controller CRC-4 which adds dilution water to the suction side of the stock pump. Pulp flow to the first-stage chlorination tower is measured by a magnetic-type flowmeter, FT-5, and controlled by controller FRC-5. Chlorine gas is mixed with water in an eductor and then added to the stock in a mixer. Temperature of the stock is recorded on TR-6. The chlorine flow is controlled by controller FRC-7 whose control point is remotely set from the output of an oxidation-reduction potential recording controller, ORPRC-8, operating from a measurement taken by an oxidation-reduction potential probe, ORPE-8. The probe is located in the pulp line or tower in such a manner that it represents residual chlorine. This loop adds chlorine based on the chlorine demand of the pulp. Chlorine addition is also being made on the basis of an optical-type measurement such as a colorimeter or brightness instead of the oxidation-reduction potential.

For safety reasons, a low stock flow interlock system, which will shut off the flow of chlorine if there is any accidental or inadvertent disruption of the pulp supply from the storage chest, is always included.

Figure 8-16 is a diagram showing the remaining two stages of the typical three-stage sequence for bleaching sulfite pulp, i.e., E-caustic extraction and H-hypochlorite treatment. The instrumentation required includes level recorders LR-11 and LR-22, used to maintain the proper retention time in each tower because they are downflow-type towers. If the towers used were upflow, they would operate full at all times like the chlorination tower and, therefore, would not require a level adjustment to maintain proper retention time. Level transmitters LT-11, LT-22, and LT-31 are of the diaphragm sealed, flanged type because measurement is made on pulp stock. Caustic and hypochlorite solutions going to the pulp are measured and controlled by flow loops FT-8, FRC-8, FV-8 and FT-16, FRC-16, FV-16, which are set by the operator to obtain the desired pulp whiteness (referred to as brightness).

Steam used for heating the pulp to the temperature for best reaction of chemicals on the impurities is controlled by temperature controllers TRC-10 and TRC-21.

The approximate interface level between the heavy stock and the diluted stock in the bottom of the towers is detected by dual-temperature recorders TR-12 and TR-23, whose upper thermal element will sense the warm pulp and whose lower element senses the cooler diluted stock under normal conditions. Dilution water is added by manually operated hand stations HIC-20 and HIC-29.

Steam, hot water, and fresh water usage by the bleaching operation is recorded and totalized by integrating flow recorders FRQ-32, FRQ-33, and FRQ-34.

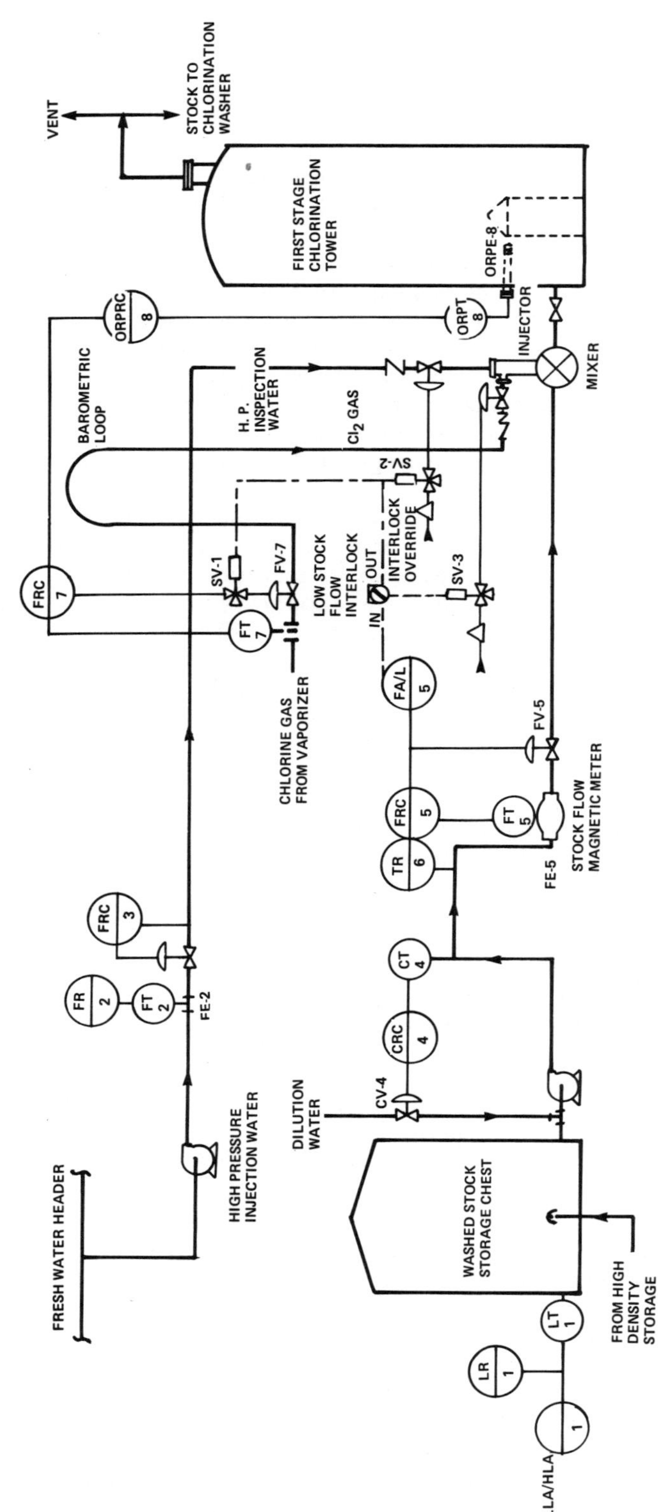

Figure 8-15. First-stage chlorination.

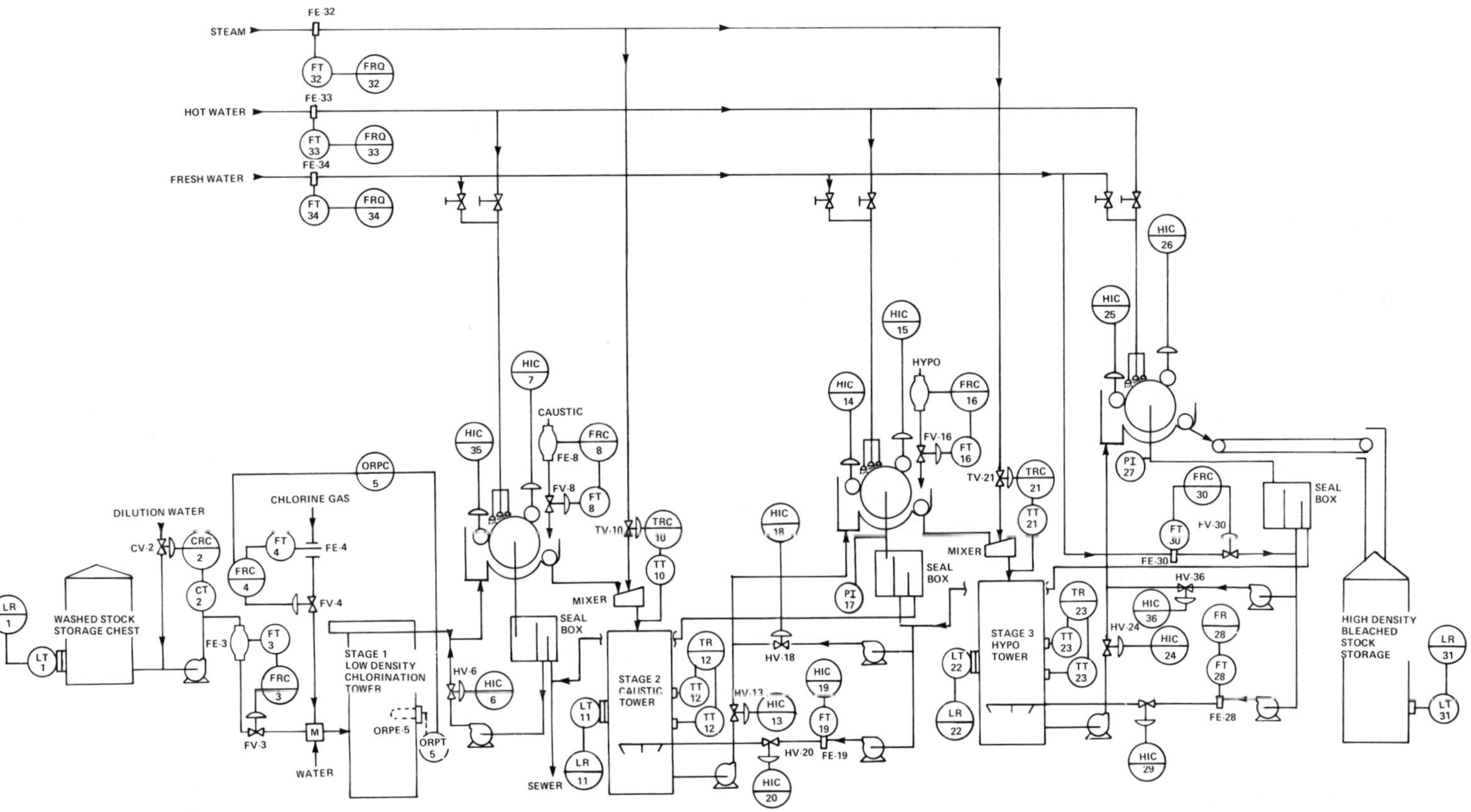

Figure 8-16. Three-stage sulfite pulp bleaching.

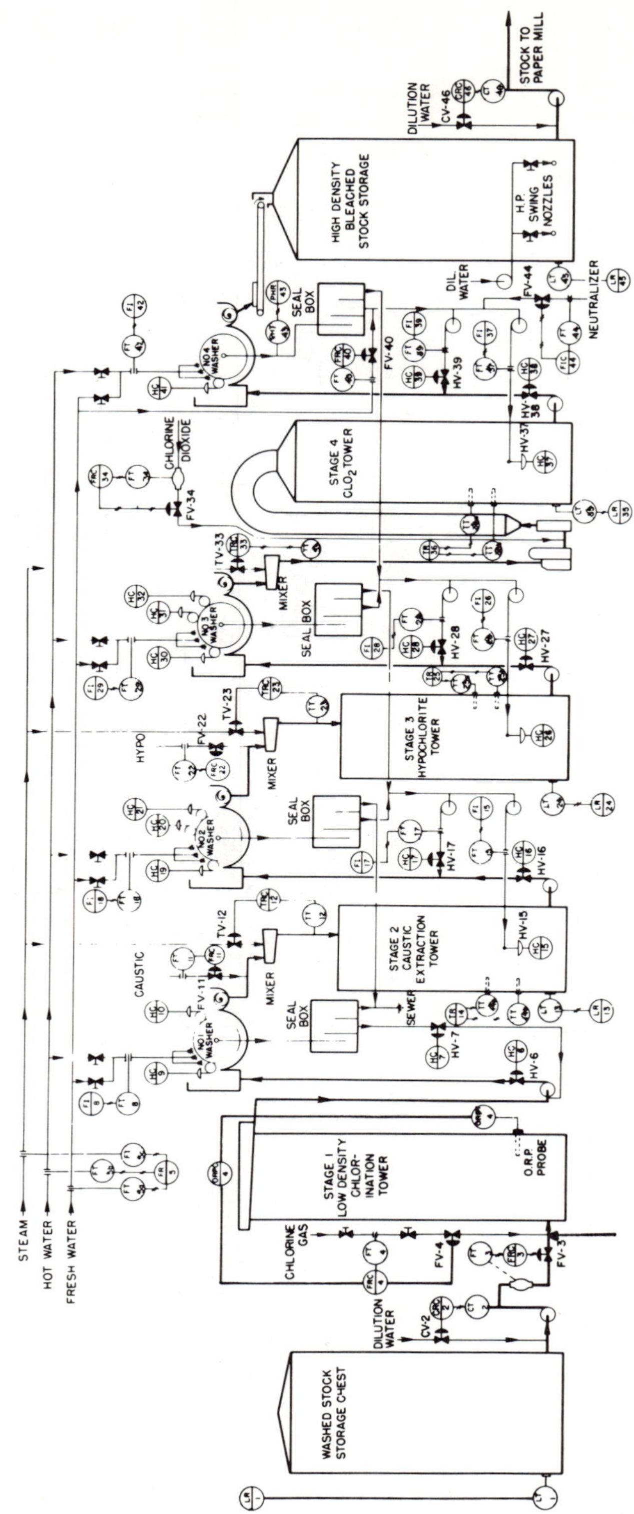

Figure 8-17. Four-stage sulfate pulp bleaching.

Sulfate Pulp Bleaching

As with sulfite pulp bleaching, sulfate bleaching systems are not all exactly alike in design and operation. The particular configuration used by a mill results from research and planning to fulfill the objective of a given mill. Although the number of stages used can vary from three to seven, the basic principles are the same and can be found in a typical four-stage bleach plant diagrammed in Figure 8-17. Other system arrangements are usually a variation or combination of the stages shown here, with a P-peroxide stage being added to or substituted for the D-chlorine dioxide stage.

Four-Stage Bleaching. Four-stage bleaching of sulfate pulp is also accomplished by a series of chemical reactions. These reactions must be closely controlled in order not to excessively destroy the desirable fiber component and affect its strength. Various bleach chemicals are mixed with the wet pulp, heated to the desired temperature with live steam, and retained for a period of time as they pass through a tower. On leaving the tower, the chemicals are washed out by dilution and dewatering, using suction on interstage washers similar to those used in the pulp washing operation. The washed pulp then goes to the next treatment stage.

The first stage of modern multiple-stage pulp bleaching is the low-density chlorination where chlorine mixed with the pulp slurry is allowed to react with residual lignins. After washing, the pulp enters the second caustic extraction stage where chlorinated fiber residues and other alkali-soluble constituents are dissolved in sodium hydroxide. The pulp is washed again and enters the third hypochlorite stage. This is the whitening stage where coloring matter is destroyed by a hypochlorite solution, usually sodium hypochlorite but sometimes calcium hypochlorite. After another washing, the pulp is further whitened by the use of chlorine dioxide (ClO_2) in the fourth stage, followed by a final washing. It is then stored in storage towers prior to being sent to the paper mill to be prepared for papermaking.

Level recorder loops LT-1, LR-1 and LT-45, LR-45 on washed and bleached stock storage tanks are used as operator guides to stock supply and paper mill demand. The desired pulp throughput on which chemical additions and retention time in reaction towers depend is controlled by flow control loop FT-3, FRC-3, FV-3 and consistency control loop CT-2, CRC-2, CV-2. Chlorine dosage is based on the residual chlorine (active chlorine left at the end of chlorination). An oxidation-reduction potential (ORP) measurement made at the proper location in the chlorination tower is related to the residual chlorine. Therefore, chlorine addition can be controlled by an oxidation-reduction potential controller, ORPC-4, cascading its output to the set point of chlorine flow recorder controller FRC-4. This controller regulates chlorine control valve FV-4 in the chlorine supply line.

Level recorders LR-13, LR-24, and LR-35 are used to maintain proper retention time in each tower. The chlorination tower operates under full condition at all times; hence no level recorder is required. Flow loops FRC-11, FT-11, FV-11 and FT-22, FRC-22, FV-22 and FT-34, FRC-34, FV-34 measure and control the exact amounts of caustic, hypochlorite solution, and chlorine dioxide solution going to the pulp. Settings are made by the operator periodically, depending on the efficiency of the chlorination stage and degree of pulp brightness desired.

Temperature controllers TRC-12, TRC-23, and TRC-33 control the amount of steam used for heating the pulp to the temperature necessary for optimum action of the chemical on the impurities.

Temperature controllers TR-14, TR-25, and TR-36 are used to detect the approximate interface level between the heavy stock and the diluted stock in the

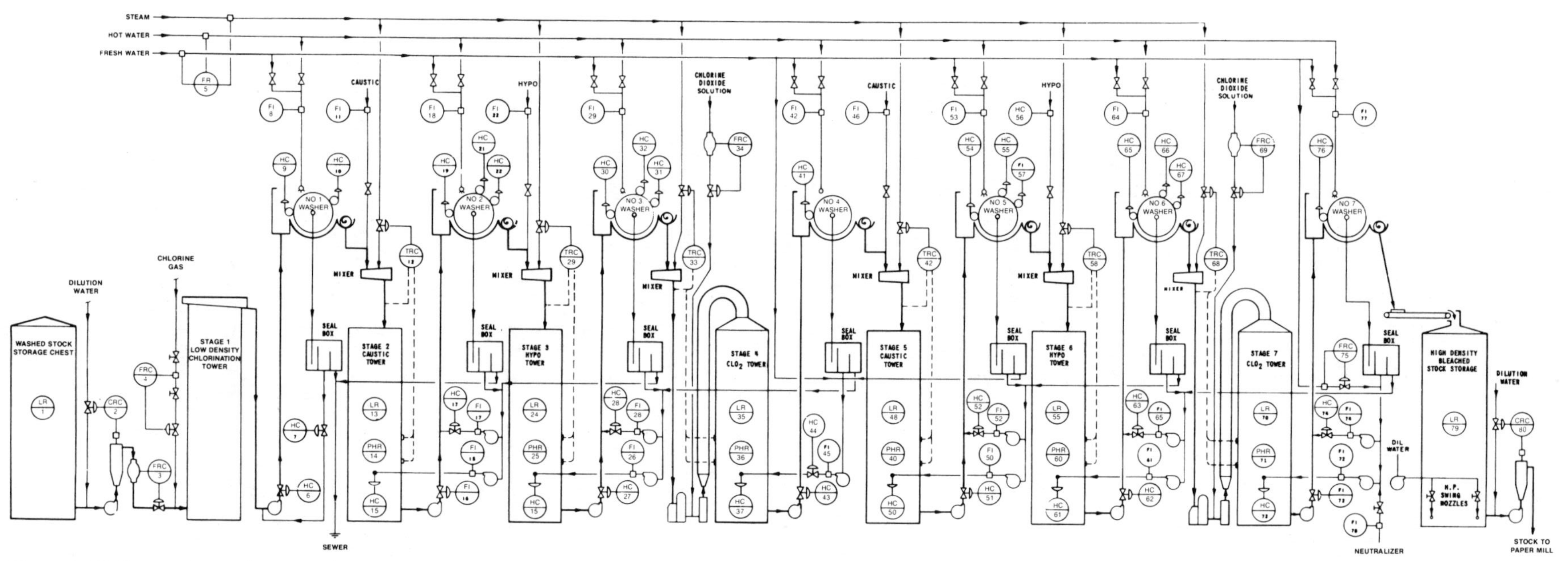

Figure 8-18. Seven-stage sulfate pulp bleaching.

bottom of the towers. Modern bleach plants operate the caustic extraction, hypochlorite, and chlorine dioxide stages at high consistencies. However, the pulp must be diluted to be pumped to the washer. Manual dilution is achieved by the operation of valves HV-7, HV-15, HV-17, HV-26, HV-28, HV-37, and HV-39 by hand-control stations HC-7, HC-15, HC-17, HC-26, HC-28, HC-37, and HC-39. Under normal conditions the upper bulb will sense the warm pulp and the lower bulb will sense the cooler diluted stock. Dilution water is added to maintain the interface level between the two bulbs.

FR-5 records the steam and water flows to the bleach plant and FRC-40 regulates the amount of fresh water added to the system. Recorder pHR-43 supplies a record of the acidity or alkalinity of the ClO_2 washer effluent, which is used as a guide in determining the set point of neutralizer flow controller FRC-44.

The electromagnetic flow transmitters for pulp slurry and chemical flow measurements are selected to meet specific plant requirements. Tower temperature measurements use special heavy-duty thermal wells to withstand the stress and strain produced by the heavy, moving pulp. All instrument materials are selected to meet the severe corrosive services found in bleach plants.

Seven-Stage Bleaching. A typical seven-stage system for bleaching sulfate pulp is shown in Figure 8-18. This figure illustrates that when more than four stages are used the overall arrangement is a combination and/or repetition of the four basic stages of C-chlorination, E-caustic extraction, H-hypochlorite, and D-chlorine dioxide, with P-peroxide stage added to, or substituted for, the D-chlorine dioxide stage. The instrumentation for each stage of bleaching is also similar to that used for the four-stage configuration.

Bleaching With Diffusion Washing. There are some bleaching systems in operation that use diffusion-type washers between stages instead of vacuum rotary-type washers. The diffusion washer system consists of two concentric rings of vertically oriented double-sided screens which are separate but interconnected and supported in the lower extremities with three drainage arms. These arms are connected through linkages to three overhead series-connected hydraulic cylinder units which provide an up and down cyclic operation to the bank of screens. A vertical lift movement is followed by a rapid downward thrust. With this rapid plunge, a simultaneous momentary cutoff of the filtrate flow is triggered, and this combination of events wipes the screen perforations free of fiber accumulation. The cycle is continually repeated. The cutaway drawing in Figure 8-19 shows how the diffusion washer is constructed and how it works.

Figure 8-19. Diffusion washer.

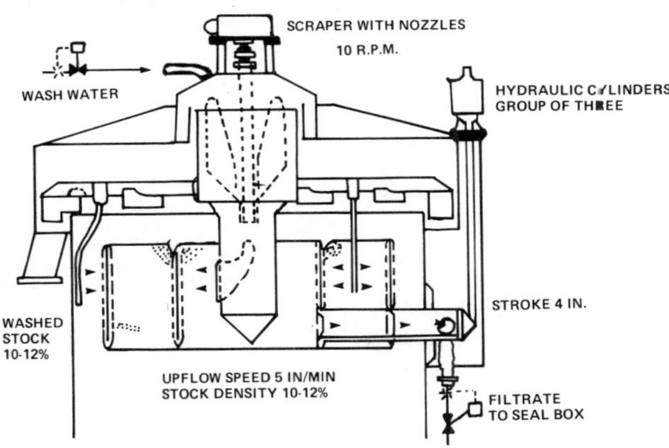

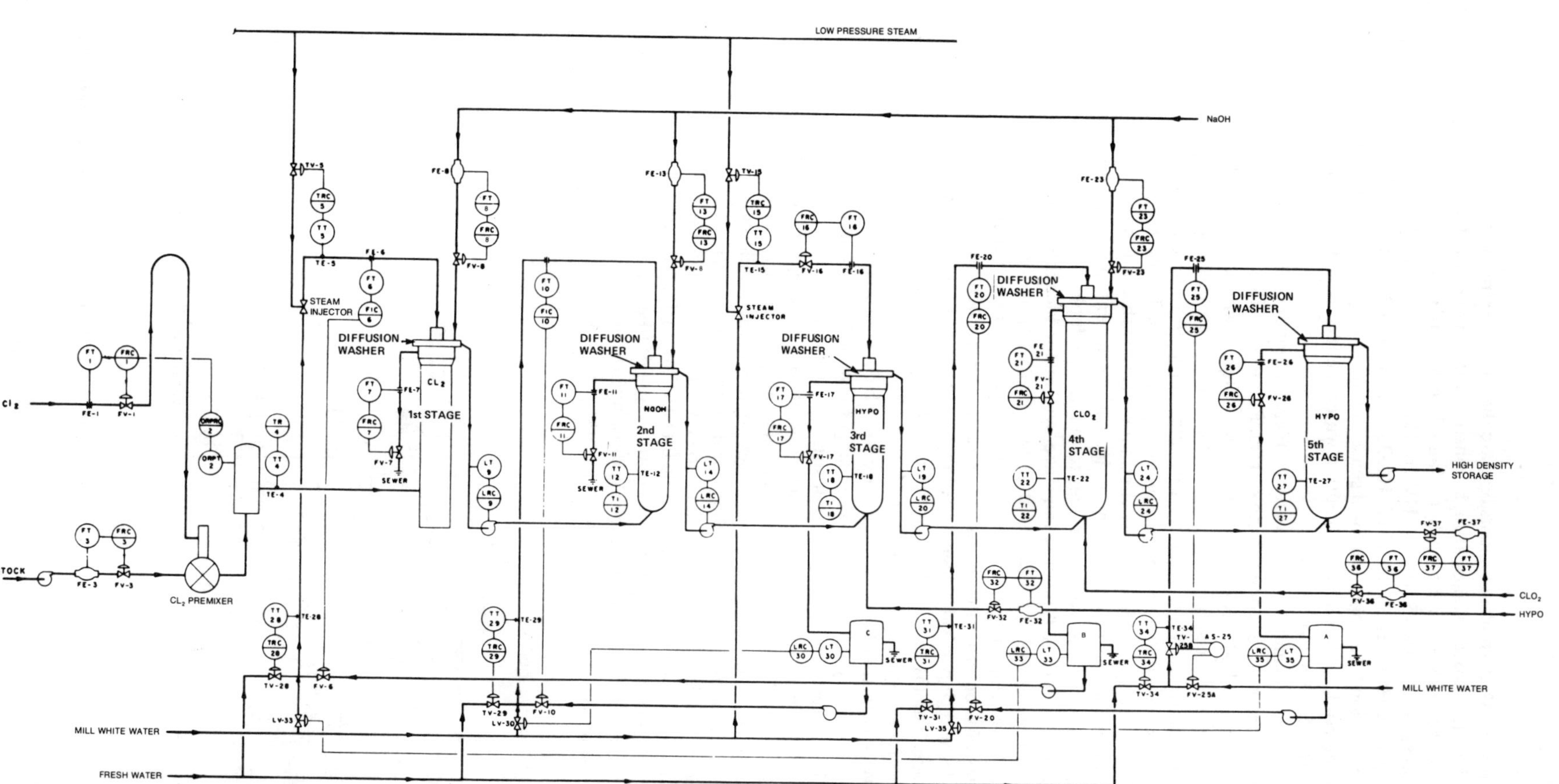

Figure 8-20. Five-stage bleaching with diffusion washers.

When used in the pulp bleaching process, diffusion washers are usually installed on the top of the bleaching towers of the various stages instead of separately between the stages as with vacuum rotary washers. Figure 8-20 illustrates a typical arrangement with a five-stage bleaching system. The instrumentation used is similar to that used when vacuum rotary washers are employed.

Displacement Bleaching. A new pulp bleaching concept recently introduced to the industry has been designed to provide savings in space, energy, water, and effluent volume. This has been used in several mills. The concept takes advantage of the principle of the diffusion washer in which the wash media diffuses through the pulp while the diffuser itself is submerged within the plug. In displacement bleaching, a number of diffusers are mounted in a single tower to accomplish almost any bleaching sequence. The basic flow sheet of a four-stage displacement bleaching process, depicted in Figure 8-21, shows a system where chlorination is performed in a separate tower and drum washer. Two E-caustic extraction and two D-chlorine dioxide stages are accomplished in displacement bleaching in one tower.

Oxygen. As shown above, most sulfate bleaching processes involve C-chlorination, E-extraction, and H-hypochlorite as the first three stages. It is in these

Figure 8-21. Basic four-stage displacement bleaching process.

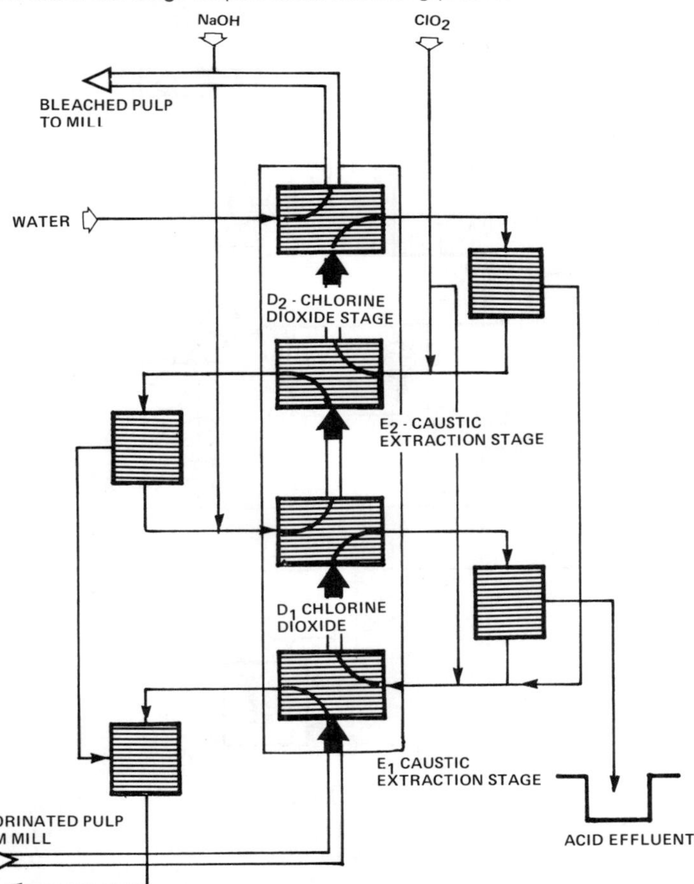

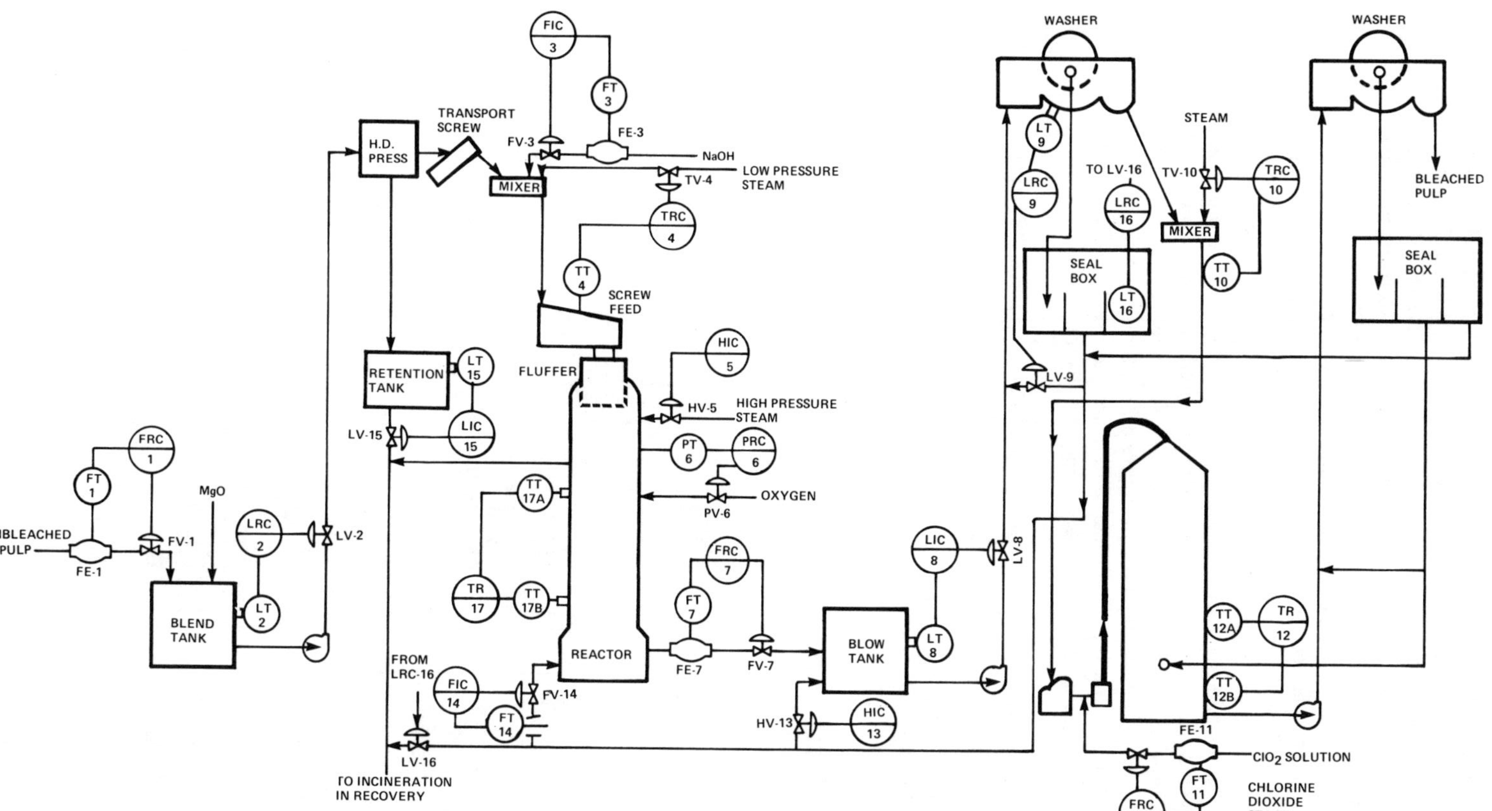

Figure 8-22. A typical oxygen pulp bleaching process.

stages that most of the coloring impurities and residual lignins are removed. However, in removing these impurities the desirable portion of the fiber is also attacked and, therefore, it is in these stages that most of the fiber and fiber strength loss occurs. Consequently, the effluent from these stages contains a high degree of contaminants such as dissolved solids, color, and BOD (biochemical oxygen demand) load, making it a large potential source of effluent stream pollution from the bleaching operation.

It was found that elemental oxygen, under alkaline conditions at elevated temperatures and pressures and in a specially designed reactor, would preferentially degrade lignins and make other coloring matter soluble without attacking the desirable fiber cellulose. The effluent was also suitable for recycling through evaporation and incineration in the recovery boiler to reduce the pollution load from the process.

Although several design versions involving the replacing of the C-chlorination, E-caustic extractor, and H-hypochlorite with an O-oxygen have been developed and tried, they are basically similar in process objectives. A typical configuration showing the instrumentation is illustrated in Figure 8-22.

The flow of unbleached pulp to the bleaching process is controlled, by recording controller FRC-1, to a blend tank where magnesium oxide is added as an inhibitor of undesirable degradation during oxygen bleaching. This is sometimes referred to as a *protector*. The level in the blend tank is maintained by level control loop LT-2, LRC-2, LV-2. After passing through a high-density pulp press which raises the consistency to about 30 percent bone dry, the pressed stock is transported by a screw conveyor to a single-shaft mixer where sodium hydroxide is added by flow indicating controller FIC-3. The temperature is regulated by temperature recording controller TRC-4 from a temperature element located in the screw feeder. This moves the pulp into the top of a reactor, forming a gas-tight plug in the feed line to prevent oxygen loss.

Steam, under manual control of HIC-5, is used to heat the pulp as it drops from a *fluffer* in the top of the reactor where it is finally disintegrated. It is then dropped into the body of a reactor where pressure is maintained at a desired level by controlling the admission of oxygen with pressure recording controller PRC-6. Reactor top and bottom temperatures are recorded by two-pen recorder TR-17.

The pulp bed formed in the reactor descends slowly until it reaches a rotary plate feeder. Bleach stock flow is ejected by the hydraulically driven feeder and actually "blows" the pulp into a blow tank. The blowing rate is controlled by a magnetic-type flow element, FE-7, and recording controller FRC-7.

The pulp is then pumped from the blow tank, whose level is controlled by throttling valve LV-8 through level indicating controller LIC-8, to a washer. Level control loop LT-9, LRC-9, LV-9 regulates the level in the washer vat by addition of filtrate to the incoming pulp. The remainder of the process and instrumentation is the same as the bleaching stages previously discussed. In this case, the last stage is a D-chlorine dioxide stage. The final effluent from the washer seal box is sent to the recovery boiler for incineration together with effluent from the high-density press. The level in the seal box is held constant by level recording controller LRC-16 which operates control valve LV-16 in the effluent line to the recovery area. Effluent is also recycled for dilution purposes under flow controller FIC-14 to the reactor and under a remote hand control, HIC-13, to the blow tank.

Mechanical Pulp Bleaching

The bleaching of pulp made by the mechanical and semimechanical process differs from the bleaching of pulps manufactured by chemical processes, not only

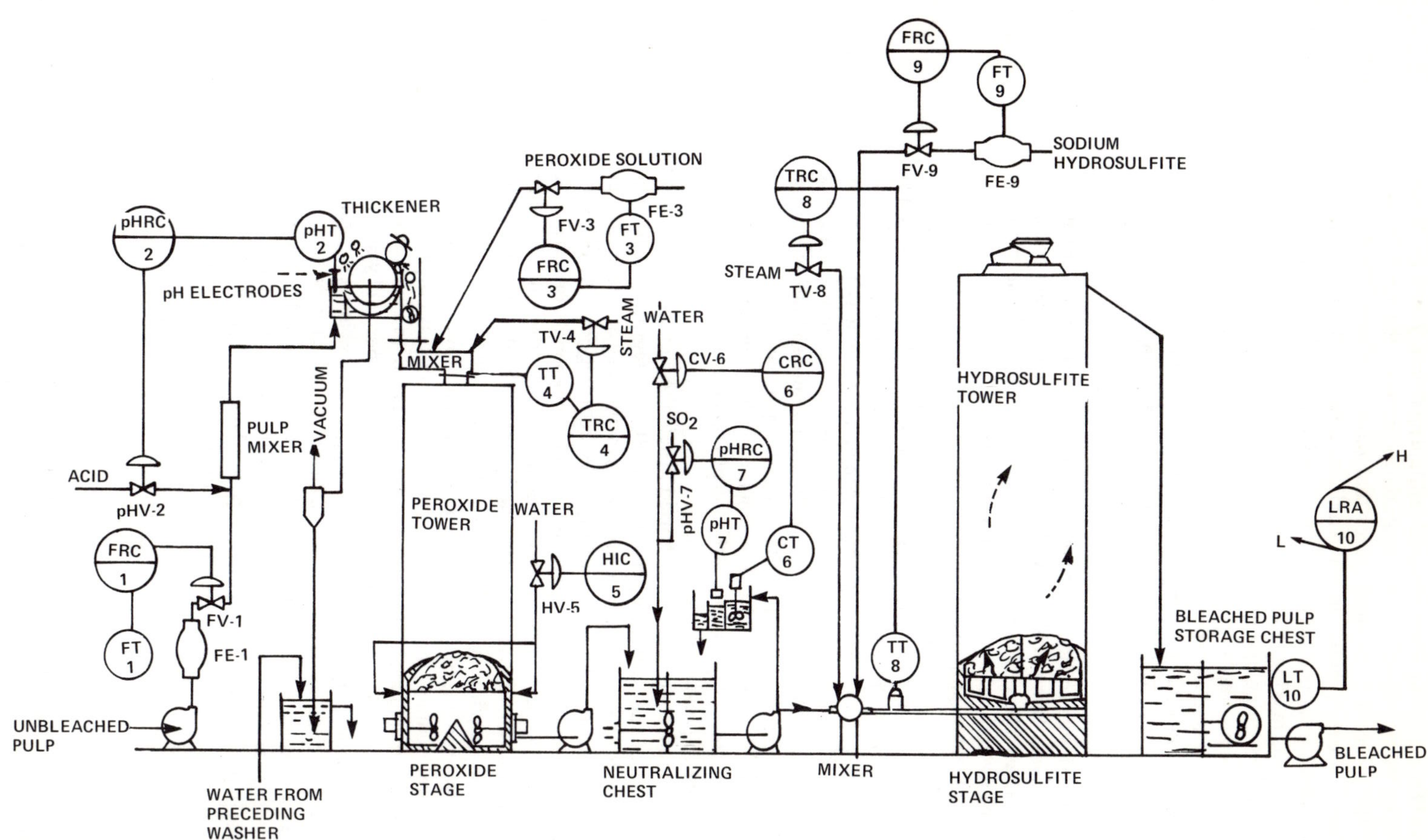

Figure 8-23. Two-stage peroxide-hydrosulfite bleaching process.

in the process equipment used but also in the type of bleaching chemicals used. Figure 8-23 shows a typical equipment plan, chemicals, and type of instrumentation used for a two-stage peroxide-hydrosulfite process for bleaching mechanical or semimechanical pulp made from softwoods or hardwoods.

In operation, the flow of 4 to 5 percent consistency b.d. unbleached pulp to the bleaching unit is controlled by a flow control loop consisting of a magnetic-type flowmeter, FE-1, and flow recording controller FRC-1. It is acidified by the addition of 2 percent sulfuric acid to obtain a pH of 2.5 to 3.0 by pH recording controller pHRC-2, making a pH measurement with submersion-type electrodes in the vat of thickener after mixing. In the thickener, the pulp is washed with fresh water and thickened to about 15 percent consistency. Peroxide bleaching solution is added to the pulp by flow controller FRC-3 in a mixer which is heated by steam to raise the temperature of the pulp to 180°F where it is maintained by temperature recording controller TRC-4. The pulp, with the bleaching solution, is then dropped into a retention tower equipped with a bottom discharge where it resides for about one and a half hours.

After bleaching is completed, the pulp is diluted at the bottom of the tower to 4 to 5 percent consistency by adding water with control valve HV-5, remotely operated by hand control HIC-5. The pH is adjusted to 6.0 in the neutralizing chest by adding sulfur dioxide solution with pH recording controller pHRC-7. Consistency of the pulp is trimmed by the use of consistency control loop CT-6, CRC-6, and CV-6.

Pulp is then pumped to a mixer where it is heated to 140°F by steam controlled by temperature recording controller TRC-8. Sodium hydrosulfite bleaching solution, under flow control of FRC-9, is also added to the mixer. It is then transferred to the low-density upflow tower which has sufficient size to retain it for one hour at 4 to 5 percent consistency. The bleached pulp is discharged continuously from the top of the low-density tower into a pulp storage chest whose level is measured by level transmitter LT-10 and recorded on a high and low alarm recorder, LRA-10.

9 Stock Preparation

Bleached pulp, as produced in the pulp mill, generally is not suitable as such for the manufacture of most grades of paper today. A sheet made only from this pulp would result in a sheet having low strength, high bulk; open, irregular texture; and wild, uneven formation; and it would disintegrate readily when wetted by water. Therefore, in order to make the pulp suitable for forming into a sheet of paper, various characteristics must be imparted to the fibers. This is done by adding other types of pulp, dyes, chemical additives, and fillers with a blending proportioning operation; mechanical treatment by working on the fibers with beaters and refiners; and other operations such as repulping. After various types of pulps are mixed together with dyes, chemicals, and/or fillers and necessary work is done by refiners, etc., it is then referred to as *stock*, although *pulp, stock,* and *furnish* are used interchangeably in many paper mills. The operations used to accomplish this actually are preparing the stock for papermaking and are, therefore, collectively categorized as stock preparation.

CONSISTENCY CONTROL

As in other phases of pulp and paper making, the control of the pulp slurry consistency is especially important. Without uniform consistency the process is practically impossible to control. Whether the uniformity is achieved with instrumentation or by other means, it must be obtained. The manual method of measuring consistency consists of selecting and weighing a representative sample, removing the water, and weighing the remainder. This method is satisfactory for spot checking but it is not acceptable for continuous control. Changes in consistency must be monitored continuously and on-line if satisfactory control is to be achieved. For control purposes, it is important to be able to sense the variations in consistency rather than to determine the absolute value of consistency.

It is usually desirable to store the slurry in a high-consistency solution to allow the use of smaller storage tanks. However, it is difficult and inefficient to pump the high-consistency slurry through pipelines. Consequently, it is quite common to use consistency control systems to add dilution water to the suction side of the pulp from the storage tank.

The term *consistency*, as used in the pulp and paper industry, is defined as the percentage by weight of dry, fibrous material in any combination of pulp and water or stock (pulp and additives) and water. Consistency is expressed in the following terms:

$$C = \frac{F}{W} \times 100$$

Where: C = consistency of sample (percent)
W = total weight of sample
F = weight of fibrous material in sample

Consistency is usually expressed on a bone dry (b.d.) basis, wherein the percentage of fibrous material is determined by comparing the total sample weight with a stabilized, oven-dry sample weight. Occasionally, consistency is expressed as air dry (a.d.) with the assumption that the fibrous material contains 10 percent water. However, since it is the practice of the industry to use the bone dry basis, this basis should be assumed if no specific reference is given.

The designations pulp and stock are sometimes used interchangeably. However, pulp slurries primarily consist of fibrous material and water with additives such as fillers, dyes, and chemicals. The amount of nonfibrous additives can be determined in the laboratory using the ash test. The results of this test are sometimes used to correct the weight of the fibrous material for use in the expression given above.

Single Dilution

A typical single-dilution control loop, shown in Figure 9-1, is commonly used to reduce the consistency of pulp slurries by no more than 0.5 to 1 percent. It consists of a consistency measuring element, CE-1, a consistency transmitter, CT-1, and a consistency recording controller, CRC-1, with a second pen recording the controller output (valve position), VPR-1, and control valve CV-1 located in the dilution water line.

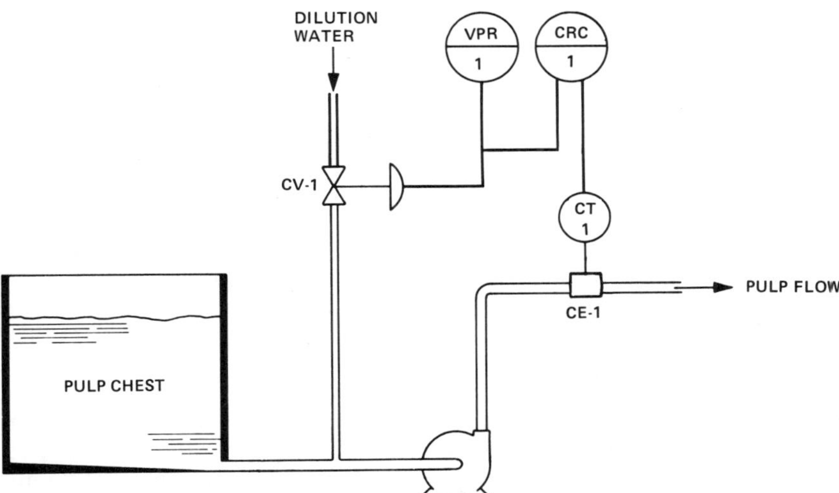

Figure 9-1. A typical single-dilution consistency control loop.

Double-Dilution Control

When it is desired to dilute a slurry more than 0.5 to 1 percent, it is normally done in two water addition steps by the use of a double-dilution control system, illustrated in Figure 9-2. This is very common on high-density storage chests where stock is stored at consistencies of over 6 percent. For ease of pumping, stock is diluted to about 4 percent in the lower or dilution zone of the chest. Final or secondary consistency control is obtained by a consistency loop similar to that shown in Figure 9-1. First, or primary, dilution is accomplished by adding the major part of the water with another controller, CIC-1A, and valve CV-1A.

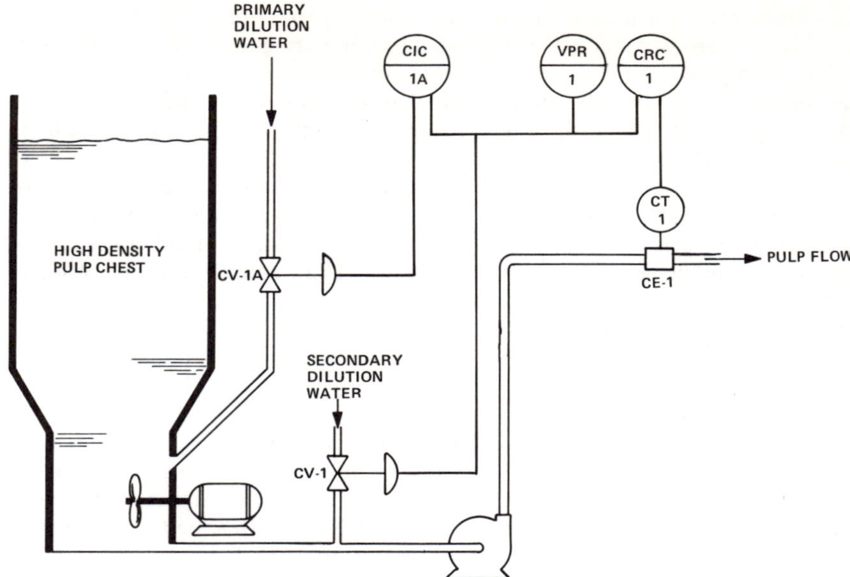

Figure 9-2. Double-dilution consistency control system.

This primary controller is usually an indicating controller with either a standard proportional control mode or a gap-action floating control mode. The measurement for this controller is taken from the output of the consistency controller, CRC-1, to the valve, CV-1, in the secondary loop.

Flow Compensated

The double-dilution consistency control system works satisfactorily under most mill operating conditions. However, there are occasions when there are large changes in pulp chest throughput or pulp flow rate. In Figure 9-3 the double-dilution consistency control is accomplished by the addition of a flow measurement signal from the magnetic flow element, FE-2, being combined in a computing relay (1:1 repeater) with the output of the secondary controller, CIC-1A. This computing relay acts as a 1:1 repeater with a positive bias. The bias is supplied by the flow transmitter, FT-2, and its effect can be manually adjusted. Now, any large swings in throughput of pulp flow rate are compensated for by readjusting the primary dilution control valve, DV-1A, to add more or less water for the variation and for maintaining the consistency at the desired value.

Dilution Control Valve Sizing

The size of the dilution control valve used is determined by the amount of water required to dilute from the maximum incoming consistency to the minimum outgoing or controlled consistency. For example, assume that in Figure 9-1 the maximum consistency that could exist in the storage chest is 4 percent and that the minimum consistency that would ever be required for control is 3.5 percent. Production rates or pumping rates from storage chests are expressed in tons per day of pulp or stock slurry. Therefore, again assume that the flow rate is 100 tons per day. The flow rate in gallons per minute should be known at 4

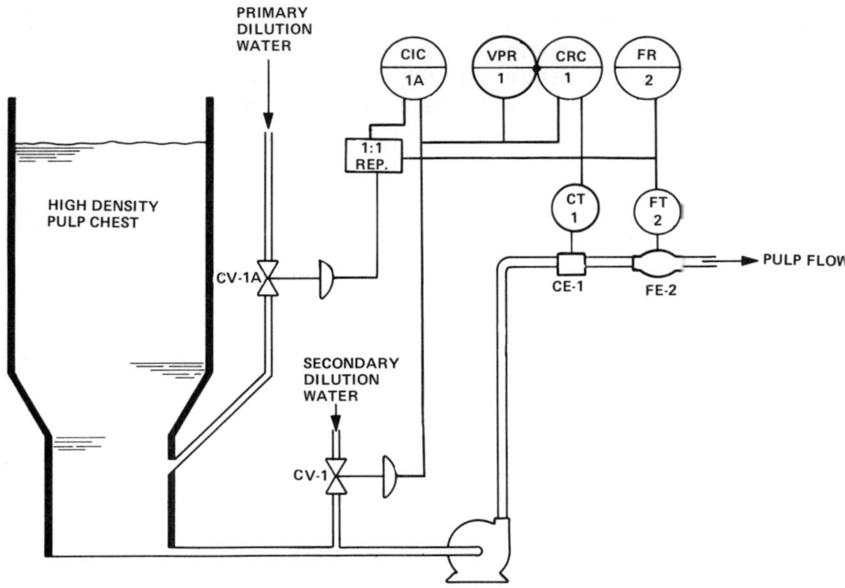

Figure 9-3. Double-dilution consistency control with flow compensation.

percent and 3.5 percent consistency in order to determine the amount of dilution water required. To convert the flow from tons per day to gallons per minute, the following formula can be used:

$$\text{gpm} = \frac{16.65}{\text{percent consistency}} \times \text{tons per day}$$

Where: 16.65 is a factor based on the weight of one cubic foot of slurry which equals 62.4 pounds.

then, at 4 percent consistency and 100 t/d:

$$\text{gpm} = \frac{16.65}{4} \times 100 = 416 \text{ gpm}$$

At 3.5 percent consistency and 100 t/d:

$$\text{gpm} = \frac{16.65}{3.5} \times 100 = 476 \text{ gpm}$$

Therefore, 476 gpm − 416 gpm = 60 gpm of water to be added to dilute 4 percent consistency slurry to 3.5 percent. This flow rate, V, is used in the formula given below.

Sound engineering practice has shown that a minimum of 5-psi pressure drop should exist across the dilution valve at all times. The C_v of a valve to meet these requirements would then be:

$$C_v = V\sqrt{\frac{G}{\Delta P}} = 60\sqrt{\frac{1}{5}} = 26.8$$

Therefore, a 1½-inch Foxboro VI Series valve is one of the valves that can be used to meet the C_v requirements of this particular installation.

A shorter method of determining dilution valve size is to use Table 9-A, which shows the consistency and volume flow relationship of slurries. From Table 9-A:

At 3.5 percent:
Flow of slurry per minute per t/d = 4.76

At 4 percent:
Flow of slurry per minute per t/d = 4.16

Dilution water:
Gallons needed per minute per t/d = 0.60

Table 9-A. Consistency-Volume Flow Relationship of Slurries

Consistency* (C)	Volume† (16.65/C)	Consistency* (C)	Volume† (16.65/C)	Consistency* (C)	Volume† (16.65/C)	Consistency* (C)	Volume† (16.65/C)
0.000		1.10	15.1	6.00	2.78	20	0.833
0.012	1388	1.20	13.9	6.25	2.66	22	0.757
0.024	694	1.30	12.8	6.50	2.56	24	0.694
0.036	463	1.40	11.9	6.75	2.47	26	0.640
0.048	347	1.50	11.1	7.00	2.38	28	0.595
0.060	278	1.60	10.4	7.25	2.29	30	0.555
0.072	231	1.70	9.79	7.50	2.22	32	0.520
0.084	198	1.80	9.25	7.75	2.15	34	0.490
0.096	173	1.90	8.76	8.00	2.08	36	0.463
0.108	155	2.00	8.33	8.50	1.96	38	0.438
0.120	139	2.20	7.57	9.00	1.85	40	0.416
0.15	111	2.40	6.94	9.50	1.75	42	0.396
0.20	83.3	2.60	6.40	10.00	1.67	44	0.378
0.25	66.6	2.80	5.95	10.50	1.59	46	0.362
0.30	55.5	3.00	5.55	11.00	1.51	48	0.347
0.35	47.6	3.25	5.12	11.50	1.45	50	0.333
0.40	41.6	3.50	4.76	12	1.39	52	0.320
0.45	37.0	3.75	4.44	13	1.28	54	0.308
0.50	33.3	4.00	4.16	14	1.19	56	0.297
0.55	30.3	4.25	3.92	15	1.11	58	0.287
0.60	27.8	4.50	3.70	16	1.03	60	0.276
0.65	25.6	4.75	3.51	17	0.979	65	0.256
0.70	23.8	5.00	3.33	18	0.925	70	0.238
0.75	22.2	5.25	3.17	19	0.876	75	0.222
0.80	20.8	5.50	3.03			80	0.208
0.85	19.6	5.75	2.90			85	0.196
0.90	18.5					90	0.185
0.95	17.5					95	0.175
1.00	16.7					100	0.167

* Percent consistency—lb of bone dry fiber in 100 lb of slurry
† Gallons of slurry per minute per ton of bone dry fiber per 24 hours

therefore:

0.60×100 t/d = 60 gpm needed.

Using a 5-psi pressure drop and Foxboro valve slide rule, a C_V of 26.8 is obtained, indicating that a 1½-inch VI Series valve is one of the valves that can be utilized for this purpose.

PULP SAMPLING

Studies of sampling methods for manual analysis of stock consistency to check operation of consistency measuring devices have been made. It has been found that for best results the sampling point should be at the point of highest turbulence, where the main stream flows over a weir immediately following the point of installation of the consistency detecting device. (See Figure 9-4.)

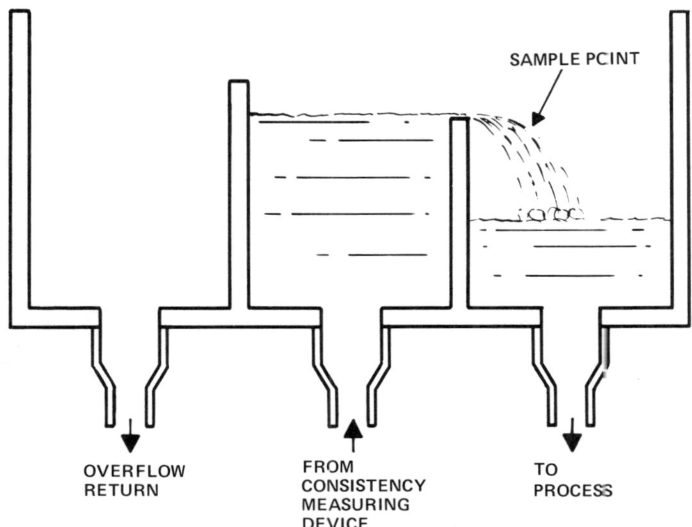

Figure 9-4. Consistency sample location.

However, such a location is not always available and other methods have been used. Figure 9-5 illustrates some of the other methods used. They are:

1. Pipe wall sampling point. A 1-inch sample pipe is installed at a 30-degree angle in the side of a pipe wall. An orifice plate is sometimes installed upstream of the sample line to provide turbulence for a good sample.
2. Elbow sampling point. A right-angle elbow is equipped with a sampling tap of reduced diameter (not less than 2 inches). The center line of the sampling tap is located at an angle of 90 degrees along the outer radius from the face of the elbow. The elbow causes the slurry to mix thoroughly before the sample takeoff point.
3. Tee sampling point. A full-size tee is located in the main flow line in such a manner that the full stream impinges directly on the sample takeoff point.
4. 45-degree sampling point. A full-size 45-degree fitting is installed so that a sample is taken from the side wall of the main stream pipeline. This

method is not considered as good a method as No. 1, the pipe wall sampling point, and is seldom used on long fiber stocks or on consistencies that will run over 3 percent.
5. Y sampling point. This consists of a full-size Y fitting with an angle of 60 degrees between the arms. This arrangement has also been used mostly on short fiber stocks and consistencies of less than 3 percent.

Although these methods have been used satisfactorily, none are ideal because studies have shown that the magnitude of the sample flow should be at least one-fourth of the main flow in order to obtain a good representative sample. It is quite doubtful that many sampling systems are operated to meet this requirement for representative sampling.

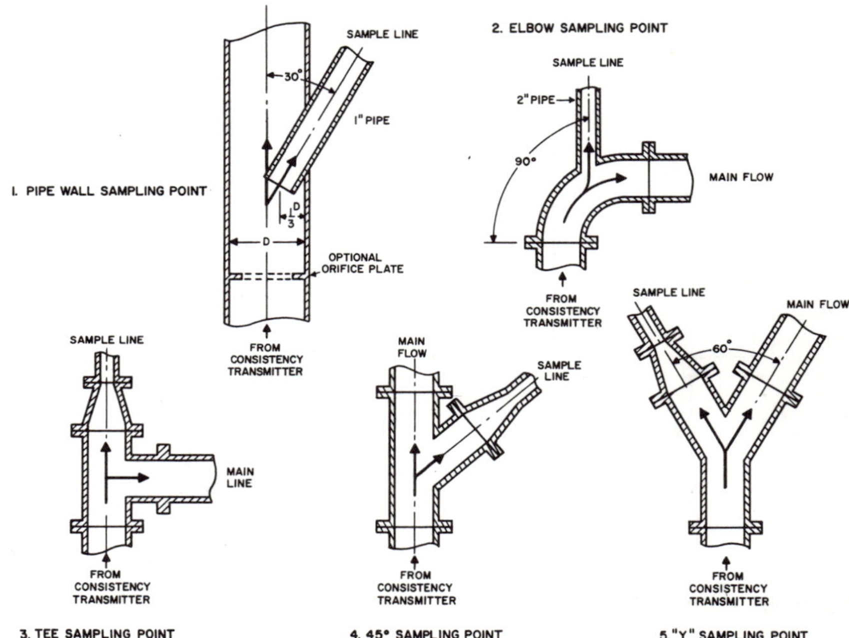

Figure 9-5. Pulp sampling methods.

STOCK PROPORTIONING

An early method used to mix pulps with varying characteristics was to batch-blend predetermined amounts in mixing tanks or machine chests prior to running stock on the paper machine. Chemical additives and dyes were also added to the mixing tanks in the same manner in order to impart certain interior and exterior properties to the final sheet of paper. Subsequent continuous mechanical proportioning systems, many of which are still in use today, consisted of individual positive displacement metering compartments, one for each pulp, dye, and additive, as required. The compartment rotors are driven by variable-speed drives which, in turn, are connected to the variable-speed shaft of a variable-speed master drive. Thus, the speed of each individual drive can be adjusted to give the required proportions. The speed of the master drive is controlled from the mixing chest level; the speed recorders on the compartment rotors are used to record the proportions. More recently, the continuous flow metering method of stock proportioning has been developed. With the magnetic flowmeter now being used as the flow measuring device, it has become the predominant method used for blending pulps, additives, and dyes in the paper industry today.

Continuous Flow

Figure 9-6 shows that a typical continuous-flow-type stock proportioning system blends stocks, dyes, and additives by using a magnetic flowmeter, a flow rate controller, and a flow valve. A level controller on the mixing chest automatically adjusts the ratio flow controller to regulate amounts of blend components to the mixing chest. Level recording controller LRC-9 maintains the level in the machine chest by adjusting valve LV-9 in the stock line from the mixing chest to make up for changes in paper machine demands. The quantity is recorded by FR-8 from magnetic flow transmitter FT-8 in the flow line. Level recording controller LRC-1 maintains the level in the mixing chest in response to stock drawn from it by the action of level controller LRC-9. Therefore, the mixing chest level will, in turn, reflect demand changes by the paper machine. As the mixing chest tries to vary in response to the paper machine demand changes, level controller LRC-1 varies its output to the ratio mechanisms located in the flow ratio recording controllers on pine, FrRC-2, hardwood, FrRC-3, and broke, FrRC-4, pulp flows. This signal is also transmitted to similar mechanisms associated with dye, additive, and starch flows, FrRC-5, FrRC-6, and FrRC-7.

The ratio devices adjust the set points of each controller in accordance with the desired amount of each ingredient relative to the total demand flow, which is previously determined and set on the ratio devices. Flows of pulp components are measured by magnetic flowmeters FT-2, FT-3, and FT-4; and dye, additives, and starch flows are measured by magnetic flowmeters FT-5, FT-6, and FT-7. These flow signals are transmitted to their respective flow ratio controllers. Level transmitters LT-1 on the mixing chest and LT-9 on the machine chest are usually of the flanged diaphragm type although the bubble type can be used.

Continuous Flow With Pacing

Periodically in a stock proportioning system, the availability of a particular pulp flow may diminish to the point where its control valve will go wide open. A loop in this condition is then out of control and the system will be producing off-specification blend.

Some electronic-type instrumentation is equipped to inherently solve this problem by operating features of its electrical design construction. Generally, and specifically with pneumatic-type instruments, this condition can be taken care of automatically by the addition of an autoselective system which consists of a valve position controller, a number of high signal selector relays, depending on the number of stock flows involved, and a low signal selector relay as shown in Figure 9-7. VPC-4 is a reverse acting (to increase measurement signal decreases output signal) valve position controller whose set point is fixed at some high value, such as 95 to 98 pecent of full scale. The set point of the mixing chest level (demand) controller, LRC-3, is set to the desired tank level. Pulp flow control valves FV-1, FV-2, and FV-3 are of the direct action type, i.e., an increase in signal to the valve opens it.

Outputs from flow ratio recording controllers FrRC-1, FrRC-2, and FrRC-3 go to the control valves and also through high signal selector relays HSS-1 and HSS-2. The highest output signal is selected as the measurement signal to valve

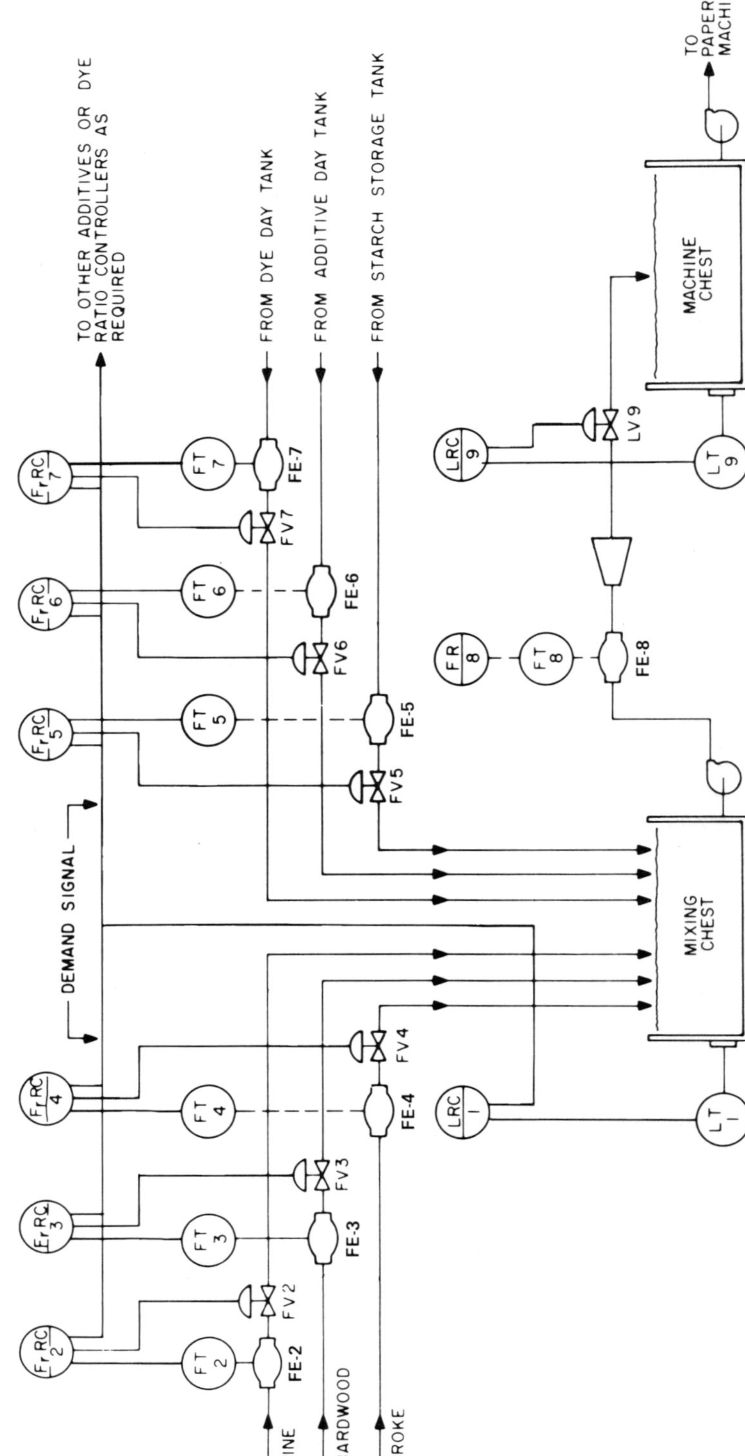

Figure 9-6. Representative scheme for instrumentation of a stock proportioning system.

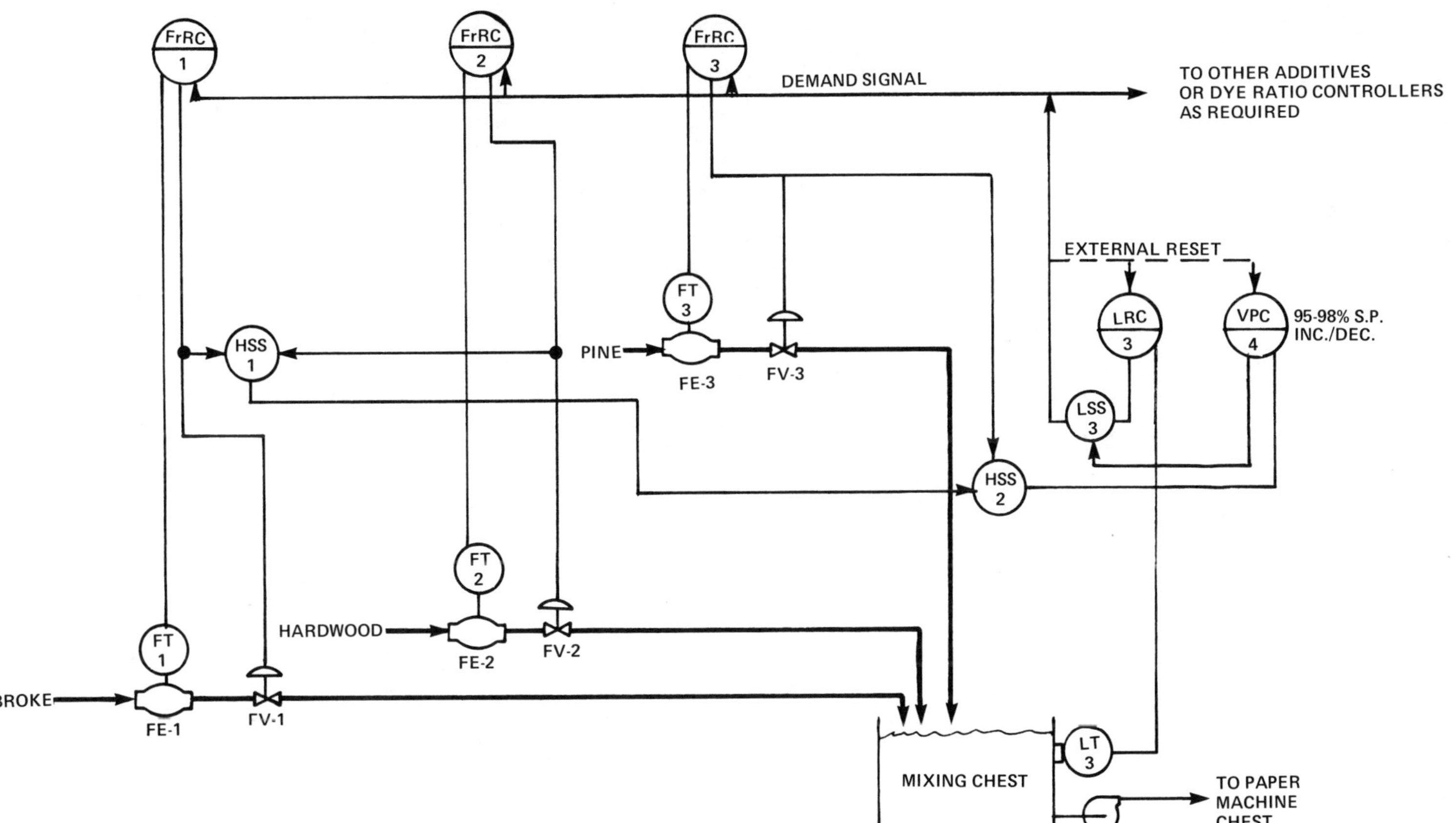

Figure 9-7. Stock proportioning system with pacing.

position controller VPC-4. As long as this signal is below the controller's set point (95 to 98 percent of scale), the controller output will be "high" and, therefore, rejected by low signal selector relay LSS-3. While the output of VPC-4 is higher than that of LRC-3, the output to the pulp flow ratio controllers, FrRC-1, FrRC-2, and FrRC-3, stay under control of LRC-3, whose output represents the demand of the system because the lower of the two signals passes through the low signal selector relay, LSS-3.

Should any of the pulp flow control valves, FV-1, FV-2, or FV-3, try to open all the way (over 95 to 98 percent), the output from VPC-4 decreases. When the output from LRC-3 becomes higher than the output from VPC-4, the output from VPC-4 is selected by LSS-3 and this signal begins to decrease. This action reduces the signal to flow controllers FrRC-1, FrRC-2, and FrRC-3 until the valve that tries to go beyond 95 to 98 percent is brought under control. VPC-4 continues to reduce the signal to the ratio controllers should the supply of broke, hardwood, or pine pulp continue to be unavailable. Although the total blend rate is reduced, the system's action preserves this blend ratio and, therefore, maintains on-specification production. This feature is referred to as *pacing* the system to the lowest ingredient flow. However, the mixing chest level is not being maintained while the system is operating in the pacing phase.

When the flow of pulp in short supply is restored, its valve will start to close (move from wide open position). Next, the output to the ratio controllers will switch from VPC-4 back into the normal condition where LRC-3 controls the output to the ratio controllers and brings the mixing chest level back to normal. To prevent windup of LRC-3 and VPC-4 when their output is not controlling the system, they are normally provided with an antiwindup feature or the output to the ratio controller is fed back to the controllers as an external reset signal.

With this feature, as much specification blend as possible will be produced at all times. This system can also be designed to shut down automatically if one of the pulp flows becomes completely unavailable.

System Sizing

The successful operation of a stock proportioning system depends on the proper selection of flowmeter ranges, ratio ranges and scales for flow ratio controllers, and ratio relays. In order to do this, the maximum and minimum paper machine demand rate, consistency of stock components, maximum percentage ratios of stocks required for furnish being run, and maximum additive and dye rates must be known.

Assume that the maximum machine demand of a certain consistency stock will be 942 gallons per minute. This can be rounded off to a total range of 0 to 1000 gallons per minute. Next, assume that the desired percentage adjustment range for each stock stream to cover all grades of paper to be made will be:

Pine flow	30% to 70%
Hardwood flow	20% to 60%
Broke flow	10% to 20%

The maximum percentage values and maximum paper machine demand of 1000 gpm establishes the selection of the following flowmeter ranges:

Pine	0 to 700 gpm
Hardwood	0 to 600 gpm
Broke	0 to 200 gpm

Dye and additive ranges are selected in the same manner except that, instead of ratio ranges being expressed in terms of percentage, they are expressed in terms of gallons per 1000 gallons of total flow because of their usually low flow ranges. For instance, if an additive's flow is to be 3 to 15 gallons per 1000 gallons, the flowmeter range will be 0 to 15 gpm.

The individual proportioning control loops use the output of the level controller on the mixing chest, which is proportional to the flow demand of the paper machine, as its master, or *primary*, signal. This signal is transmitted to ratio relays which remotely set the set points of flow controllers on each individual stock or additive flow controller, or directly to ratio units located in the individual flow ratio controllers. These ratios are operator-set for automatic proportioning of the flows. The ranges of these ratios can be determined by a number of methods. One way is by the use of the following calculations:

$$\text{Instrument ratio} = \frac{\text{Stock flow \%}}{\text{Stock flow range/total machine demand}}$$

For pine stock:

$$\text{at minimum \%, I.R.} = \frac{0.30}{700/1000} = 0.43\%$$

$$\text{at maximum \%, I.R.} = \frac{0.70}{700/1000} = 1.0$$

The ratio unit setting dials are then marked 30 percent at the minimum ratio of 0.43 and 70 percent at the maximum ratio of 1.0. The remainder of the ratio ranges of the individual stock loops are calculated in the same manner. The individual flow rates are then controlled at a flow proportional to output of the mixing chest's level controller as set on the ratio devices by the operator. The stock ratio percentage settings must be set so that their sums equal 100 percent.

The dye and additive ratio ranges can be determined in the same manner:

$$\text{at minimum additive \%, I.R.} = \frac{0.003}{15/1000} = 0.2$$

$$\text{at maximum additive \%, I.R.} = \frac{0.015}{15/1000} = 1.0$$

The ratio unit setting dials are then marked 3 at the minimum ratio of 0.2 and 15 at the maximum ratio of 1.0.

Some mills proportion the pine and hardwood based on percentages of total virgin stock (pine + hardwood) and broke as a percentage of total virgin stock. In such cases, the system remains the same but the calculations of ratio ranges and percentages are modified to reflect this difference in system operation.

STARCH COOKING

Most chemical additives and dyes used in the blending of pulp during the stock proportioning operation are shipped to the mill in powder form or in concentrated solutions. All that is required by the mill is to mix or dilute these materials before using.

Other additives like starch must be further processed before they can be added to the pulp mixture. Thick, high-viscosity, raw starches are converted to thinner, lower viscosity types by additional refining and treating processes—either by the starch producing industry or the paper manufacturers, who usually use enzymes to break down the starch for use in the mill.

Figure 9-8 shows a typical enzyme starch conversion system designed to convert raw starch to a desired viscosity on a batch basis. Enzyme conversion usually goes through three phases during the operation of the system.

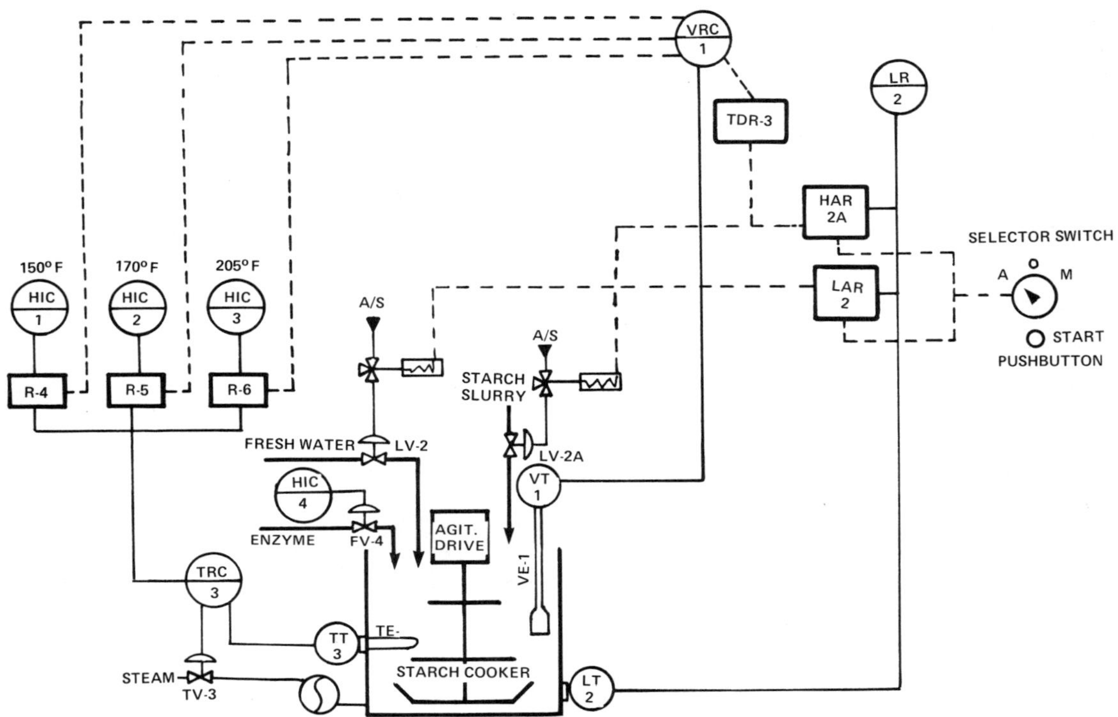

Figure 9-8. Typical starch cooking system.

In starting a batch cycle with the starch cooker kettle empty and the agitator running, the selector switch is turned from *off* to *automatic* and the *start* pushbutton is enabled. This starts to fill the cooker with water to the level set on low level alarm relay LAR-2 on the output signal of level transmitter LT-2 by operating a solenoid which opens and then closes fresh water valve LV-2 at the desired level. After the water valve closes, the starch slurry valve, LV-2A, is opened by high level alarm relay HAR-2A, also on the output signal of level transmitter LR-2. Meanwhile, enzyme is added manually, locally, or remotely by the operation of a manual control station, HIC-4, and valve FV-4.

When the proper level in the cooker kettle is reached, as detected by high alarm relay HAR-2A, the starch slurry valve is closed and viscosity recording controller VRC-1 activates relay R-4, which allows the set point generated by manual set station HIC-1 to set the control point of temperature recording controller TRC-3 to 150°F. This opens steam valve TV-3 and the first or preconversion phase is started as the temperature of the starch cooker is raised to 150°F. There it is maintained by temperature controller TRC-3 for the predetermined period of time set on timer relay TDR-3. At the end of the time period, the control set point on TRC-3 is raised to 170°F by shutting off the output from HIC-1 through relay R-4, and applying the output from HIC-2 to set the control point of TRC-3 through relay R-5. This activates the enzyme and puts the system in the second or conversion phase. The temperature is then controlled by TRC-3 until the conversion is completed to obtain a proper viscosity as measured by a viscosity device, VE-1, and set on viscosity recording controller VRC-1. When the proper viscosity value is attained, VRC-1 raises the control set point of TRC-3 to 205°F by activating relay R-6 and allowing the output of HIC-3 to go to TRC-3. As the temperature is raised rapidly to 205°F, the process enters into the third, or inactive, phase where it is maintained for a sufficient time to inactivate the enzyme as set on timer relay TDR-3. At the end of this period, the control set point of TRC-3 is set to the bottom of its range. The starch batch can now be pumped out and is ready for a new batch cycle which is repeated by pressing the *start* button again.

There are also continuous systems for cooking starch. The basic phases of the process and instrumentation used are similar to the batch process.

MECHANICAL TREATMENT

In the early days of papermaking, the mechanical treatment of pulp was accomplished manually in order to improve its matting or felting properties. One of the first machines used was known as a *beater* or Hollander. It consists essentially of an oval tub with a partition or midfeather. On one side of the midfeather is a cylindrical motor-driven roll whose surface is fitted with steel bars or knives aligned parallel to the axis. Rotation of the roll causes the pulp to circulate around the tub. Under the roll is a *bed plate*, which consists of knives similar to those imbedded in the top of the roll. The roll setting determines the amount of work done on the fibers in the pulp. For many years, the use of beaters was the predominant practice to accomplish the desired disintegration of fiber in the pulp. Most of them are operated in a batch manner and many are still in use today.

Advancing technology toward continuous processing of pulp has led to the modern continuous pulp refining equipment used today. Two major designs of refiners have eventually evolved in the paper industry: conical and disc. The basic elements are a stationary housing and movable, rotating members. The instrumentation for these is similar.

The conical-type refiner consists of a cast iron tapered shell, fitted inside with metal bars running full length of the inner surface. A conical plug fits into the shell, which is also fitted with metal bars running lengthwise on its surface. The plug revolves inside the shell and it can be moved back and forth so that the distance between the bars in the plug and those in the shell can be varied. The pulp slurry is fed into the small end and discharged at the large end.

Disc refiners are characterized by a flat, disc-shaped movable rotating element mounted in similar shaped housings with facing surfaces consisting of alternate bars and grooves called *tackle*. Two basic disc refiner designs have evolved in the paper industry: single and double disc. There are several variations of this design but the main difference is that in the single-disc refiner only, one disc is driven in a rotating fashion while the second disc is stationary. With the double-disc design, both discs are driven and rotate in opposite directions. In both cases the degree of refining is controlled by varying the distances between the disc faces as the pulp passes between them. This is accomplished by moving one disc toward or away from the second disc.

Refiners were first controlled by a handwheel, working through gear reducers to manually position the movable element of the refiner in or out. To enable remote operation, the handwheel was supplemented or replaced with a reversing motor. Automatic operation has been achieved by using refiner drive loading, plug pressure, couch vacuum, differential temperature of the stock across the refiner, a freeness measurement, or a combination of factors to control the automatic moving member driving mechanism.

Figure 9-9 illustrates a typical system using a combination of drive motor power and temperature difference across a conical-type refiner to control plug position.

The primary objective of the control scheme is to regulate the amount of mechanical work done on the pulp by positioning the rotating component (referred to as *plug* or *disc*).

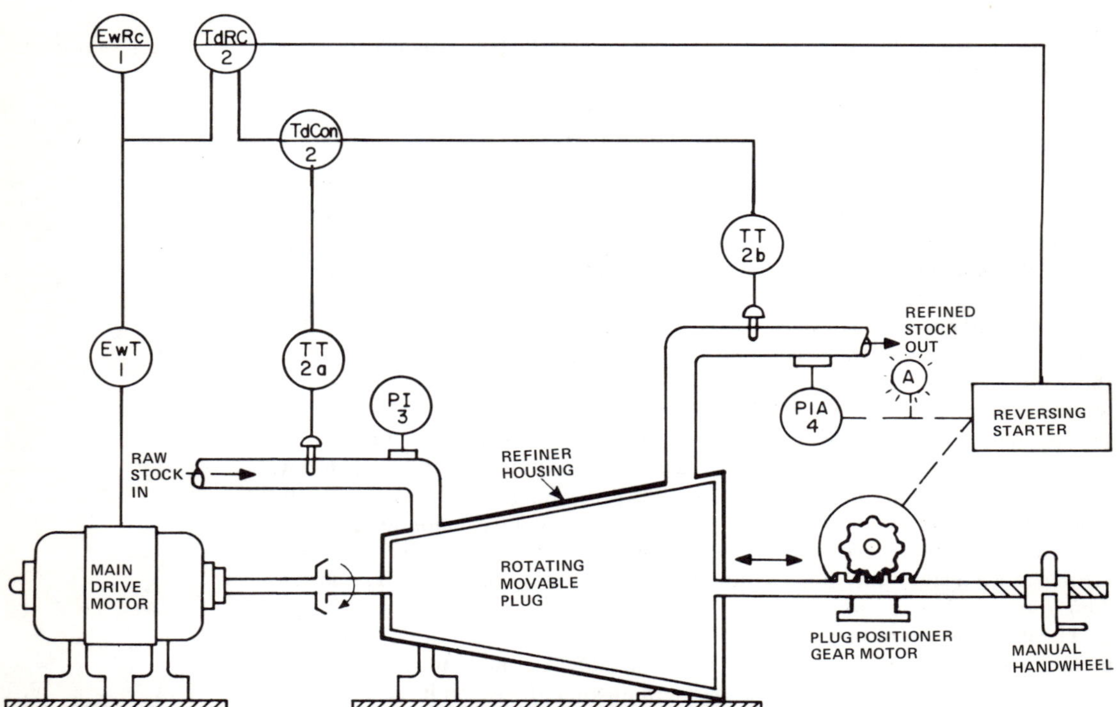

Figure 9-9. Typical refiner control instrumentation.

Temperature bulbs are located at the inlet and outlet of the refiner. These temperatures are transmitted by TT-2a and TT-2b to temperature difference converter TdCon-2, which subtracts the inlet temperature from the outlet temperature and transmits a signal proportional to temperature rise across the refiner to temperature difference recording controller TdRC-2. This is a pulse duration type of controller, which emits a pulse-type signal to energize the appropriate relay in the reversing starter of the plug positioning motor. This action moves the plug in or out (or varies the distance between discs on a disc-type refiner) as required to maintain the proper level of mechanical action (work) on the plug.

Main drive motor load power is measured and transmitted by transmitter EwT-1 to recorder EwRC-1. This same signal is fed into the temperature difference recording controller, TdRC-2, and used to supplement its control action. When an upset occurs, a time lag exists until its effect is detected by the differential temperature measurement, and the plug is repositioned. The more sensitive motor load power measurement detects upsets more rapidly and sends a corresponding signal to the temperature difference controller, which uses it to make compensating adjustments to its control action. A manual handwheel is also provided.

Pressure indicator PI-3 is used as a guide to possible plugging or loss of flow to the refiner. Pressure indicating alarm PIA-4 is also used for this purpose but, in addition, it will automatically withdraw the rotating plug on low flow by overriding the control signal from the temperature difference controller to the reversing starter. At the same time, it energizes a visible and/or audible alarm to call the operator's attention to this condition.

When the freeness measurement of the refined stock is used as the basis to control the operation of the refiner, the control can be made to directly operate the plug adjustment motor of the refiner, or to act as a master controller to adjust the set point of a secondary controller in a cascade fashion, as shown in Figure 9-10. The freeness measurement, QE-5, can be made either in-line or by bypassing a sample of the stock stream through an off-line freeness measuring device. The output of the freeness recording controller, QRC-5, adjusts the control set point of temperature difference recording controller TdRC-2 so that it will move the refiner plug in and out to achieve the desired freeness characteristics in the refined stock.

Figure 9-10. Typical refiner control instrumentation with freeness control.

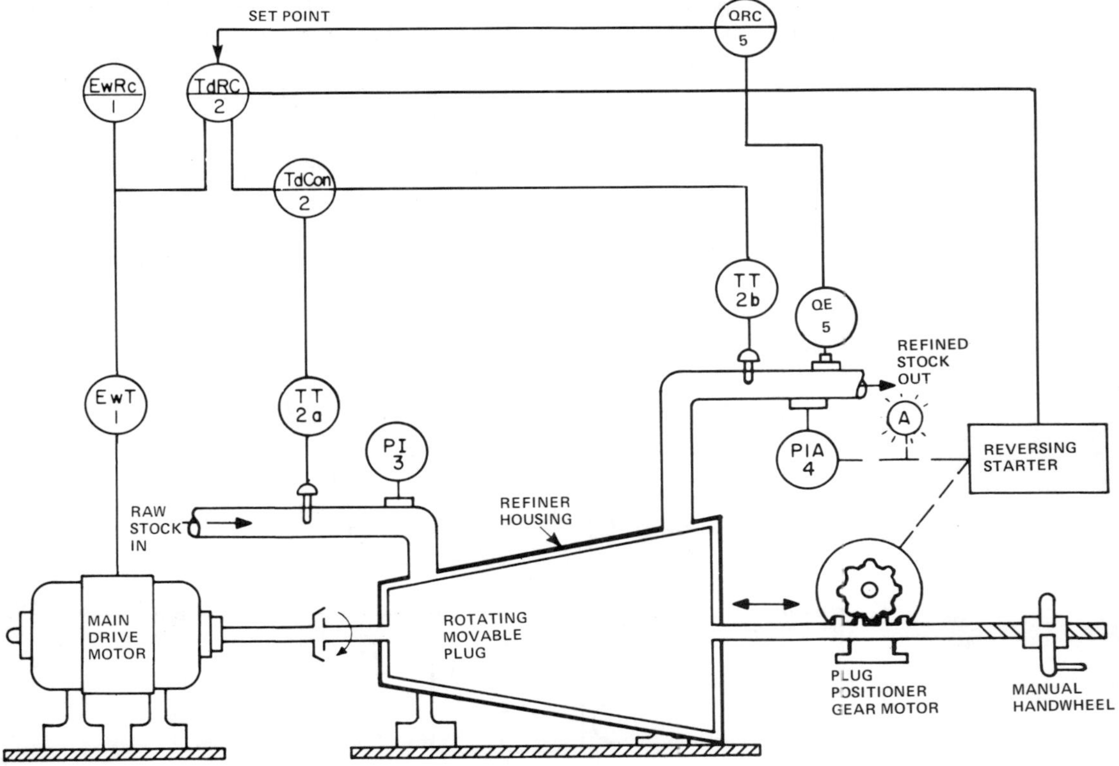

PULPERS

During the stock preparation phase of papermaking, it is necessary in many mills to break up, defiber, and disperse dry pulps of various forms, commercial waste papers, or other fibrous materials in slush form preparatory to further processing and conversion into paper or paperboard. This operation is often referred to as *repulping* and the machine used to do this is, therefore, a *repulper*, commonly called a *pulper.* Although there are various designs available, they basically consist of a tank or chest with suitable agitation to accomplish the dispersion with a minimum consumption of power. When the pulp originates as recycled product due to paper machine breakdown, it is known as *broke* and the equipment used to break it up is called a *broke pulper.* Even though they may be located at the paper machine they are considered as a form of stock preparation.

Accordingly, pulpers can be placed in three general categories, depending on their use. The first category is the pulper, which is located off the paper machine and is required to handle hand-fed broke, bales, and slabs of dried pulp. The second category is the pulper, which is located under the dry end of the paper machine and is required to handle winder trim, slabs, and loose broke as well as the full machine production which occurs during a paper machine breakdown. The third category is the pulper, located under the paper machine and away from the dry end. This pulper is required to handle only the full machine production when a break occurs, and does not handle the trim and loose broke.

Off-Machine Pulper

Figure 9-11 illustrates the basic instrumentation used on a typical pulper to control pulper tub level, pulper tub consistency, and discharge stock consistency.

Level is shown controlled by means of a three-way pulp valve, LV-2. Level controller LRC-1 receives the level signal from transmitter LT-1 and positions the pulp valve so as to direct a portion of the flow back to the pulper for level control. The pulp not required for level control is directed to the process.

The discharged pulp consistency is measured by a consistency detecting element, CE-2. The measurement is transmitted by CT-2 to a consistency recording controller, CRC-2, which functions to modulate dilution water valve CV-2 to maintain the desired consistency value for the process. If, for some reason, the consistency in the pulper increases to the point that the dilution water valve

Figure 9-11. Basic pulper instrumentation.

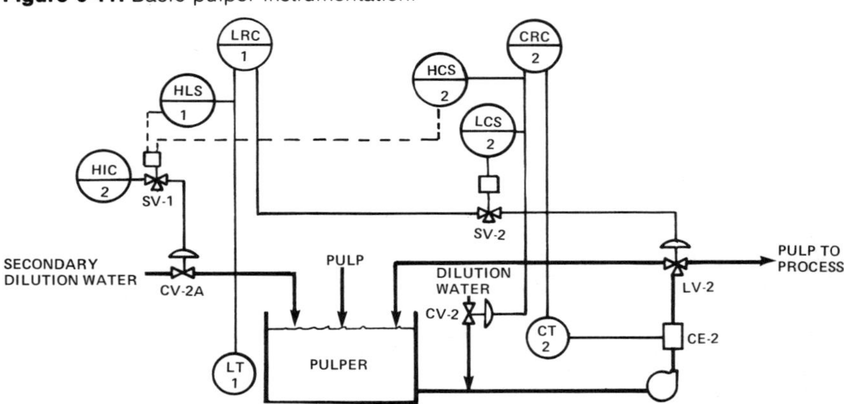

approaches a wide open position, high-consistency switch HCS-2, located in the output signal line of consistency controller CRC-2, senses this condition. It then energizes a solenoid valve, SV-1, in the supply line to the secondary dilution water valve, CV-2A, opening it and adding water until the consistency in the pulper allows dilution water valve CV-2 to start closing to its normal operating position within the range of the consistency controller. The high-consistency switch, HCS-2, then closes the secondary dilution water valve, CV-2A. The amount of the secondary dilution water valve opening required to maintain this desired consistency in the pulper is adjusted by a manual control station, HIC-2, in the supply line to the valve. In case the level of the pulper gets too high during the secondary dilution water addition, high level switch HLS-1, located in the measurement signal line from level transmitter LT-1, will de-energize solenoid valve SV-1 and shut secondary dilution valve CV-2A.

A low-consistency switch, LCS-2, is also connected to the output of the consistency controller to detect when the pulper consistency is so low that the signal to the control valve is at a minimum and the system is out of the control range. Under these conditions, the switch operates solenoid valve SV-2 in the output of the level controller to three-way valve LV-2 to hold it in the recirculating position until such time as the consistency has built up to an acceptable value.

Under Machine Pulper

Figure 9-12 shows a typical repulping system as it would be applied under the dry end of a paper machine. This system is basically similar to that shown in Figure 9-11, with additions to handle trim and slab broke pulp from the paper

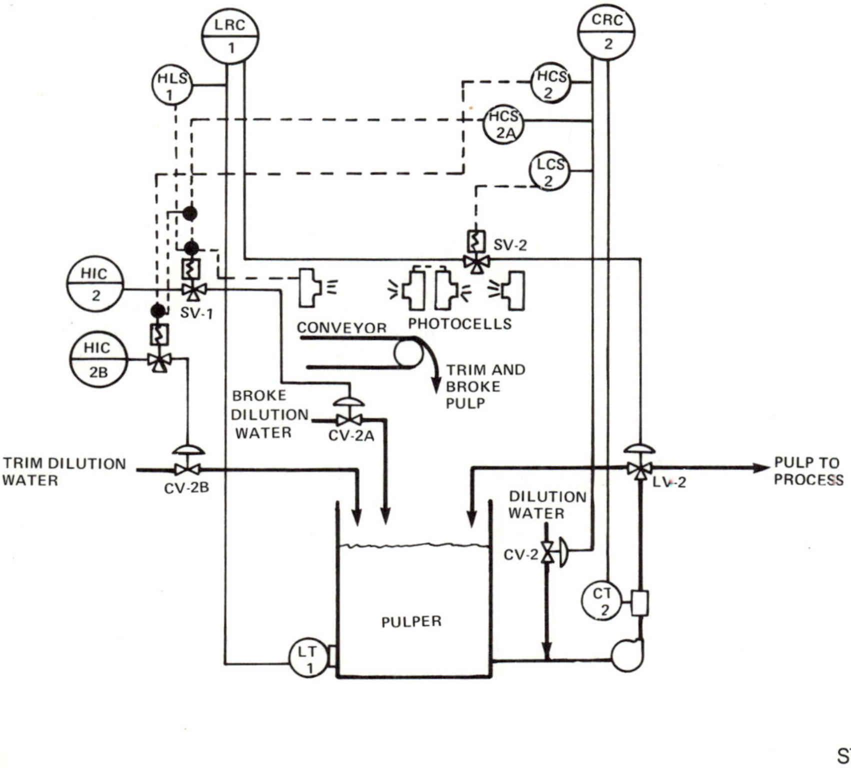

Figure 9-12. Typical instrumentation for under paper machine pulper

machine automatically. The secondary dilution water line becomes the broke dilution water line and a trim dilution water line is added.

During operation on the trim water dilution cycle, the trim dilution water valve, CV-2B, and the broke dilution water valve, CV-2A, are closed. The only dilution will be through the consistency control valve, CV-2. This valve is controlled in the normal manner to maintain discharged consistency. As the consistency in the pulper increases, the consistency control valve, CV-2, will continually open to maintain control of the discharged pulp slurry. High-consistency switch HCS-2, mounted in the signal line from consistency controller CRC-2 to this control valve, will open the trim water dilution valve, CV-2B, when valve CV-2 is more than half open. This tends to lower the pulper consistency back to the range of the consistency controller. When the consistency control valve drops below one-quarter open, switch HCS-2 closes the trim dilution water valve. The cycle is continually repeated during normal trim operation. If, for some reason, the pulper should be fed excessively heavy, the dilution water furnished by the consistency control valve and the trim dilution water valve may not be sufficient to maintain the desired discharge consistency. If high-consistency switch HCS-2 functions to open trim dilution water valve CV-2B, and consistency control valve CV-2 continues to open, high-consistency switch HCS-2A is so set that it opens the broke water dilution valve through solenoid valve SV-1 to aid in the dilution of the pulp in the pulper. This extra dilution will result in a throttling down of the consistency control valve and, when it reaches its one-half open position, the broke dilution water valve will automatically close. The consistency control valve has final control under all conditions. This system also uses low-consistency switch LCS-2 to throw three-way valve LV-2 into the recirculating position. This occurs if the consistency drops to a value which is not acceptable for pumping to the process.

When a paper machine break occurs, the full sheet drops onto the conveyor or into the pulper. The sheet interrupts a photocell light beam, energizing solenoid valve SV-1 which opens broke dilution water valve CV-2A to a preset position set on manual loading station HIC-2. This water is introduced through a shower which wets the sheet so that it falls into the pulper. The volume introduced is adjusted so that it is just less than that required to hold the consistency for full machine break. This results in increasing the consistency in the pulp, and dilution water valve CV-2 continues to open. At about one-half open, the trim dilution valve, CV-2B, will automatically open to aid the broke dilution water in controlling consistency. As the consistency then tends to decrease, the dilution water valve will be throttled down toward the one-quarter open position where the trim water dilution valve will close. This sequence is automatically repeated as necessary during full machine break. When the machine break is over, the light beam to the photocell is again restored and the system automatically goes back to trim operation.

Master And Slave Pulpers

Two pulpers are sometimes utilized under one machine. Figure 9-13 shows a typical configuration of such a system. The master pulper is normally located under the dry end of the paper machine and handles the dry end trim as well as the full machine break. The slave unit is located near the wet end of the machine and it handles only the full machine break. The master pulper is similar to that depicted in Figure 9-12.

The slave pulper uses very little instrumentation. It discharges into the master pulper for final control. The broke dilution water valve, HV-4, is set by a remote

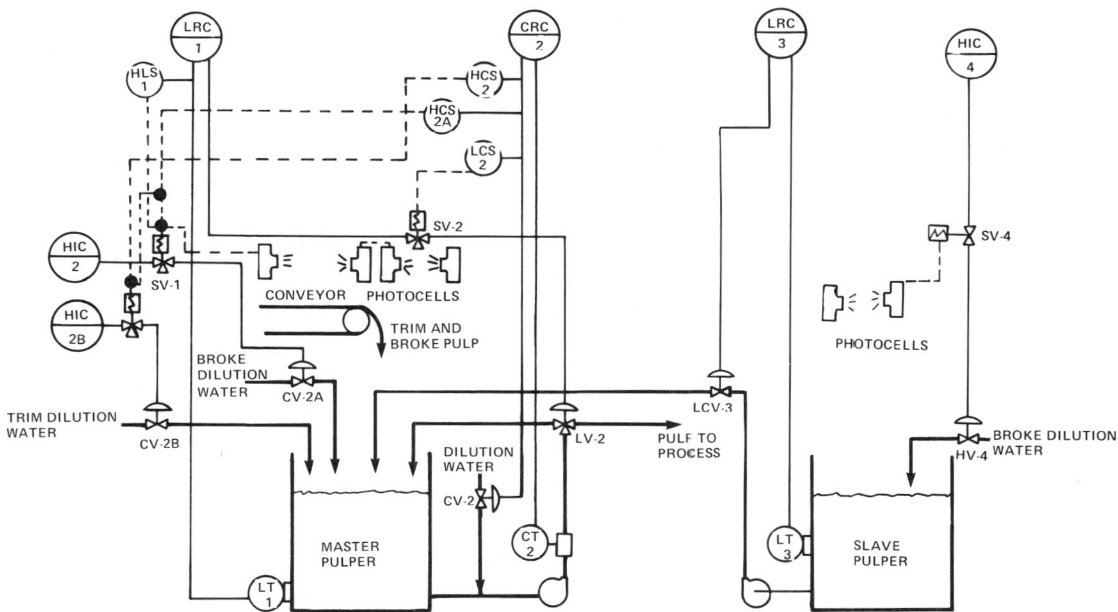

Figure 9-13. Typical master and slave pulper for under paper machine.

manual station, HIC-4, so that the discharged consistency will be high enough to permit the consistency controller for the master pulper to control it.

The slave pulper will normally be shut down. When a paper machine break occurs and the photocell light beam is interrupted, the pulper drive and stock pump will start up. The discharge rate from the pulper is controlled by level controller LRC-3, which modulates the pulp discharge valve LCV-3 to hold the pulper level constant. The broke dilution water valve, HV-4, opens to a preset position which is determined by manual loading station HIC-4. This should be set so that the consistency of the pulp being delivered to the master pulper is high enough for final consistency control. When the break is over and the light beam is restored, the broke dilution water valve, HV-4, is closed by solenoid valve SV-4, and the stock pump shuts down after the pulper level controller is satisfied.

SAVE-ALLS

Another paper mill function that occurs at the paper machine, but is sometimes thought of as a stock preparation activity, is the save-all system operation. Save-alls are used to recover fiber and filler from paper machine white water. The fiber or fillers are reused in papermaking while the water effluent is utilized for dilution and wash water in stock preparation and in other pulp and paper mill operations.

There are seven general types of save-alls:

1. Stationary save-all which is essentially an inclined screen.
2. Revolving screen or cylinder, consisting of a wire-covered cylinder which revolves in a vat containing white water.
3. Rotary vacuum type which employs a wire-covered drum or wire-covered discs which revolve in a vat containing white water.

4. Settling or sedimentation tank which allows white water to flow slowly in a tank or reservoir; suspended matter settles as water reaches the discharge point.
5. Flotation unit where white water is aerated in such a manner that the water contains minute bubbles of air and, after suitable defoamers are added, the white water passes to an area of low velocity where the fiber and filler float to the surface and are removed. The clarified water leaves at the bottom of the unit.
6. Filter press type is a two-stage process in which alum assists coagulation.
7. Upflow clarifier combines the functions of chemical addition, flocculation. and sedimentation, including sludge removal, in a single unit.

The instrumentation shown in Figure 9-14 is typical of that used on cylinder- and rotary vacuum-type save-alls. The flow of white water to the save-all is controlled by flow control loop FE-1, FT-1, FRC-1, and FV-1. The proper level in the vat for efficient operation of the save-all is maintained by level controller LIC-2 which adjusts the speed of the floating drum, thereby increasing or decreasing the rate of filtration through the screen. Warm water is added to the save-all's discharge side for dilution purposes and to aid the filtration by remote manual station HIC-3. Consistency of the stock is regulated by the addition of return water to the outlet section of the save-all with consistency controller CRC-4.

Figure 9-14. Typical save-all instrumentation.

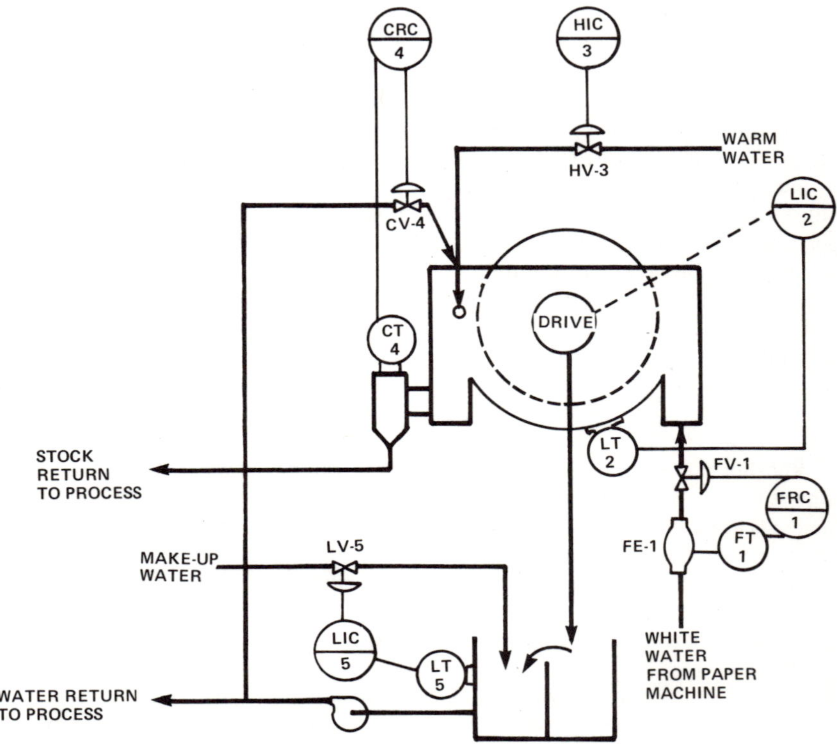

Paper Machine: Wet End 10

After the stock is suitably prepared in the stock preparation area by mechanical treatment and the addition of the required dyes, chemicals, and additives, it is furnished to papermaking machines located in an area commonly referred to as the *machine room*, where the final formation and drying of the sheet of paper or paperboard is accomplished.

There are a number of types of paper machines used to manufacture the many and varied grades of paper and paperboard. However, the general method they use to make paper is essentially the same. First, the formation of a web of paper from an aqueous suspension of fibers (stock) and further dewatering on what is called the *wet end* of the paper machine. Second, the drying and finishing or smoothing of the sheet by the section of the paper machine called the *dry end*.

The most common types of paper machine wet ends can be classified into general categories such as: (1) the fourdrinier, (2) the cylinder, and (3) the more recently developed line of wet ends known as sheet-formers. Additionally, a combination of the fourdrinier and the cylinder has been used to manufacture a particular product.

THE FOURDRINIER

The fourdrinier is by far the most extensively used type of paper machine wet end. In operation, the stock is deposited on a continuous belt-like moving wire screen as a free jet. The width of the jet is fixed in accordance with the weight of the sheet being made, the characteristic of the stock, and drainage capability of the wet end at any given speed. Older, smaller wet ends were designed to operate on paper machines designed to run at speeds up to 300 feet per minute, producing a sheet of paper between 72 and 100 inches in width. Today, machines are being designed to run at speeds approaching 5,000 feet per minute, and making sheets of over 380 inches in width. Although the general configuration of these wet end systems is similar, the details of the components have been modified in order to operate at the higher speeds.

The principal components of a typical fourdrinier wet end designed to operate with higher speed paper machines are depicted in Figure 10-1 and include the machine storage chest, refiner, stuff regulating box, fan pump, cleaners, screens, headbox, wire screen, white water tank, wire pit, couch pit, presses, save-all, broke pulper, and all associated equipment.

Beginning with the machine chest, instruments LT-1, LRC-1, and LV-1 measure and maintain normal operating stock level in response to machine demand or production. Consistency loop CT-2, CRC-2, and CV-2 ensures uniform stock consistency to the refiner by automatically adding water to the suction side of the pump. The degree to which the stock is refined is controlled by refiner motor load controller EwRC-11A, operating the motor for the refiner plug positioner in response to couch roll vacuum transmitter VT-11, which remotely sets its control point accordingly.

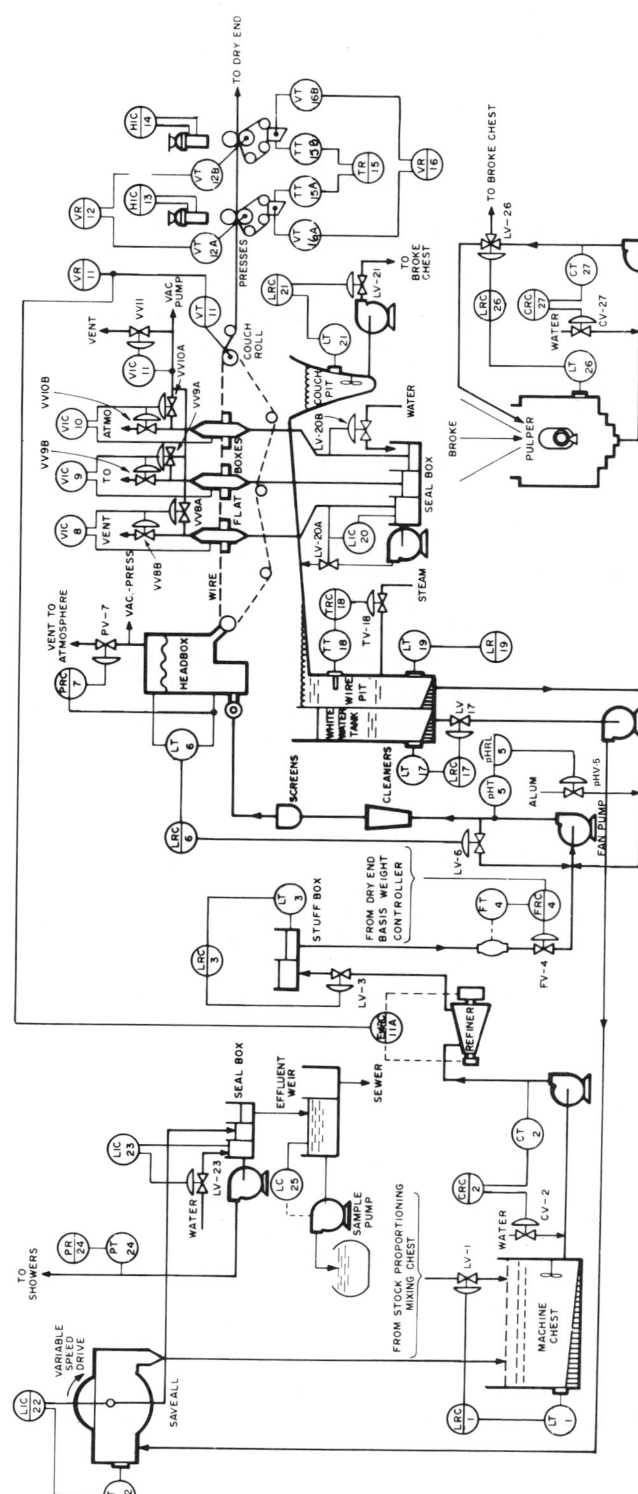

Figure 10-1. Instrumentation for the wet end of a paper machine.

Temperature difference and freeness measurements have also been used for this control. Depending on the grades of paper being made, a paper machine may or may not have a stuff or regulating box whose level is measured and controlled by instruments LT-3, LRC-3, and LV-3. The rate of stock delivery to the machine is automatically controlled by having a basis weight controller at the dry end to cascade-set the control point of the magnetic-type stock flow metering control system, FT-4 and FRC-4.

Many grades of paper are sized for resistance to water penetration by adding rosin to the stock before it goes to the machine. Alum in generally used to "set" the size on the individual fibers. This is best accomplished at a definite pH level. A pH recording control loop, pHT-5, pHRC-5, pHV-5, regulates the addition of alum to the paper stock. To attain a uniform flow through the slice onto the wire and to ensure good sheet formation, both total head (pressure) and level must be controlled in the headbox. Level controller LRC-6 controls headbox level by adjusting valve LV-6 in the bypass stock line around the fan pump. The total head is controlled by PRC-7 operating vent valve PV-7 in a vacuum pressure exhaust system attached to the headbox. Efficient and safe water removal on the wire is further enhanced by controlling the vacuum in the individual flat boxes with vacuum loops VIC-8, VV-8A, VV-8B, VIC-9, VV-9A, VV-9B and VIC-10, VV-10A, VV-10B, with loop VIC-11 and VV-11 controlling the vacuum in the header to the vacuum pump.

The couch roll drives the wire and is evacuated to create a vacuum under the wire. This measurement is recorded on VR-11 and also transmitted to EwRC-11A to cascade-set the load on the refiners. Vacuums on the presses are recorded on a two-pen vacuum recorder, VR-12, while remote loading of the press rolls is accomplished by hand indicating control stations HIC-13 and HIC-14. Loops VT-16A, VT-16B, VR-16 and TT-15A, TT-15B, TR-15 measure and record vacuum and temperatures of felt conditioners as a guide to proper operation.

The vacuum system seal box level is maintained at a uniform level by level indicating controller LIC-20, operating level control valve LV-20A, which is split-ranged with control valve LV-20B on the fresh water line to ensure a minimum water level in the seal box to protect the pump suction. Loop LT-21, LRC-21, and LV-21 controls the level in the couch pit while LIC-22 controls the level in the save-all vat by adjusting a variable-speed drive on the drum. Shower pump suction on the save-all seal box is protected by a locally mounted level indicating controller, LIC-23, and the shower water pressure is recorded on PR-24. A locally mounted level controller on the mill effluent weir is used to operate a sample purge to provide a means of checking fiber losses in the paper mill effluent.

Basic level and consistency control loops used on broke pulpers in paper mills are illustrated by instruments LT-26, LRC-26, LV-26, and CT-27, CRC-27, CV-27.

Screens and Cleaners

On most paper machines the diluted stock is screened ahead of the wet end. Quite often, cleaning is also employed as a complementary operation and is shown in Figure 10-1 as auxiliary equipment such as screens and cleaners.

Screens. The purpose of screening is to remove shives, strings, lumps, and foreign objects. Several screen designs have been developed and used, but the underlying principle of the process is that it is a straining operation. The most common type of paper machine screen is of the vibrating, slotted plate, rotary configuration, with flow direction from the inside out. High-pressure water showers inside the screen drum impinge on the slots and flush the dirt which collects on them into an outside tray, from which it usually goes to a sewer. The

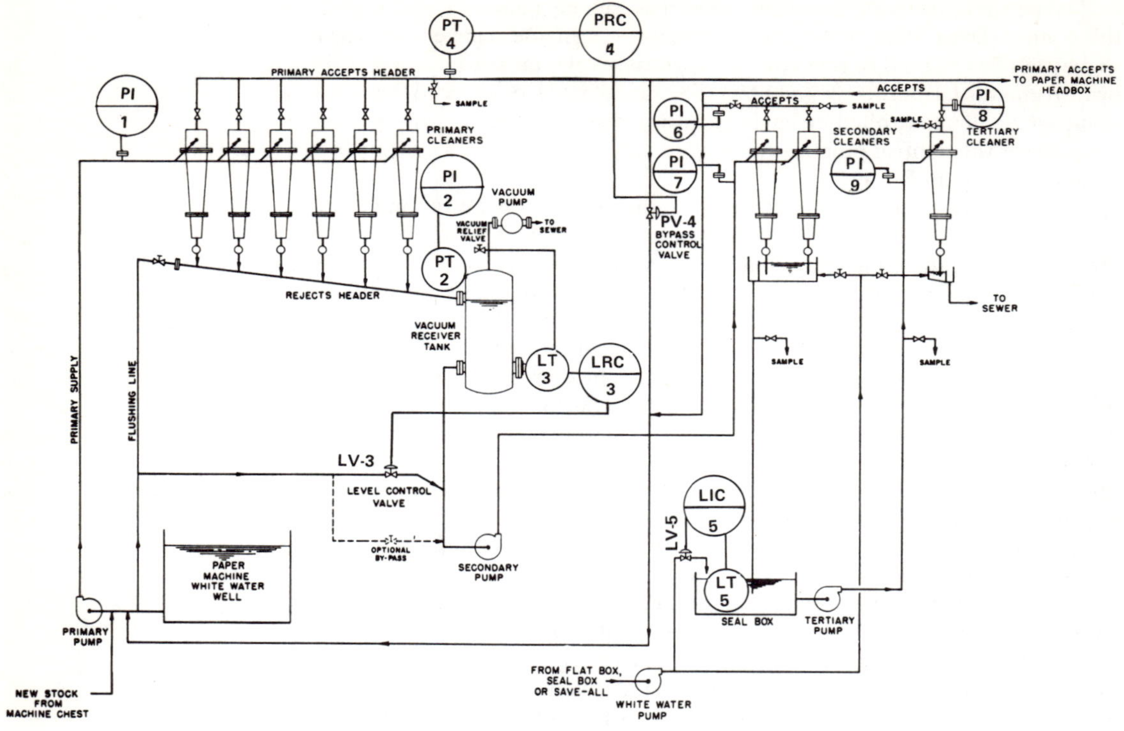

Figure 10.2 Centrifugal stock cleaning.

instrumentation used is minimal and can consist of a simple pressure controller on the flush water line and a level indicator for the operator's guide.

Cleaners. Grit, sand, bark, etc., are removed from the diluted stock by cleaners. Here also, several types of cleaners have been used successfully. A very common one is the centrifugal cleaner. Figure 10-2 shows a modern installation of nine centrifugal cleaners arranged in three stages.

A cleaner consists of a hollow-cone section joined at the top by a cylindrical body with cone-shaped connections. As the stock suspension enters under pressure, controlled by pressure recording controller PRC-4, it swirls downward and inward, and forms an upward-spiraling column extending the length of the cleaner to accept the outlet connections. Increasing velocities and centrifugal forces occur as the stock suspension moves downward to smaller and smaller radii. Heavy dirt such as sand, together with some fiber, is discharged continuously at the bottom of the cone. The cleaned stock, traveling upward along the axis of the cleaner, is discharged at the top center of the head. Because some good fiber is always rejected with the dirt, several stages of cleaning are normally used. Only the accepted furnish is delivered to the paper machine, accepts from the second and third stages being recycled back to the primary stage.

The vacuum created in the core of the cleaner by the rotational speed of the liquid can cause air to be drawn in through the reject orifice. To prevent this, a vacuum is sometimes applied to the reject side of the primary cleaners. This vacuum is indicated on PI-2 and the liquid level in the vacuum receiver tank is controlled by level recording controller LRC-3. Primary accepts header pressure is maintained by controller PRC-4 throttling automatic control valve PV-4, located in a bypass return line to the suction side of the primary pump. The level in the secondary cleaner's seal box is maintained by indicating level controller LIC-5. Supply and accepts pressures are indicated on gauges PI-1, PI-6, PI-7, PI-8, and PI-9, conveniently located for operator guidance.

Deaeration

Free air in the stock to the wet end is liable to cause objectionable foaming, slime troubles, and consistency variations that affect drainage and sheet formation. When this problem exists, the air must be removed from the stock prior to the headbox. Several deaerating devices, not shown in Figure 10-1, utilizing various methods have been used.

A typical deaerating process, which operates at high vacuum and is located in the dilute stock system between the wet end fan pump and the headbox is shown in Figure 10-3. The stock from the fan pump is sprayed into the deaerator receiver through large nozzles which whirl the stock at high velocity and discharge it in the form of a spray against impingement areas, thereby separating individual air bubbles from the fiber and water. A constant liquid level is maintained in the receiver by means of an overflow weir. This level is recorded on level recorder LR-1 from a measurement signal made by transmitter LT-1. A differential pressure type of level transmitter is used with the low side connected to the top of the deaerator receiver to compensate automatically for variations of tank vacuum. A sight glass is installed on the lowest point of this piping to detect any condensate buildup. Buildup can be periodically drained by the use of a shutoff valve. The result is that, regardless of changes in deaerator vacuum, the output of the level transmitter is proportional to changes in level only. The vacuum in the main section of the deaerator is measured and indicated by PT-2 and PI-2.

Figure 10-3. Typical deaerator instrumentation.

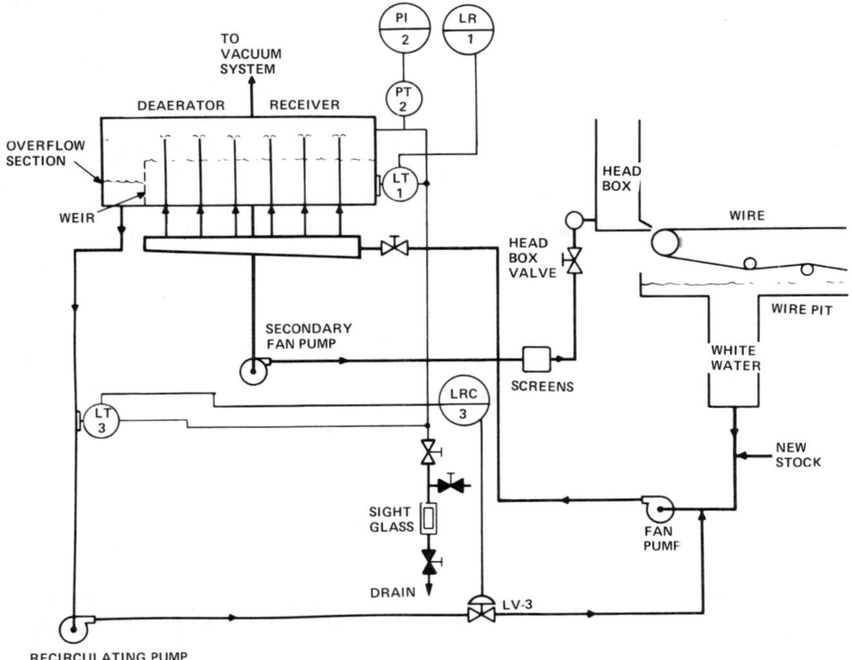

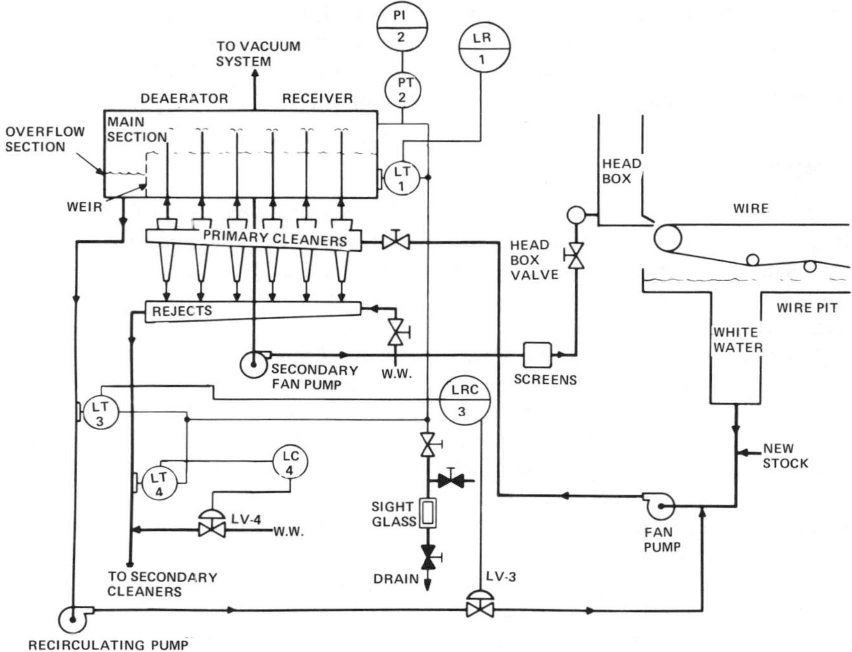

Figure 10-4. Instrumentation for a typical deaerator with cleaners.

The deaerated stock is pumped from the main compartment of the receiver through the outlet to the headbox of the wet end of the paper machine. The air is removed continuously from the receiver and indicated on PI-2. The overflowed portion of stock is removed continuously by a separate overflow recirculating pump and is returned to the suction side of the deaerator supply fan pump, where it is combined with new incoming aerated stock. A constant head is maintained on the suction side of the recirculating pump by level recording controller LRC-3.

The deaeration process may also be used in combination with centrifugal cleaners. Such an arrangement is depicted in Figure 10-4. The spinning action of the stock column from the cleaners provides the spray needed for deaeration in the presence of vacuum.

The instrumentation and control are essentially the same except for the addition of the primary cleaner rejects level control loop, LT-4, LC-4, LV-4.

Headbox

Control systems for the wet end headboxes of the paper machine operate on the basic principle represented by the formula:

$$V^2 = 2gh$$

Where: V = spouting velocity (ft/sec)
g = a constant of gravity (32.2 ft/sec^2)
h = total head (ft)

This states that the head required to maintain the necessary spouting velocity on the wire portion of the wet end from the box varies as the square of the wire velocity.

Two general types of headboxes have been used: the open to atmosphere and the enclosed or pressurized. The open headbox was generally used on earlier paper machines running at slower wire speeds of 600 to 800 feet per minute and requiring a variation of approximately 14.5 inches in level to change the corresponding spouting velocity. As faster machines were developed to operate at speeds of 3000 to 5000 feet per minute, a level change of hundreds of inches of water was necessary to produce the required spouting velocity. This made it physically impractical to use open headboxes. The enclosed headbox, which is more common today, was developed for use on these faster operating machines.

Level in the enclosed headboxes has been controlled while regulating the total head with an air pad produced with a pressure vacuum pump, by two general forms of systems: the older mechanical and the more recent instrument types.

Mechanical. The two most common variations of the mechanical types are the float and tube.

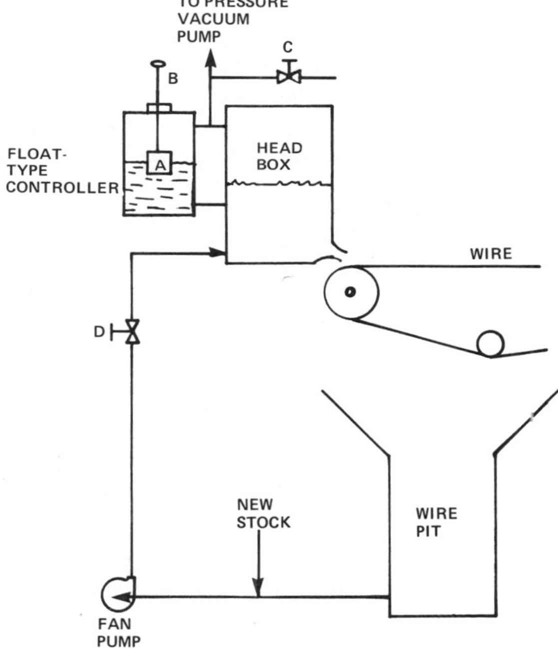

Figure 10-5. Pressure-level control system for a mechanical float-type headbox.

The mechanical float type of headbox control system shown in Figure 10-5 uses an adjustable float that operates an air release valve to control total head and liquid level. The relationship of the valve assembly to the float is changed when the machine speed is changed. To change from pressure to vacuum operation, or conversely, the valve-to-float relationship must also be changed.

In operation, the float indicates the level of stock in the headbox. Connections are made to both the submerged section and pressure pad of the headbox in order that the float read the true level of stock. When properly adjusted for pressure operation, the float, on its upward movement, causes the valve to close; on its downward movement, it causes the valve to open. For vacuum operation, the movements are reversed.

The pressure vacuum pump is valved for either operation. The system is adjusted so that when the float is at its operating level, valve B is closed and valves C and D are partially throttled. The float then establishes the proper level and total head. Valves C and D are adjusted to provide an equilibrium between the pump head and padding pressure, thus producing the proper spouting velocity.

During operation in the pressure mode, an increase in level causes the float to rise. This closes the valve and increases pressure in the headbox. As pressure increases, the fan pump experiences higher head and delivers fewer gallons to the headbox. When the level decreases, the float falls and exhausts air through valve B. Because of the lesser head due to this drop in pressure, the fan pump delivers more gallons to the headbox, thereby restoring the level.

For vacuum operation, the starting procedure is similar to the pressure operation, except that valve B is adjusted to be open when the float is at its operating position. When the level rises under these conditions, valve B opens, causing less vacuum or more pressure, and the fan pump delivers fewer gallons. For a falling level, valve B closes, increasing vacuum and the pump delivers more gallons to restore the liquid level.

Tube Type. In the tube-type headbox pressure-level control system, the float is replaced by a tube as shown in Figure 10-6. This system works on the principle that, when the tube is either fully or partially submerged, the amount of air bleeding in or out is controlled. The pump and valve arrangement can be identical to the float system. With valves C and D properly adjusted, the level and pressure will come to equilibrium.

When the level in the headbox rises, the tube becomes submerged allowing less air to bleed through the hole. This causes an increase in pressure in the headbox, thereby increasing the head on the pump which causes it to deliver fewer

Figure 10-6. Pressure-level control system for a mechanical tube-type headbox.

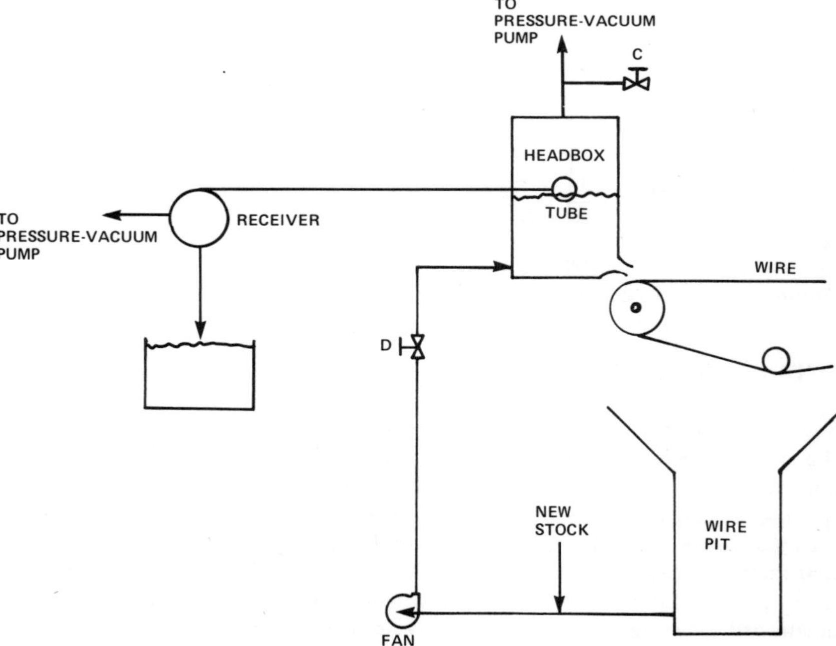

gallons and drive the level down. A decrease in level has the opposite effect; it allows more air to be exhausted as the level drops due to the increased area of the tube. This loss in pressure causes the pump to deliver more gallons, causing the level to rise.

Vacuum operation of the tube type is similar in principle to that of the float type. In this case, a separator is employed, allowing a reverse function of the system. Air passes from the tube through the pump to atmosphere. Any dry stock that spills over goes to the wire pit. Valve C is used to trim the capacity of the vacuum pump. As the level rises, the tube becomes increasingly more submerged, and less vacuum is in the pad because the pump is exhausting less and valve C is allowing the head box to come to atmospheric pressure. Thus, the fan pump experiences an increasing head, and the level falls. Should the level fall below the set point, the vacuum pump pulls more vacuum, the fan pump experiences a lesser head, and the level rises.

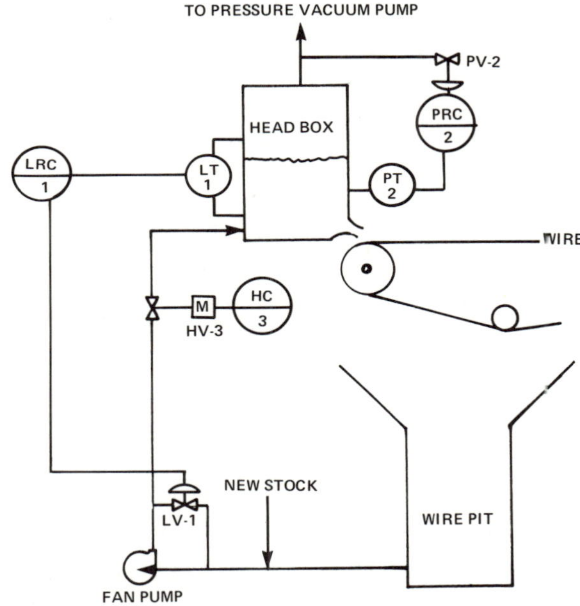

Figure 10-7. Instrument-type head box pressure-level control.

Instrument Type. When instrumentation is used to control the operation of the head box, it accomplishes control by the throttling of a valve in the main flow, throttling of a shunt valve across a manually controlled valve in the main flow, controlling the recirculation to the suction side of the pump, and/or controlling the fan pump speed. Figure 10-7 illustrates a typical instrument system for controlling both liquid level and total head.

The sensing element, LT-1, on the liquid level control loop is a differential-pressure type transmitter. The final control element, LV-1, is a positioner-operated valve on a pipeline that bypasses the fan pump. This valve is regulated by recording controller LRC-1. A pump head regulating valve, HV-3, is located in the stock line in series with the fan pump discharge. If a variable-volume pump is used, the head regulating valve is not used. In operation, the control valve bypasses only a small percentage of the fan pump output (approximately 10 percent maximum). If this valve approaches the full open or closed positions, valve position switches actuate pilot lights or alarms to alert the operator so that he can readjust the pump head valve, HV-3, to bring the bypass valve back into the proper operating range.

The total head control loop measurement, which includes both the liquid level and applied air pressure, is made with pressure transmitter PT-2. The final control element, valve PV-2, controls total head by exhausting air to atmosphere.

Many variations of instrument systems have been used to control pressure-level in headboxes, such as controlling level by throttling the air bleed and controlling the total head by throttling the fan discharge valve.

Rush/Drag Control

The formation of the paper sheet on the wet end wire is affected by the relationship between the rate at which it is deposited on the wire, the slice jet velocity, and the speed of the wire. Some grades of paper require that the slice jet velocity be slightly faster than the wire speed for best formation. In essence, it is "rushing" the stock on the wire and, therefore, referred to as *rush*. Other grades of paper require that the wire speed be greater than the slice jet velocity, resulting in a "dragging" effect on the stock by the wire. This phenomenon is called *drag*. The relationship or ratio between them is expressed as rush/drag or efflux ratio, and its relationship to the slice jet velocity and wire speed can be stated as:

$$\frac{R}{D} = \frac{S.V.}{W.S.}$$

When the ratio is equal to 1, the slice jet velocity is the same as the wire speed. When the ratio is less than 1, the wire speed is greater than the slice jet velocity and the stock is being dragged by the wire. Conversely, when the ratio is greater than 1, the slice jet velocity is greater than the wire speed and the stock is rushing the wire.

There are several ways to implement rush/drag control on the wet end of a paper machine. Figure 10-8 shows the instrumentation involved in a typical version.

It operates in accordance with the basic principle of slice jet and headbox pressure or the total head relation of $V^2 = 2gh$. The total head can then be expressed as $h = \frac{V^2}{2g}$. Since $2g$ will always be constant, the relationship can be simplified and expressed as $H = V^2$.

As the slice jet velocity of the stock to the wire is uniform and made proportional to the wire speed, the theoretical head required is automatically computed from the speed of the wire as measured by a tachometer generator type of speed transmitter, ST-2. This signal is recorded on speed recording controller SRC-2 and squared in a computing relay to obtain the theoretical head corresponding to the wire speed. This signal is combined in a multiplying computing relay with a signal from manual loading station HIC-2 as a bias calibrated in rush/drag ratio. This biased signal from the multiplying computing relay is proportional to the biased theoretical total head to give the rush/drag ratio desired between slice jet velocity and wire speed, and is used to remotely adjust the set point of the actual total head controller, PRC-2, to establish the correct rush/drag relationship.

In other versions of this control system, the biased theoretical total head signal is back-loaded into total head transmitter PT-2, which, in this case, would be of the pressure differential type. Total head controller PRC-2 would then be a deviation type with the set point set at midscale.

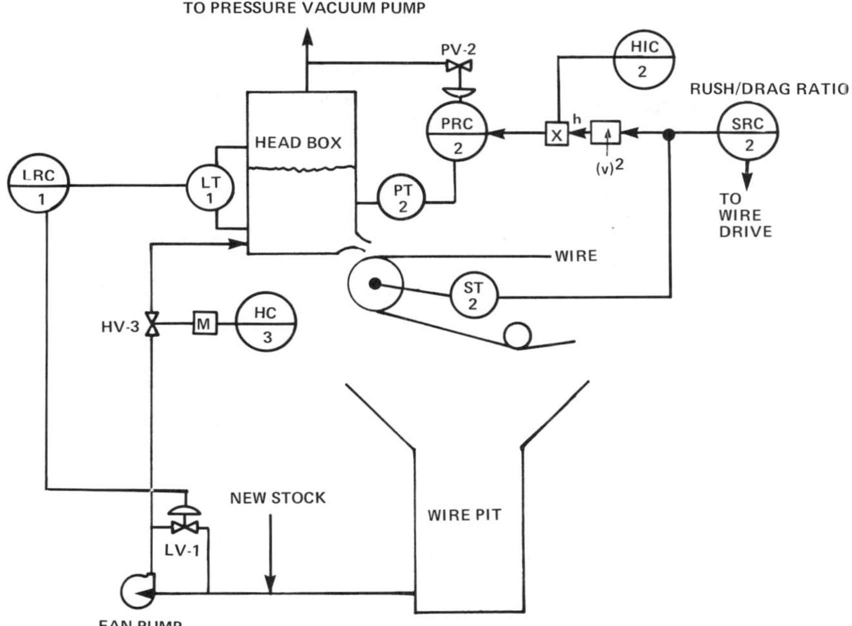

Figure 10-8. Wet end rush/drag control.

pH Control

Ordinarily, the stock coming to the wet end of a paper machine has a higher pH than desired, and it is reduced by adding alum or sulfuric acid, or both. An alum solution is the most common acidifier, particularly when sizing is used in the furnish. The aluminum ion performs an important function in binding size particles to paper fibers when pH is held at the correct value, generally around 4.5 pH. However, the pH may be adjusted and controlled at whatever value is required by local conditions to secure optimum overall results.

The pH of the stock is measured by the potential difference between a glass electrode and a reference electrode. Some types of electrode assemblies can be immersed directly in stock lines. Other types are more delicate and must be installed in a sample box or flow-through electrode chamber, through which a sample of stock is continuously run. All electrodes require periodic servicing. The frequency of this servicing can be decreased by the use of ultrasonic cleaning devices with the electrodes.

Depending on the type of wet end configuration, the pH measurement is made at several locations. Figure 10-9 schematically shows three such points. pHE-1 makes its measurement from the stock to the headbox. This is the recommended location. An alternative location, pHE-1A, measures the stock in the discharge from the fan pump, and another alternative, pHE-1B, is located in the white water from the trays. Of the two alternatives, pHE-1A reflects changes in pH sooner, but the installation is more difficult. The other alternative, pHE-1B, reacts to changes in pH later, but is preferred by some from the standpoint of convenience.

The pH measurement is transmitted to recording controller pHRC-1 which adjusts valve pHV-1, regulating the amount of alum solution being added to the stock at the suction side of the wet end fan pump.

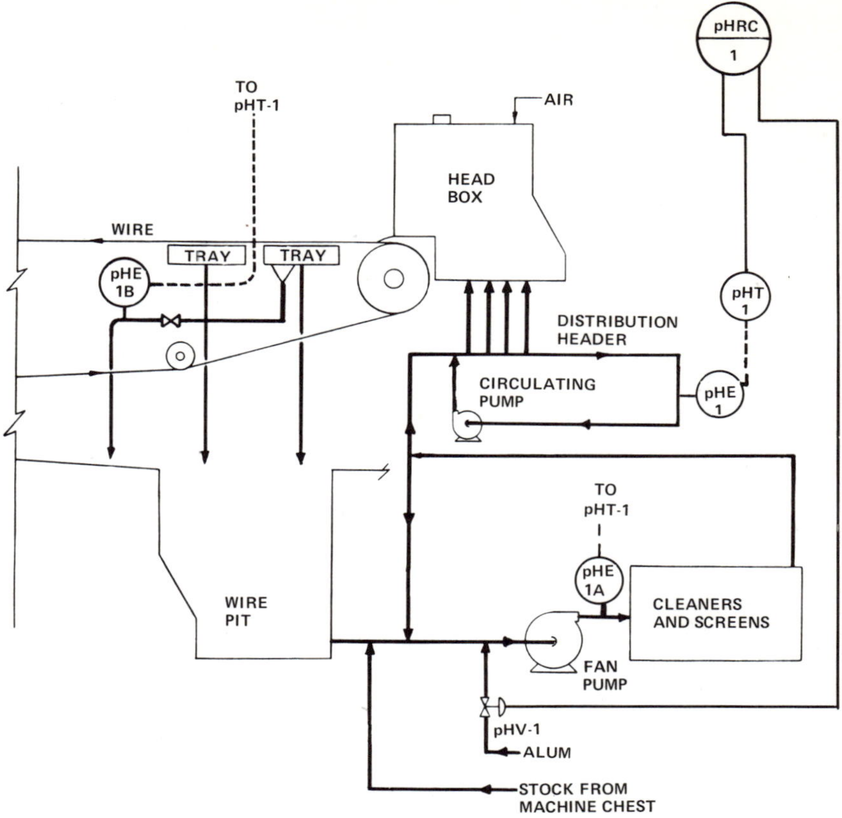

Figure 10-9. Wet end pH control.

Flat Box Vacuum Control

During the formation of the sheet of paper on the wet end wire, water is removed by suction flat boxes located below it. Here, water is removed by a difference in pressure between the top and bottom surfaces of the sheet with atmospheric pressure above and vacuum of 4 to 10 inches of Hg (mercury) underneath. This pressure difference caused by the use of a vacuum pump removes the water suspended in the sheet against the resistance of the fiber mat, the wire, and the openings in the suction flat box. The resistance of the mat increases progressively. The amount of water extracted varies with the freeness of the stock and with the vacuum in the boxes. An increase of vacuum increases the drag on the wire and, therefore, decreases its life. Figure 10-10 illustrates the instrumentation used to permit operation at optimum vacuum levels for maximum water removal and minimum wire wear due to surges in the vacuum system.

The system consists of a master vacuum controller, PIC-1, operating a vent valve, PV-1, in the main header, plus controllers PIC-2, PIC-3, and PIC-4 on each individual flat box vacuum line.

The output of each flat box controller operates a pair of valves in sequence. If the flat box vacuum falls below the controller set point, valves PV-2A, PV-3A, and PV-4A to the main vacuum header open more to increase the vacuum at the box. Should the flat box vacuum exceed the controller set point, the controller

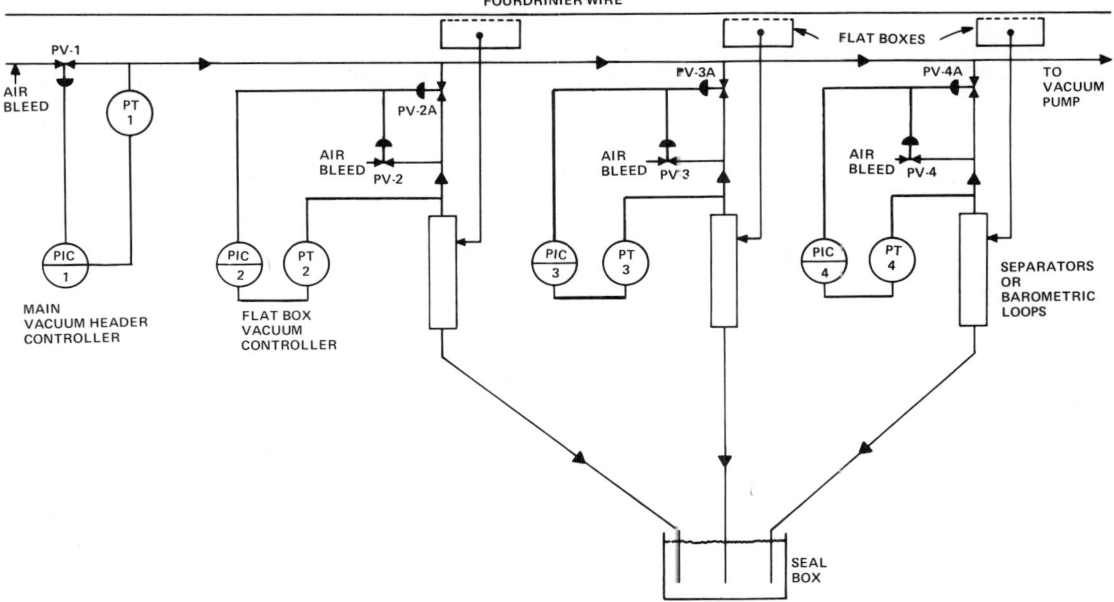

Figure 10-10. Wet end flat box vacuum control.

outputs will throttle valves PV-2A, PV-3A, and PV-4A to the main vacuum header. If the vacuum is still above the set point with this valve closed, the controller outputs will open atmospheric vent valves PV-2, PV-3, and PV-4 to bleed air into the individual flat box systems.

CYLINDER MACHINE

The cylinder machine was the first papermaking machine in America. It was the counterpart to the fourdrinier-type wet end of the paper machine.

In the cylinder machine a revolving cylinder mold, covered with wire mesh, turns partially submerged in a vat into which the stock is fed. The water passes through the face of the cylinder mold under static head and out the ends, leaving the fibers as a sheet on the face of the cylinder. The wet sheet is picked up or *couched* from the top of the cylinder mold by means of a traveling woolen felt pressed into contact by the couch roll and, thus, is carried to presses and the drying section. When multiply sheets such as liner boards are made with cylinder machines, two or more cylinder vats are used with each cylinder forming a liner of the multiply sheet. It is also more suitable for producing heavier grades such as various types of paperboard.

The two most used vat types on cylinder wet ends are the direct or parallel flow and the counterflow. In the counterflow, the stock enters the cylinder vat and flows in the opposite direction of the pickup felt travel. The flow of stock in the parallel-type cylinder vat is in the same direction as the pickup felt movement. Figure 10-11 shows a schematic arrangement of a typical direct-flow cylinder wet end.

The instrumentation used is normally not as extensive as in other wet end types. The stock chest is maintained at a satisfactory level to provide a steady supply of stock to the machine by level indicating controller LIC-1. High and low alarms are usually included to warn the operator of loss of stock supply or an

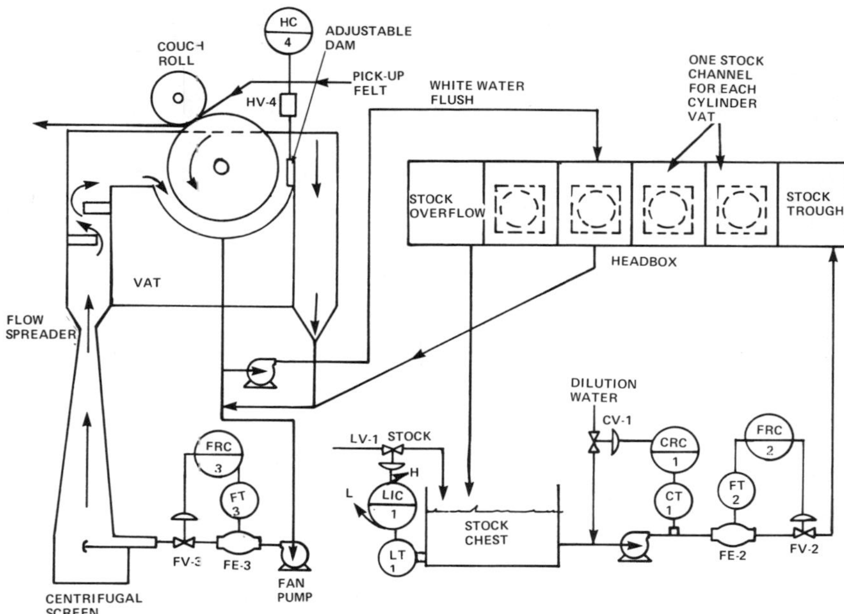

Figure 10-11. Direct-flow type cylinder wet end.

overflowing chest. The consistency of the stock to the cylinder machine is regulated by consistency control loop CT-1, CRC-1, CV-1, while flow is controlled by flow loop FE-2, FT-2, FRC-2, FV-2. The total stock to the cylinder vat is controlled by flow recording controller FRC-3. The height of stock in the cylinder vat is regulated at operating level by manual adjustment of dam HV-4 from remote manual station HC-4.

FORMERS

A more recent development in the papermaking industry has been a series of wet end types which can be categorized under the general heading of *formers.* They resemble cylinder-type wet ends in that a cylindrical forming roll is used to support the forming wire mesh. The difference lies in the method of admitting the stock to the wire surface with a nozzle or specially shaped inlet. Vacuum is applied to the interior of the roll by means of a suction located for drainage to a specific zone.

This type of wet end has been constructed in several different forms, characterized by different size nozzles, compartmented internal suction boxes, vertical twin wires, multiwires with primary and secondary head boxes, and others. Although the instrumentation varies somewhat for each type, all are similar to that found in the fourdrinier- and cylinder-type wet ends.

11 Paper Machine: Dry End

The formed sheet of paper, after passing through the press section of the wet end of the paper machine where further removal of free water is accomplished and the bulk is reduced, is conveyed to the dry end portion of the paper machine. Heat is applied in the dry end to drive off water by evaporation and bring the sheet to the desired state of dryness.

There are several types and arrangements used in dry end configurations, depending on the grade of paper being produced. The paper sheet comes to the dry end with about a 70 percent water content. A very common dry end consists of a series of steam-heated, cast iron drums arranged in sections called dryers.

MULTICYLINDER DRYERS

The dryers vary in number and arrangement to facilitate the passing of the sheet from one dryer to another and compensate for the effects of shrinkage on the drying sheet. Each section is driven at a slightly different speed to compensate for the shrinkage occurring in the sheet during the drying operation.

Most papermaking machines, except paperboard machines, are fitted with an endless canvas belt or felt which holds the paper sheet in contact with the surface of the drum. Heat from the steam introduced into the dryers is conducted through the dryer shell to the sheet, evaporating the water. The vapor is carried away by a ventilating system consisting of exhaust fans in tightly fitted hoods vented to atmosphere.

Some typical instrumentation used to control the operation of the dry end of a paper machine is shown in Figure 11-1.

To obtain efficient extraction from input steam, the machine is divided into sections, and the differential pressure between sections is regulated. With this arrangement, the condensate will flash from each successive separator to the following section inlet header.

Steam used in the paper drying operation is measured, recorded and totalized by instruments FT-1 and FR-1. Control loops dPRC-2, dPRC-3, dPRC-5, and dPRC-9 maintain the necessary differential pressure across each section of dryers in order to ensure good condensate and air removal for dryer efficiency. Autoselector-type pressure controllers PsC-3, PsC-5, and PsC-9 establish the necessary differential gradient between sections. These controllers are provided with auxiliary mechanisms which will automatically reduce pressure in subsequent sections if pressures approach values too close for good condensate removal. The cascade moisture control system, MT-7, MRC-7, and PRC-7, controls the sheet moisture prior to the size application in the size press. The moisture measurement is made by a variety of methods. The output of the moisture controller automatically adjusts the pressure controller set point in a cascade fashion in order to control the steam pressure at the desired level to maintain a constant moisture value. In the same manner, the moisture control system, MT-12, MRC-12, PT-12, and PRC-12, maintains uniform sheet moisture going to the reel.

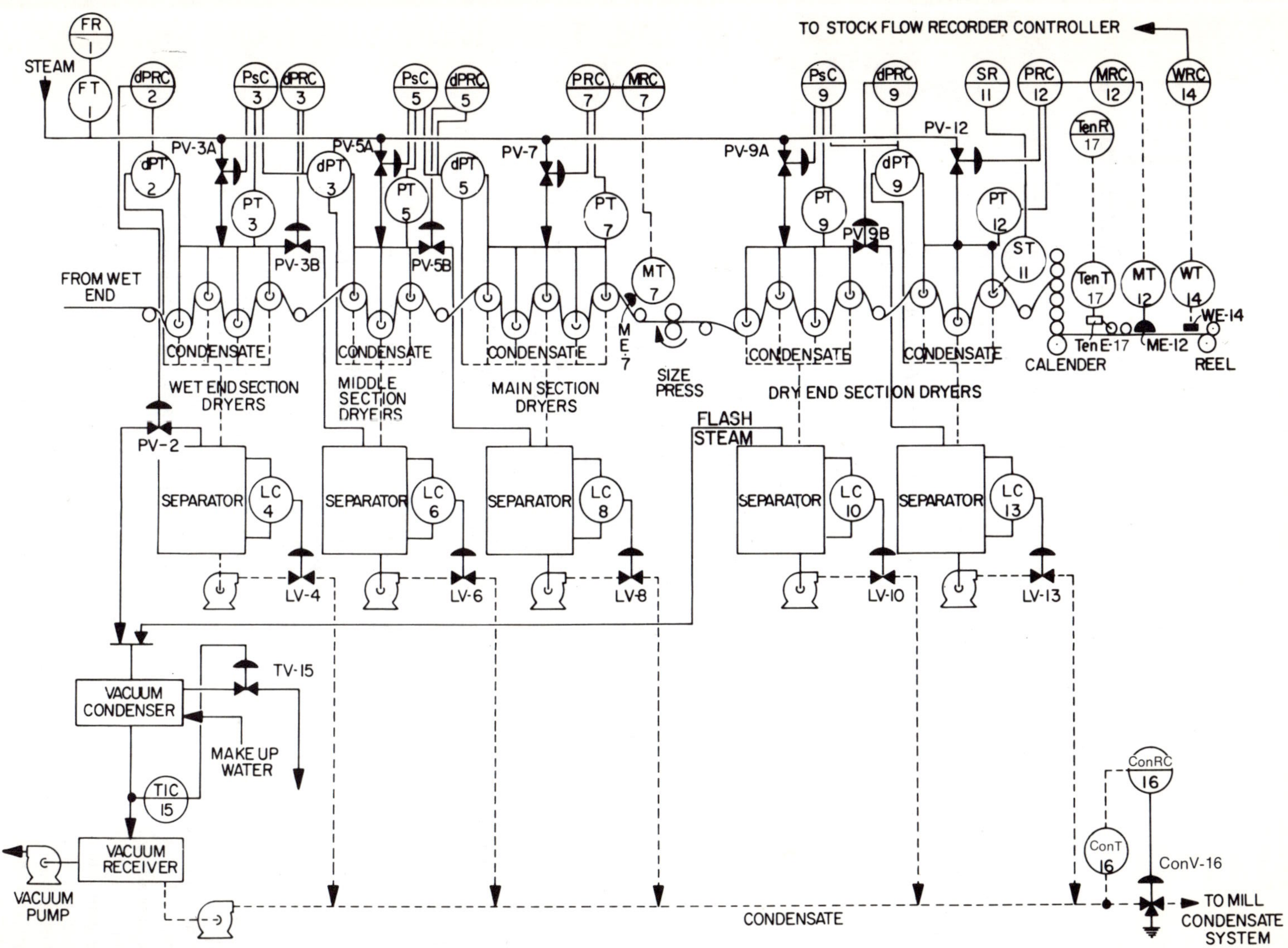

Figure 11-1. Paper machine dry end instrumentation.

Also, by cascade control, the basis weight of the sheet going to the reel is maintained at predetermined levels by instrument system WT-14 and WRC-14 which automatically adjusts the set point of the flow controller on the stock to the fan pump of the paper machine. Different methods of the basis weight measurement are employed. Sheet tension to the reel is measured by TenT-17 and recorded on TenR-17. Speed of the machine is measured by speed transmitter ST-11, which is usually a tachometer-generator, and a record is made on recorder SR-11.

Level control loops LC-4, LC-6, LC-8, LC-10, and LC-13 ensure proper level in order to achieve satisfactory condensate flashing in the separator tank and maintain a positive head of condensate to protect the pump suction.

Temperature loop TIC-15, TV-15 controls the outlet temperature of the vacuum condenser, while condensate recording controller ConRC-16 diverts condensate to the sewer when it reaches a dangerously dirty condition, allowing only clean condensate to pass back to the feedwater system for the power boilers.

Thermocompressors

Thermocompressors or steam jet ejectors have been successfully applied to dry end drainage systems. With the increase in machine speeds and consequent drying and drainage requirements, the use of thermocompressors has increased.

A thermocompressor consists of a motive-steam nozzle, a mixing chamber, and a diffuser, schematically depicted in Figure 11-2. The nozzle expands the motive steam from inlet to suction pressure and converts the pressure energy to velocity energy. In the mixing chamber, the jet of motive steam mixes with the suction stream. This accelerates the suction steam and decelerates the motive steam, producing a mixed stream at an intermediate velocity. At this point, the process of compressing the steam to discharge pressure begins. The diffuser is designed to reconvert the velocity energy into pressure.

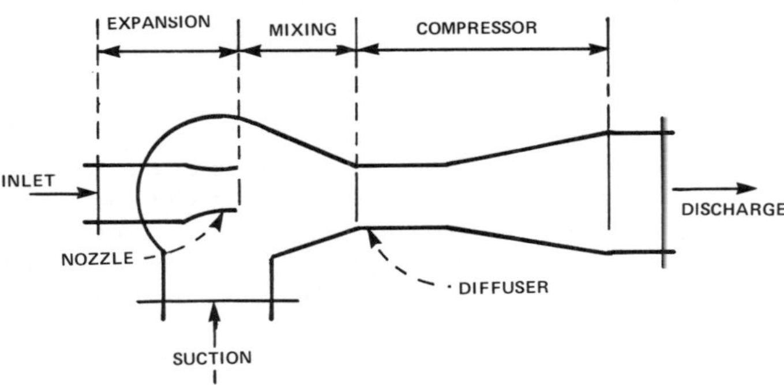

Figure 11-2. Typical thermocompressor construction.

Figure 11-3 shows the wet end section of the dryer system shown in Figure 11-1, modified for thermocompressor operation. The thermocompressor recirculates the flash steam from the flash tank separator, and is also split range with steam pressure makeup valve PV-3A. Thermocompressors are also used on the middle, main, and dry end sections in the same manner. When all sections in the paper machine dryers are equipped with thermocompressors, the improved circulation and reduction of blow-through steam to the condenser decreases the need for cascading, or the use of differential pressure control operating across each section, to a minimum. However, differential pressure control across each

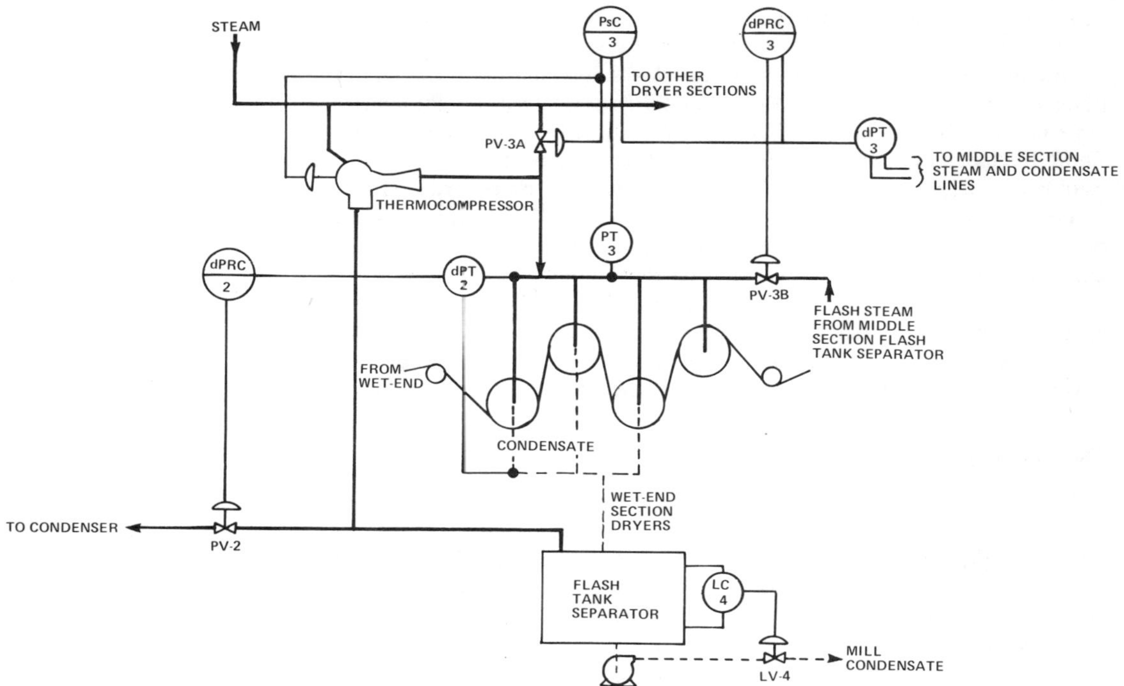

Figure 11-3. Wet end section of dryers with thermocompressor.

section, when used as shown, does add a dimension of safety by making it possible to have a minimum differential setting across the dryers.

Ventilating Hoods

In order to assist the machine dryers in evaporating water from the sheet of paper and to remove economically the resulting vapor, paper machines are equipped with ventilating hoods. The main purpose of a hood over a paper machine is to catch and confine the vapors from the machine before they have a chance to disperse in the room. In general, hoods can be classified as open or totally enclosed. The more modern machines, making wider sheets and running at higher speeds, are using the totally enclosed hood system to handle the proportionally larger exhaust volumes.

Figure 11-4 is a simplified drawing of the enclosed hood of a typical paper machine, showing the basic instrumentation normally used. The control of the exhaust is accomplished by making a wet bulb or dewpoint temperature measurement, TT-1, just ahead of the air heat exchanger and regulating the position of damper TV-1, preceding the exhaust fan, with recording controller TRC-1. It is also possible to regulate the speed of the exhaust fans by the use of a variable-speed drive.The dry bulb temperature of the supply air is measured by transmitter TT-2 and controlled by the amount of steam admitted by valve TV-2 to the booster coils of the air heater by controller TRC-2. The humidity of the air entering the enclosed hood is controlled by making a wet bulb temperature measurement with transmitter TT-3. The amount of fresh air and recirculated air from the exhaust is proportioned by controller TRC-3, driving dampers TV-3 and TV-3A in an adjustable relationship to each other.

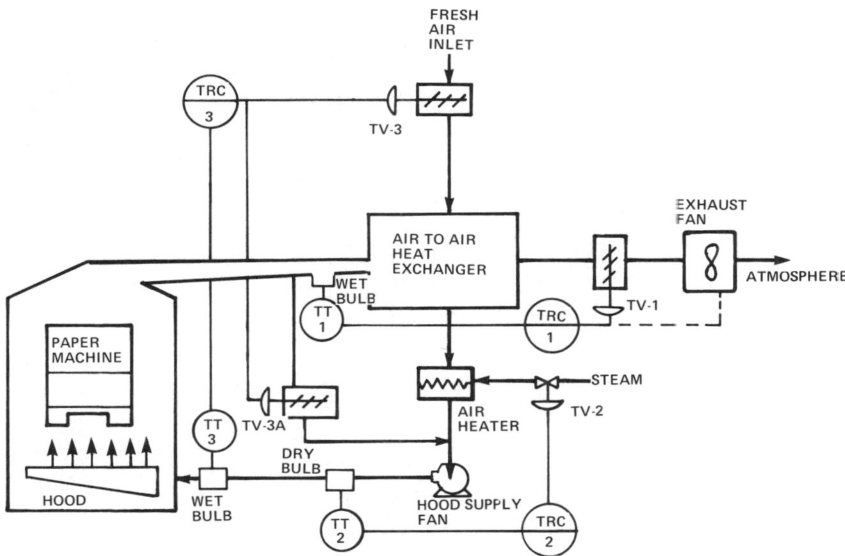

Figure 11-4. Enclosed hood instrumentation for a basic paper machine.

SINGLE-CYLINDER DRYER

Figure 11-5 illustrates a single-cylinder dry end section normally used for making light papers, such as tissue, creped wadding, and cigarette paper. It is normally referred to in the industry as a *Yankee* dryer. It has one large drum typically running 10 to 12 feet in diameter with widths of about 10 to 20 feet. Speeds range anywhere from 1200 to 5000 feet per minute. The dryer surface has a high polish. The press rolls run against this surface. This gives the produced sheet a glossy finish on one side only. Such paper is known as machine glazed (MG).

Dryer Control

The instrumentation used in controlling the steaming of a Yankee dryer is also shown. The pressure controller, PIC-1, is the basic operating control of the dryer. Where a thermocompressor is used as shown in the drawing, the high-pressure steam is admitted through it as required to maintain the dryer pressure. The thermocompressor also promotes circulation in the system by creating a reduced pressure at the bottom connection.

At times of peak dryer load, high-pressure steam may be needed through a bypass around the thermocompressor, but it is desirable to avoid the use of the bypass until the main steam valve is wide open. Positioners on the two valves are individually adjusted to accomplish this automatically.

The surface temperature of the dryer must be held within narrow limits for uniform drying of the paper sheets. Considerable variations can occur in this surface temperature due to load changes, because of the large temperature drop through the thick dryer shell. For most stable control, pressure controller PIC-1 handles fast-changing loads due to variations in steam supply, and its set point is automatically adjusted by surface temperature controller TIC-2 to take care of the slower variations at the roll surface. The roll surface temperature head, TT-2, measures roll surface temperature with negligible errors due to friction,

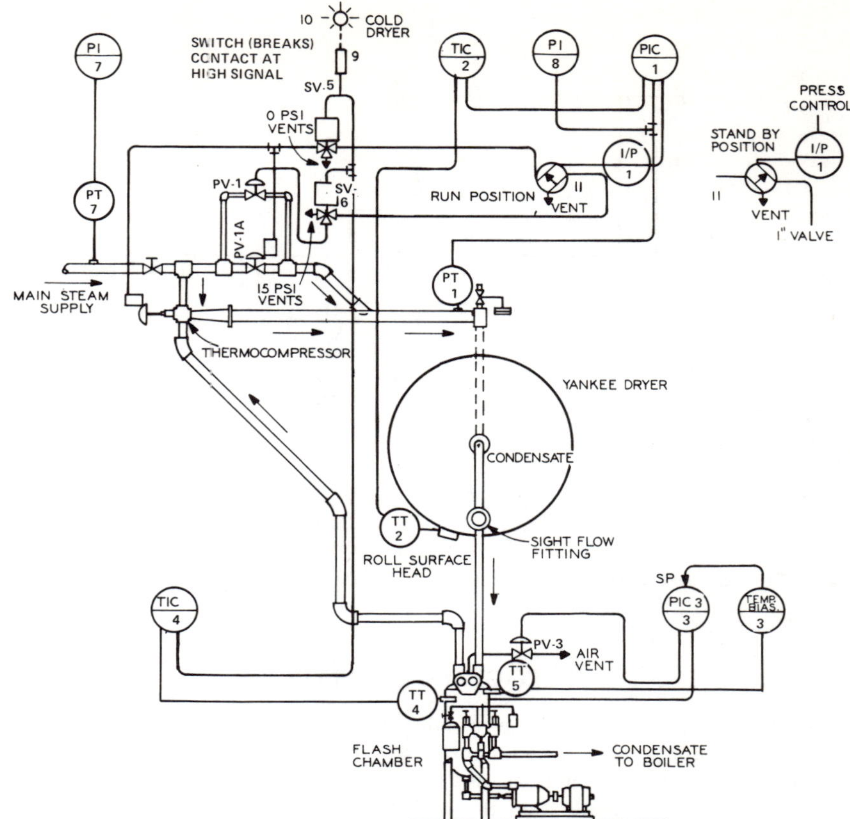

Figure 11-5. Single-cylinder dry end control system.

radiation, and air currents around the dryer. Its bearing surface is made of material which will not scratch the polished dryer surface.

During a dryer shutdown, steam condensation produces a vacuum which draws air into the dryer through the joints. It has been shown that as little as 5 percent of air mixed with steam will reduce heat transfer by as much as 20 percent. High drying efficiency, therefore, is contingent on effective air removal. This may be accomplished automatically by measuring the temperature and pressure above the condensate level in the flash chamber. If no air is present, the temperature will be that of saturated steam at the pressure at this point. If air is present, the temperature will be lower for the same pressure. The temperature-pressure relation control configuration (PIC-3 set by Temp Bias) keeps the vent valve closed as long as the temperature and pressure are in correct relationship, but opens the vent to allow air to escape when the temperature falls below the desired relationship. This method saves steam by venting the flash chamber only when necessary to remove air.

Because of the great mass of metal in the dryer, it is of utmost importance that the unit be heated gradually during startup; otherwise thermal stresses may be set up which might cause a fracture of the shell. Steam must be admitted slowly with adequate safeguards. A separate small bypass valve is used to admit steam in this warmup process. The valve is sized to heat up a cold dryer to approximately 240°F in four to six hours, even if left wide open.

Operation of the main valves is prevented by temperature controller TIC-4, which measures condensate temperature. The output of this on-off controller is zero as long as the temperature is below the set point. This output keeps the main valves closed through the action of solenoid valve SV-5, and the cold dryer light, 10, is kept on by pressure switch 9. The small bypass valve is controlled by the main pressure controller, PIC-1, by moving the switch from the *run* to the *standby* position. Should the switch be inadvertently left at or turned to the *run* position, the small bypass valve would remain closed, since solenoid valve SV-6 would then vent the signal to the small bypass valve. However, with the switch in the *standby* position, the small bypass valve will be closed when the condensate temperature reaches the set point of temperature controller TIC-4, through the action of solenoid valve SV-6. Thus, the operator cannot bring the dryer up to operating conditions until the warmup has been completed, and not then until the switch is turned to *run*.

The *standby* position on the switch permits operating the dryer safely at reduced pressure over a weekend shutdown. To do this, the surface temperature controller, TIC-2, is set to a low value to keep from raising the set point of the pressure controller. The pressure controller, PIC-1, is set to the desired lower pressure. The switch is turned to *standby* so that only the small bypass valve is used. Should the condensate temperature exceed the set point of the warmup controller, TIC-4, the small bypass valve would be shut off, overriding the pressure controller for safety.

Pressure gauges PI-7 and PI-8 can be used to show steam supply and pressure inside the dryer.

Ventilating Hood

Like the multicylinder dryers, the Yankee dryer is fitted with a ventilating hood for the purpose of drawing off and exhausting the vapor produced by the evaporation of water during the drying process. Heat is forced into the hood under pressure. A recirculation system and either a heat exchanger or economizer system are usually provided, as shown in Figure 11-6.

Figure 11-6. Instrumentation for a Yankee dryer ventilation system.

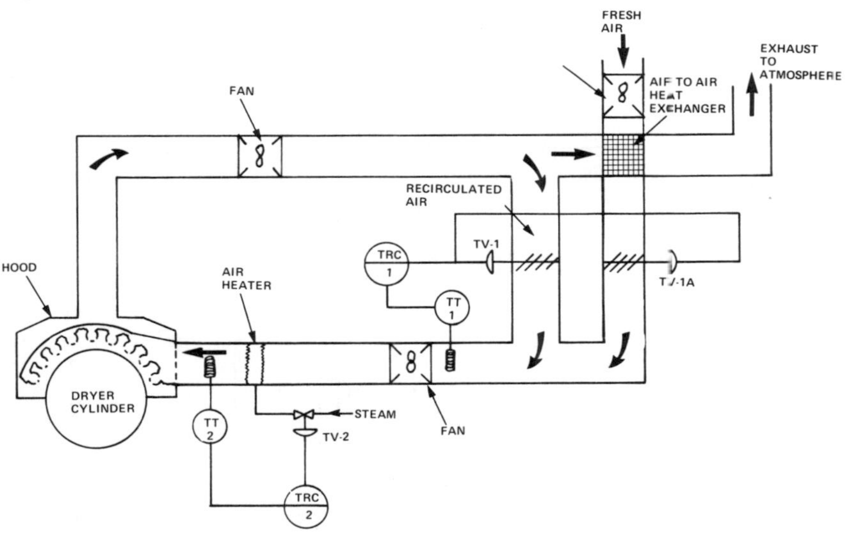

Temperature control is used on the air system to operate the heating controls and recirculation dampers. The operator can adjust the mixture of fresh and recirculated air to avoid building too high a humidity value in the hood supply. This is accomplished by varying the relationship in which the damper louvers, TV-1 and TV-1A, are operated (as they are moved in opposite directions from each other) by the output signal of air inlet temperature recording controller TRC-1. The air supply temperature is measured by transmitter TT-2 and controlled by TRC-2 through modulating control valve TV-2 in the steam supply line.

PAPER MILL CHEMICAL PREPARATION

Size and coating applied to the paper sheet on the dry end or off the paper machine are usually prepared in another section of the paper mill and pumped to the paper machine area.

Size Preparation

Figure 11-7 shows a typical size preparation system where the temperature of the size cooker is controlled by temperature control loop TT-1, TRC-1, TV-1.

The levels of the mix tank are measured by a diaphragm seal, flange-type transmitter, LT-2, and recorded on LR-2.

The storage tank levels are measured by level transmitters LT-3 and LT-4 and recorded on LR-3 and LR-4, respectively.

Level of the size measuring tank is measured by level transmitter LT-5 and recorded on LR-5. This measurement can be used to determine the quantity of size sent to the paper machine.

Coating Preparation

The coating cooker in the coating preparation system depicted in Figure 11-8 is operated on a time-temperature basis. The operator needs only to start the cooking cycle; the rate of temperature rise and the time of the holding tempera-

Figure 11-7. Instrumentation for a size preparation system.

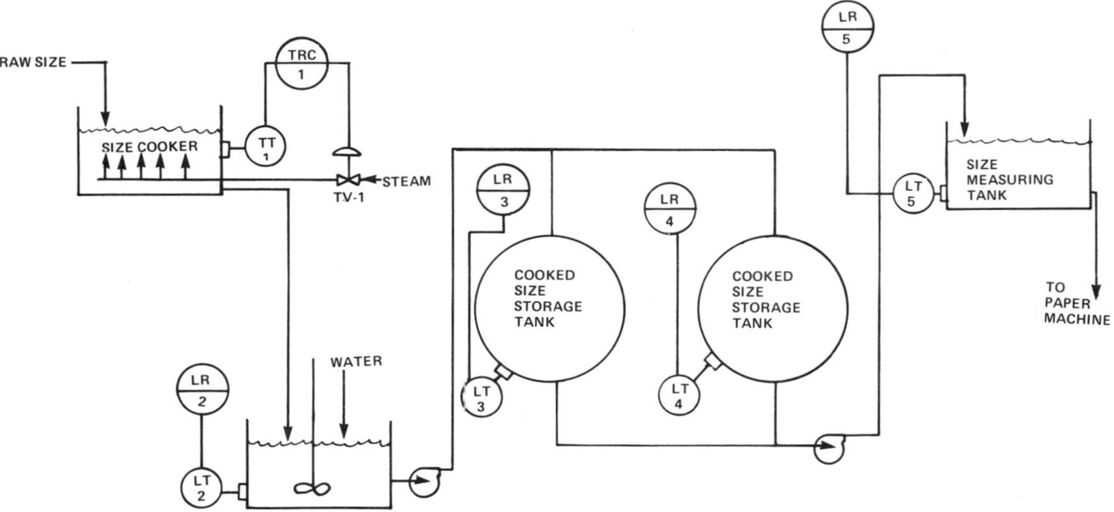

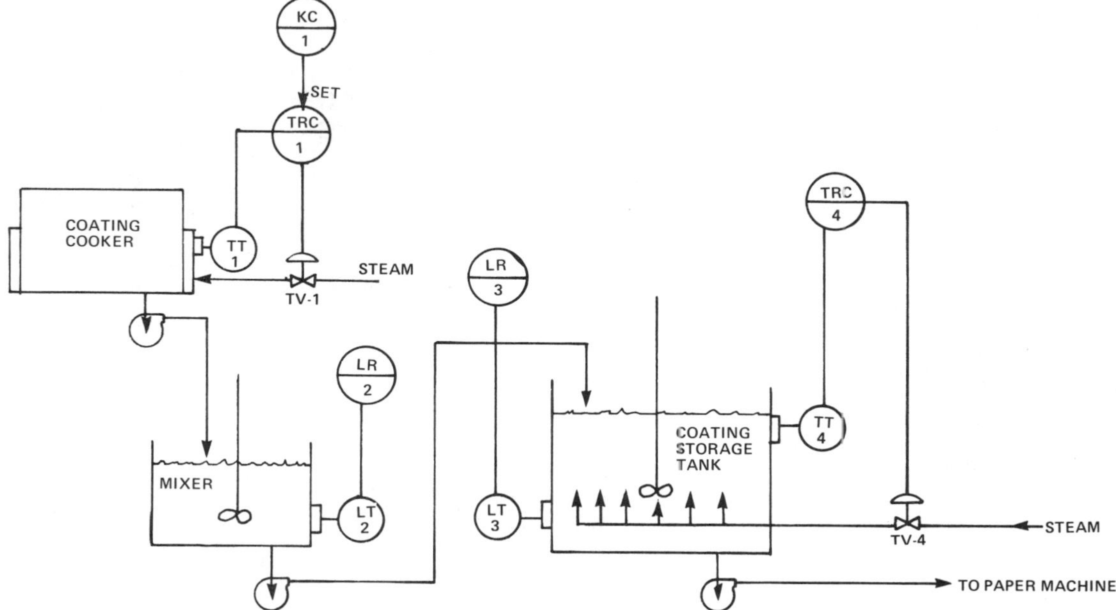

Figure 11-8. Instrumentation for a coating preparation system.

ture are governed by cam programmer KC-1 which remotely adjusts the set point of temperature controller TRC-1 in accordance with a predetermined schedule cut in the program cam. The steam valve, TV-1, is regulated accordingly to maintain the desired temperature in the cooker as set by the controller.

Levels in the coating mixer and storage tank are measured by diaphragm seal, flange-type level transmitters, LT-2 and LT-3, and recorded on LR-2 and LR-3.

The coating storage tank is maintained at proper storage temperature level by controller TRC-4 which admits steam through automatic temperature control valve TV-4.

Coating Weight Control. There are several approaches to controlling the weight of coating material as it is used in the process of coating paper. Figure 11-9 illustrates one of these methods.

Coating solution is pumped from the coating preparation area into the paper machine's level tank. From the level tank it is pumped to the color pan which coats the sheet as it passes through. Excessive coating is blown off the sheet by an air knife and drops back into the blowoff pan for return to the level tank. An increase in coating solution demand is reflected by a decrease in tank level. This change is sensed by diaphragm seal level transmitter LT-1, and is transmitted as a signal proportional to level. This signal goes to level recording controller LRC-1, which automatically adjusts control valve LV-1 to maintain the desired tank level during the coating operation.

Coating solution viscosity is sensed in a sampling chamber by VT-2 and the signal is transmitted to viscosity recording controller VRC-2, which adjusts dilution water control valve DV-2 to maintain the desired viscosity. Coating solution flow into the level tank is sensed by an in-line magnetic flow element, FE-3. The measurement signal is transmitted by FT-3 as F input to multiplying computing relay X-3 and also to flow integrating recorder FRQ-3. Tachometer speed transmitter ST-4, mounted on the backing roll, transmits its speed measurement as input V to

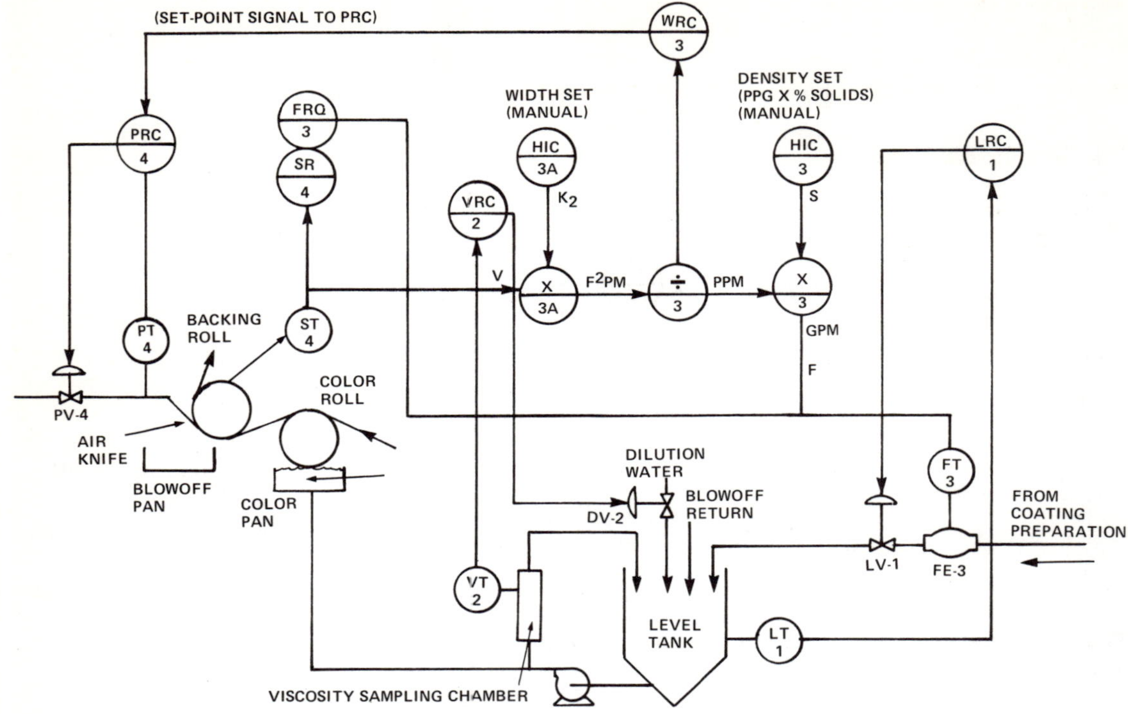

Figure 11-9. Coating weight control system.

multiplying computing relay X-3A and to speed recorder SR-4.

The operator manually sets, with manual set station HIC-3A, the paper width constant, K_2, into multiplying computing relay X-3A, whose output then represents square feet of paper per minute, F^2PM. Likewise, the coating density factor, S, set into multiplying computing relay X-3 by manual set station HIC-3. The output of relay X-3 is the mass flow of coating in pounds per minute. The two output signals from the multiplying computing relays are sent to the dividing computing relay, ÷ -3, which puts out a signal representing the dry coating weight in pounds per ream of paper. Actually, the following formula is solved by this combination of computing relays:

$$W = \frac{K_1 \times S \times F}{K_2 \times V}$$

Where: W = dry coating weight in pounds per ream
K_1 = mill ream standard (square feet per ream)
S = density factor of coating solution
F = flow rate of coating solution into level tank in gallons per minute
V = sheet surface speed in feet per minute
K_2 = sheet width in inches

The output from the dividing computing relay, ÷ -3, is transmitted to the weight recording controller, WRC-3, whose output remotely sets the control point of pressure recording controller PRC-4. PRC-4 controls the pressure of the air knife on the coater as required to maintain the desired coating weight of the sheet of paper.

The air knife pressure is sensed by transmitter PT-4 and it sends a signal proportional to the air knife pressure to pressure recording controller PRC-4.

Titanium Dioxide Slurry Control

Titanium dioxide is used in slurry form in wet ends and coatings on dry ends of the paper machine. In the past, the titanium dioxide was furnished to the paper mills in crystalline solid form, shipped in bags. Lately, the titanium dioxide has been furnished to mills by suppliers in the form of slurries of 65 to 77 percent titanium solids, supplied in tank cars. The slurry is then diluted to the concentration required in the paper mill. Figure 11-10 is a schematic of a typical layout for titanium dioxide slurry showing instrumentation and control makedown by ratio control.

The titanium dioxide is shown being shipped to the mill site by tank car as approximately 70 percent solid concentration. It is diluted by mixing with water under the control of recording ratio controller FrRC-1, and stored in a paper machine supply tank. The level in the supply tank is maintained by level recording controller LRC-2, actuating the pump and water control valve LV-2 on high and low measurement. This measurement is made with a flanged diaphragm seal type of level transmitter, LT-2. The diluted slurry is pumped into a runaround

Figure 11-10. Titanium dioxide slurry make-down system using ratio control.

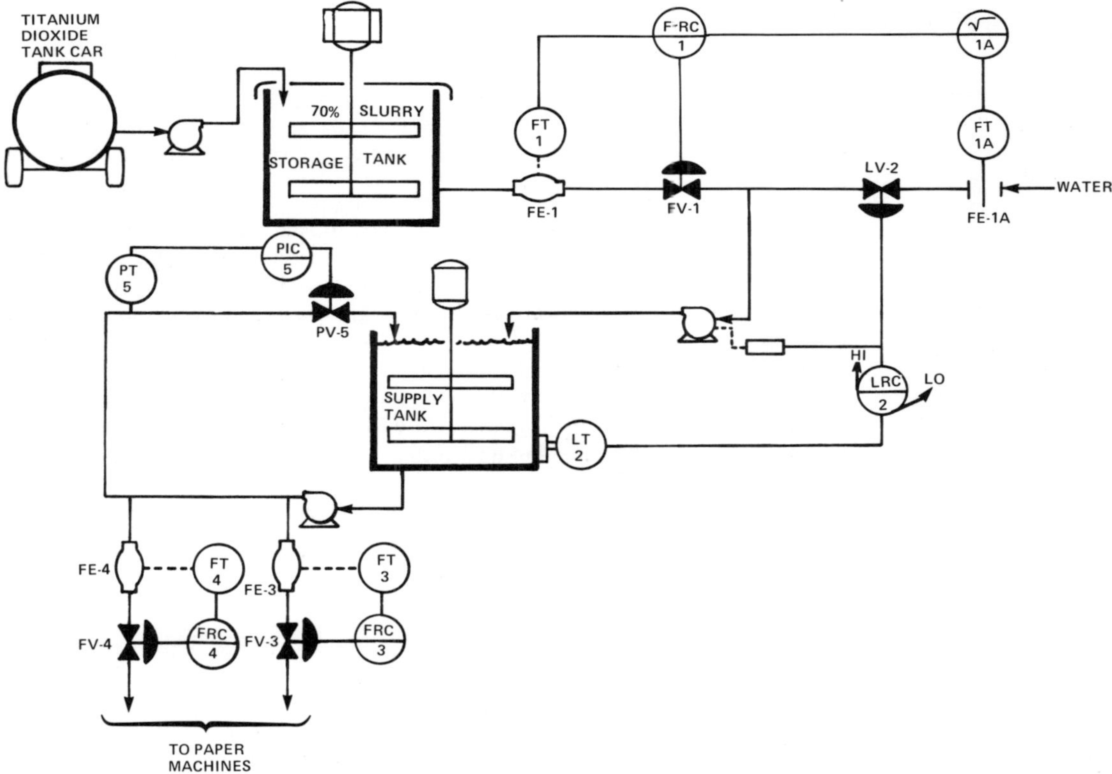

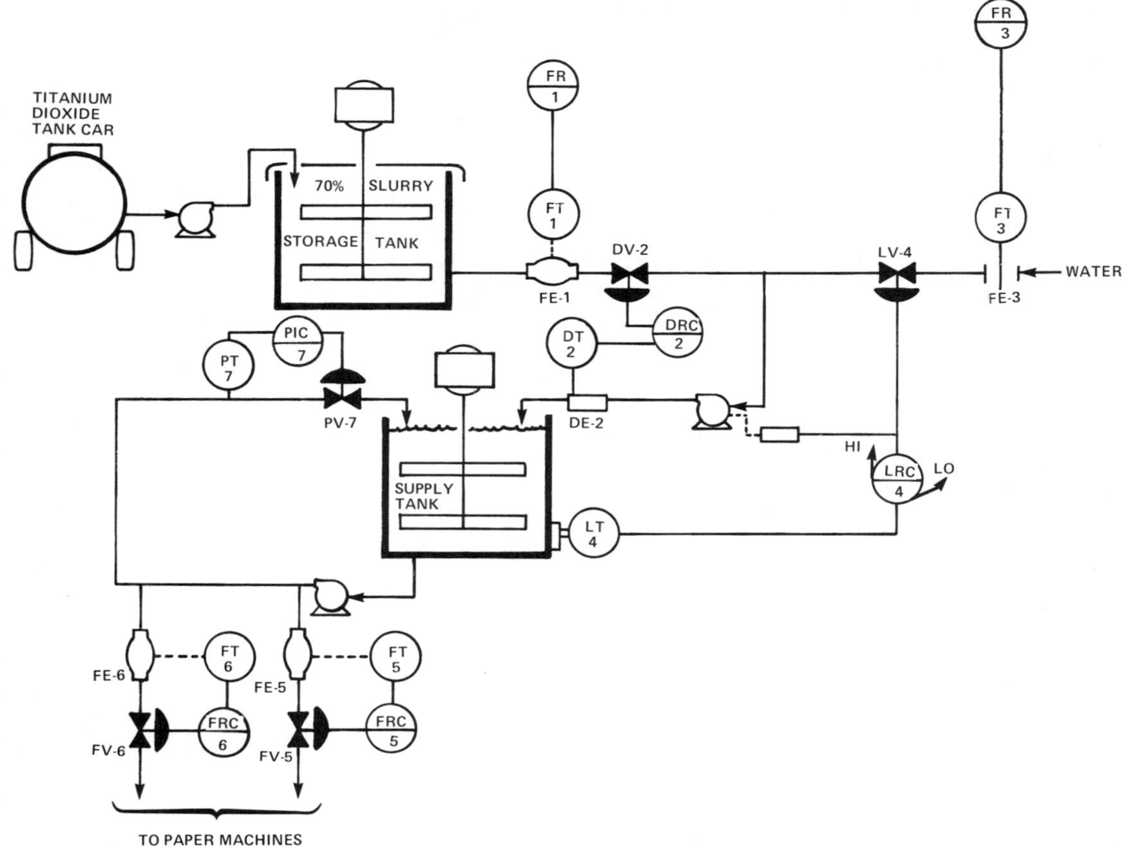

Figure 11-11. Titanium dioxide slurry make-down system using density control.

header whose pressure is controlled by pressure indicating controller PIC-5. The take-off feed lines to the various paper machines are taken off this header and the flows in the take-off lines are controlled by flow recording controllers FRC-3 and FRC-4. Magnetic flowmeters are used to measure titanium dioxide slurry flows at various points in the system.

Figure 11-11 is a schematic of the same titanium dioxide system except it uses a density measurement at DE-2 as the basis for dilution of the 70 percent slurry to the solid concentration which is satisfactory for use on the paper machine. There are various in-line and sampling-type primary elements available for measuring the density of the high-density slurry, such as radioactive nuclear electrical, electromechanical, and mechanical devices. The remainder of the instrumentation is similar to that shown in Figure 11-10.

12 Water Treatment

The pulp and paper industry demands water, not only in large volume but, in most cases, it must be of the highest quality for the type of pulp and paper being produced. Water is an integral part of the manufacture of pulp, paper, and paperboard, and it plays a critical role in virtually every step of the process from the production of pulp and the transportation of fiber throughout the mill to the generation of steam by mill power and recovery boilers.

Although some water can be used in the raw state directly from the source in such areas as turbine-generator condensers and log flumes, most of the water must be conditioned in some way to make it suitable for use in other mill areas. Generally, dissolved minerals, as they exist in most readily available supplies, have little detrimental effect. However, suspended matter, turbidity, color, and organics cause the major problems, and these must be removed to varying degrees of purity for various grades of pulp and paper. These process water impurities affect the quality of the product.

Hard waters are a source of trouble in the sizing of paper and paperboard. Suspended matter and turbidity decrease the brightness of the product, clog wires and felt on paper machine wet ends and, in general, discolor pulp and paper. Highly colored waters are undesirable for white and dyed papers, as well as for pulp. Fibers readily absorb coloring substances and stain the finished product.

Slime-forming bacteria cause slime spots, pinholes, and other undesirable results on the product. Iron in process water also causes discoloration. Fibers will absorb iron from diluted solutions which yellows white pulp or paper and dulls colored paper. pH is important from the standpoint of avoiding corrosion and maintaining optimum conditions for economical use of fillers, sizes, and dyes in process water.

Water treatment processes and equipment vary widely from mill to mill, depending on the source of water—i.e., river, lake, pond, stream or well—and also on the quality of water as dictated by the type of product being manufactured.

When treatment of the influent water to a mill is required, the objectionable characteristics of the water are removed by one or a combination of the following processes:

1. Sedimentation
2. Coagulation
3. Settling
4. Filtration

Sedimentation is the process of removing suspended matter by allowing the water to stand quiescent or to move very slowly through a long tank or reservoir, then to overflow at the far end for immediate use or for further processing. Coarse particles settle to the bottom of the tank and are removed by draining and flushing the tank. Most sedimentation units are equipped with apparatus for the continuous removal of settled matter.

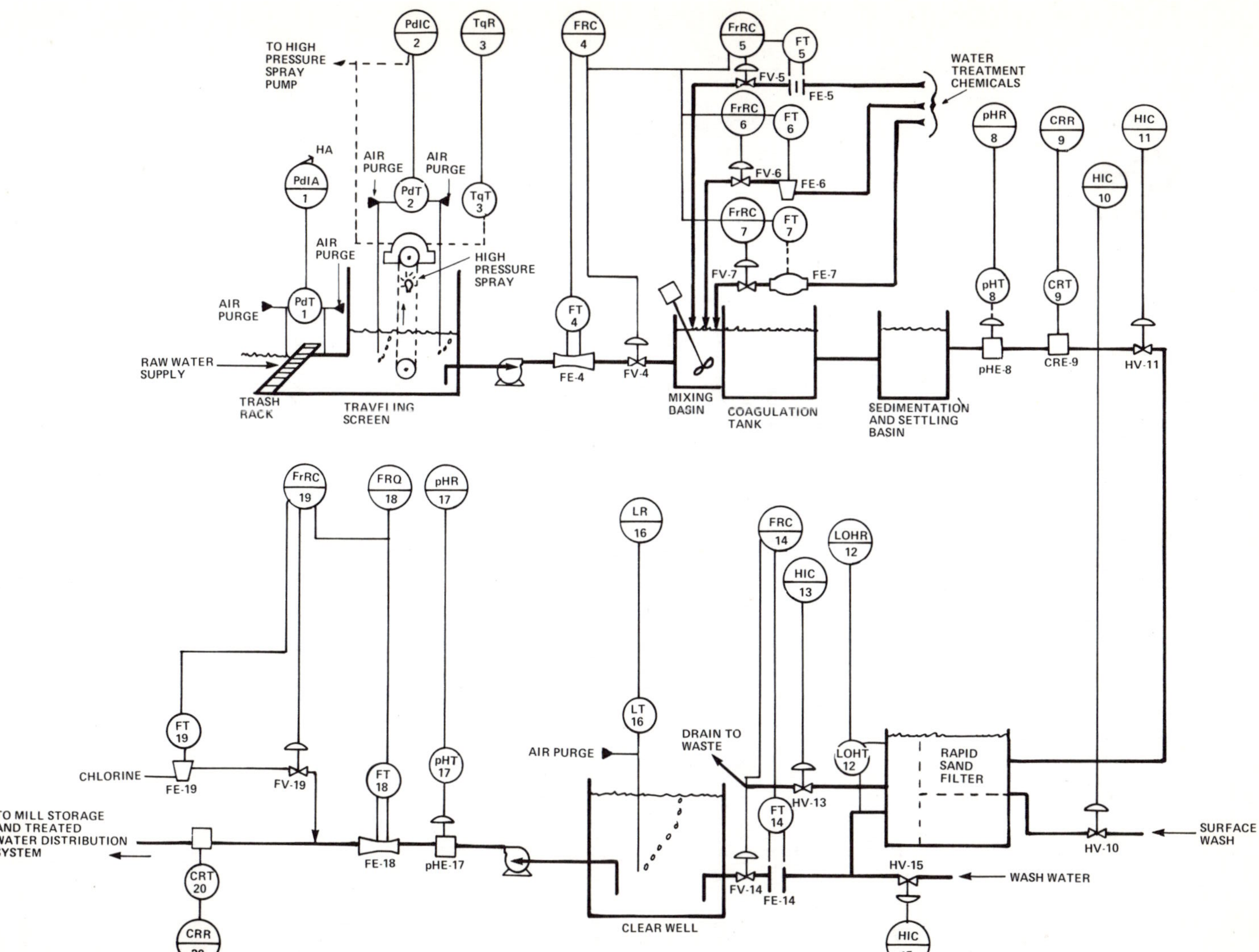

Figure 12-1. Instrumentation for a typical water treatment plant.

Coagulation is used to remove fine suspended matter and color from water, and it consists of adding chemicals to water which will react with the alkaline salts naturally present in water. This action forms a gelatinous precipitate which absorbs or enmeshes suspended matter and colors and then coalesces into coarse particles called *floc*. The floc settles relatively rapidly, carrying down the suspended impurities with it, and may be removed by settling and/or filtration.

Settling is the stage in which the floc settles rapidly in the water, carrying down the suspended matter and color with it.

Filtration is the process of further purifying the water after coagulation and settling have removed most of the suspended matter and color. This is accomplished by passing the water through a gravity or pressure sand filter.

Chlorination of water supplies is practiced in many mills to reduce the bacterial and slime growths. When chlorine is added, oxidation and chlorination of the organic and inorganic matter takes place. After the chlorine demand is satisfied, the remaining chlorine acts on the bacteria. The amount of chlorine required depends on the extent of oxidizable matter and the length of time required for oxidation, but a small excess of residual chlorine should exist.

Instrumentation

In a typical conventional sand gravity flow type of water treatment plant (shown in Figure 12-1) the raw water is pumped first to a mixing basin where chemicals are added, and then to a coagulation tank. From there, it moves to a sedimentation and settling basin.

The instrumentation found in a typical pulp and paper mill's water treatment plant is also shown. In order to prevent large debris from entering the water treatment system from the water source, the water passes through a trash rack. Any plugging at the trash rack will create a differential water level across it. Differential pressure transmitter PdT-1 senses this differential by the use of air-purged bubble tubes immersed in the water on both sides of the rack. When this value reaches a preset maximum value on indicator PdIA-1, an alarm is energized to warn the operator that the racks must be cleaned. Some installations are equipped with automatic cleaning systems.

Water is further cleaned by passing through a traveling screen. Unexpected plugging of these intake water screens can cause a temporary plant slowdown. Earlier screens were equipped with cleaning systems operated on a time basis which did not furnish protection against plugging between cycles. Today, most traveling screens' cleaning cycles are actuated by differential level measurements made before and after the screens by differential pressure transmitter PdT-2. However, in this case, the indicating pressure controller, PdIC-2, automatically starts an electric timer to operate the screen drive motor and the high-pressure spray pump to clean the screen.

The raw water flow is measured by a venturi-type flow element, FE-4, and chemicals are added to the mixing tank which are ratioed by ratio controllers FrRC-5, FrRC-6, and FrRC-7. pH recorder pHR-8 and chlorine residual recorder CRR-9 are used to monitor the effect of the treatment chemicals. HIC-10, HIC-11, HIC-13, and HIC-15 are used to regulate flow inputs to and effluents from the rapid sand filter, while LOHR-12 records the loss of head across the filter. FRC-14 controls the flow to the clear well, whose level is recorded on LR-16. pH of the water from the clear wells is measured and recorded on pHR-17. Total flow is measured, recorded, and totalized by FRQ-18. Chlorine is added by ratio recording controller FrRC-19 and its effect is checked by chlorine residual recorder CRR-20.

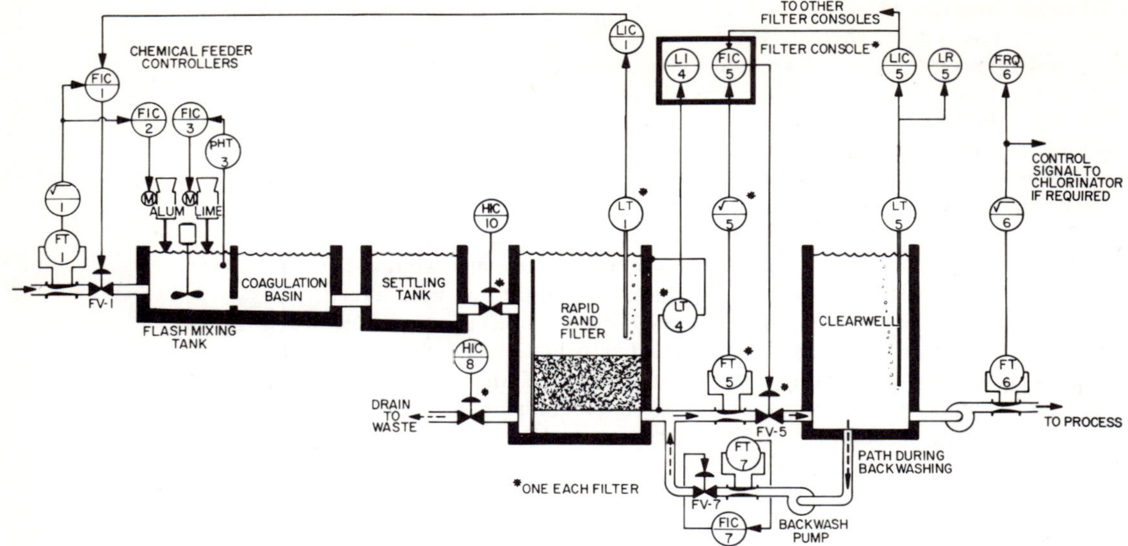

Figure 12-2. Instrumentation used to control filters in a water treatment plant.

FILTER CONTROL SYSTEMS

Another basic instrumentation system used to control filters in water treatment plants is shown in Figure 12-2. The control system required can be broken into three general categories: (1) plant throughput control, (2) backwash control, (3) chemical addition control.

In this configuration, level in the clear well is measured by level transmitter LT-5 which determines the setting of the filter effluent rate controller, FIC-5, on each filter in a cascade fashion from the output signal of level controller LIC-5. The raw water influent flow rate is also controlled by cascading the output from the rapid sand filter level controller, LIC-1, to the set point of influent flow controller FIC-1. If a change in demand occurs, the change in level in the clear well causes an increase in the filtration rate; and the lowering of the level over the filters will admit more water to the plant. Plant output is thus matched against the process demand.

Control during backwashing is dependent on proper sequencing of valves in preparing the filter. Completely automatic control consists of proper sequencing of valves on a time basis, with flow controlled by flow loop FT-7, FIC-7, FV-7.

Loss-of-Head Measurement

The backwash operation is initiated from loss-of-head measurement by LT-4 or by the operator on observation or signal from loss-of-head indicator LI-4.

Loss-of-head measurement applied to rapid sand filters provides the operator with an indication of how satisfactorily the filter is running. In normal operation of a filter, accumulation of materials filtered from the water will increase the loss-of-head (pressure drop) through the filter bed, and will reduce the efficiency of the filter. Loss-of-head measurement informs the operator of the need to backwash the filter, or it can initiate this operation automatically. Also, a sudden drop in head loss is indicative of channeling, an undesirable phenomenon whereby the water to be filtered flows through an area of least resistance and is improperly filtered.

Chemical Addition

Chemical addition is normally done by ratioing flows to influent raw water flow with controllers FIC-2 and FIC-3. A pH trim function can be accomplished by the action of pHT-3, whose output readjusts the control point of chemical flow controller FIC-3.

Filtration Rate by Filter Level

In another basic instrumentation configuration, the level of the rapid sand filter determines the filtration rate, and the level of the clear well determines the plant influent rate, as shown in Figure 12-3.

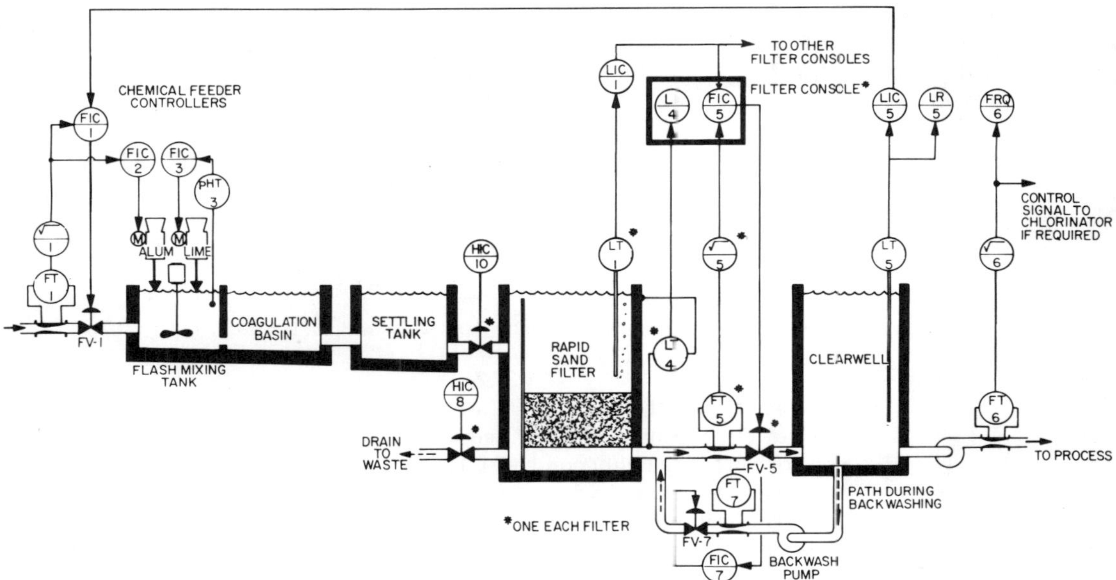

Figure 12-3. Water treatment plant control from filter level.

The basic difference from the system shown in Figure 12-2 is that the control point of the rapid sand filter's effluent flow controller, FIC-5, is set by the output of filter level controller LIC-1, and the control point of the influent flow controller, FIC-1, is remotely set from the output of clear well level controller LIC-5.

With this arrangement, the system is a type of fail-safe operation; the filter effluent valves automatically close when the water supply fails, leaving the clear well full. This diminishes the requirement for backwashing the filter on restoration of supply and prior to continuing filtration. The filters automatically return to service when the supply is restored and the plant started up.

Low Limit Selector Control

Another system (shown in Figure 12-4) is a combination of the previous two, and it operates similarly to that shown in Figure 12-2 as long as the supply rate satisfies the demand of the process. Under normal conditions, the filtration rate is set by the clear well level. The rate of raw water is controlled by the level of the filter.

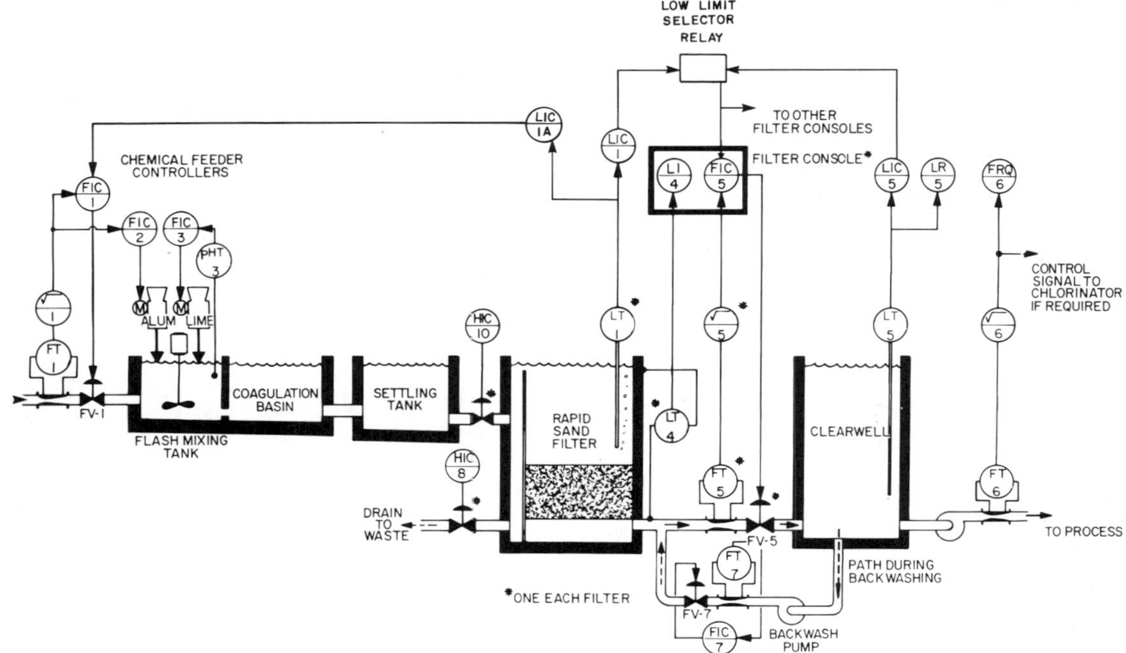

Figure 12-4. Water treatment plant control with low limit selector.

In addition to the instrumentation described in the previous system, another filter level controller, LIC-1A, and a low limit selector relay are added. The control band of the filter controller, LIC-1, is adjusted so that normal filter levels are outside and above its operating range. During normal operation, this controller, LIC-1, transmits a maximum signal, and the clear well level controller transmits a throttling signal which is less than the maximum signal. The low limit selector relay selects the lower of these two and repeats the signal to set the control index of the flow controllers. If the supply fails or diminishes, the signal from controller LIC-1 on the filter level decreases as the level drops. When the signal is less than the one being transmitted from the clear well, the filter level takes over control of filtration at a rate equal to the supply or, if the supply fails, it closes the filter effluent valve. Restoration of supply reverses the process, and the plant is back in operation. This system also provides fail-safe operation by leaving the filters full when the plant shuts down.

Backwash Control

The control during backwash is dependent on the proper sequencing of valves in preparing the filter. The techniques include remote manual operation of the valves and semiautomatic, as well as completely automatic, self-initiating systems. In remote manual operation, an operator located at a central control panel or at the filter consolette, is provided with a means of remotely actuating valves in the pipe gallery. Four-way hydraulic valves used in the past are being replaced by either pneumatic or electric control signals. It appears that this system is limited because it cannot throttle the valve—the valve is either open or closed—but normally it is not desired to throttle influent, drain, rewash, or surface wash valves. The effluent and backwash valves are the only ones to be throttled, and this is accomplished by using controllers or positioners. Semiautomatic backwash control uses a manually actuated sequencing switch to actuate valves in the proper sequence for a duration of time controlled by the operator. The operator determines how long the filter is backwashed and how long the surface washing is maintained. The system has the advantage of being mechanical, thus preventing an operator from incorrectly sequencing a valve; yet it allows the operator to exercise judgment in timing the sequencing steps. Complete automatic control of filter backwash consists of sequencing valves on a time basis with initiation either from a high loss-of-head or by an operator. High turbidity initiation could also be readily incorporated. Such systems normally incorporate a cam-type timer which actuates solenoid valves or relays to operate the valves.

13 Waste Treatment

During the early days of the pulp and paper industry, the discharge of untreated effluents from the manufacturing processes was tolerated because of the relatively small volumes involved and the ability of the receiving waterways to easily dilute and assimilate these flows. But times have changed. With the expansion of the industry and increasing competition for the use of rivers and streams, this practice has become unacceptable. Pulp and paper mills now have the potential of being one of the principal contributors to water pollution, when taken on the basis of the oxygen-consuming organic waste they are capable of discharging into the waterways. The primary reason for this is because, unlike many other industries, the pulp and paper industry uses its water largely as process water.

As previously described, an integrated pulp and paper mill consists of two basic parts: (1) the *pulp mill* which includes the fundamental processes of wood preparation, pulping, screening, washing, bleaching chemical preparation, and chemical recovery; and (2) the *paper mill* which includes stock preparation, paper machine, finishing, and sometimes converting operations.

The severity of effluent pollution potential by an integrated mill depends largely on the process used in the pulp mill and whether chemical recovery is practiced or not. It should be noted that:

1. A mechanical-type mill uses no chemicals; therefore, little or none of the wood is dissolved, causing very low water pollution loads on discharge. Average loads are between 0.5 and 0.15 lb BOD (biochemical oxygen demand)/100 lb pulp.
2. A sulfite-type mill using an acid-base cooking chemical dissolves about 40 to 50 percent of the wood, which causes severe water pollution loads on discharge. If no recovery of chemicals is practiced, average loads are between 30 and 40 lb BOD/100 lb pulp.
3. A sulfate mill uses an alkaline-base cooking chemical. About half of the wood is dissolved, but chemicals are recovered in this process and dissolved wood is burned to CO_2 and water. Pollution loads are, therefore, generally low, in the vicinity of 1 to 2 lb BOD/100 lb pulp.
4. Semichemical (NSSC) pulp mills dissolve less wood than the chemical but more than the mechanical mills. Without recovery units, the pollution loads can be in the vicinity of 20 to 30 lb BOD/100 lb pulp.
5. Not all pulp mills have bleach plants. When they do, the effluent pollution potential increases another 1 to 4 lb BOD/100 lb pulp on the waste load.

It is anticipated that pulp mills of the future will be predominantly sulfate, sulfite, and semichemical types. Solid and liquid wastes are generated from transporting and debarking operations which consist of coarse and fine particles, wood slivers, small wood debris particles, dissolved solids, and soluble organic matter. The major sources of waste waters in a sulfate pulp mill are digester relief, black liquor spill, overflows, pulp cooking, sealing water, evaporators, washing of pulp,causticizing dregs, lime mud, liquor filters backwashing, and lime kiln scrubbing. The chemical nature of these waste waters is alkaline, and they contain high concentrations of BOD and COD (chemical oxygen demand), and dissolved and suspended solids.

Waste water sources from the semichemical pulping process are from digester blowdown, evaporator condensate, liquor preparation, and spent brown liquor spills. Unrecovered chemicals are sometimes wasted from the recovery furnaces.

In sulfite pulp mills, waste waters come from digester blowdown, dirty condensate, scrubber showers, acid preparation, and chemical losses when the recovery process is used. Waste water from this type of mill is composed of spent cooking liquors which are acidic and contain high concentrations of BOD, COD, and dissolved and suspended solids. Dirty condensate waters also contribute to the waste load.

In mills where pulp bleaching is practiced, waste water sources are primarily from the pulp washing operations and are characterized by high concentrations of BOD, dissolved solids, color, and unreacted chlorine; they can be either acid or alkaline in nature.

The main waste water source from the paper mill area is white water from the paper machine operations, which contains fibers, fillers, and traces of chemical additives and dyes, depending on the type of paper being made.

Encouraged by public reactions, by the establishment of stricter government regulations on effluent discharges from mills, and by enlightened management, practically all pulp and paper mills are planning or have started to employ in-plant water, fiber, and chemical reuse, along with recovery techniques which will result in large reductions in waste effluent sources, waste loads, and waste water quantities. Along with these *in-plant* operations to reduce effluent waste load, mills are also putting in more *out-plant* waste treatment operations to minimize effluent pollution of waterways.

The selection of a satisfactory waste treatment system for a mill has become just as important to its economic health as other primary factors such as adequate wood, transportation, good water supply, and qualified personnel.

Waste treatment in a pulp and paper mill is usually accomplished by one or a combination of the following processes:

1. Recovery methods.
2. Primary treatment by coagulation and flotation to remove suspended matter.
3. Secondary treatment by one of a number of biological processes, such as activated sludge and trickling filter, to remove oxygen-demanding material.
4. Chemical precipitation to remove color.
5. Lagooning for final storage settling, and polishing for equalization and BOD reduction.

Some of the systems are fairly simple and easy to operate, while others are more sophisticated and involve more elaborate technical control.

ACTIVATED SLUDGE PROCESS

The activated sludge is a biological process in which nonsettling substances occurring in dissolved and colloidal form are converted by microorganisms to a settling form. The settled material that develops is called activated sludge. A simplified flow diagram of a typical basic system is shown in Figure 13-1. Settled waste from the primary tanks is mixed with a portion of activated sludge

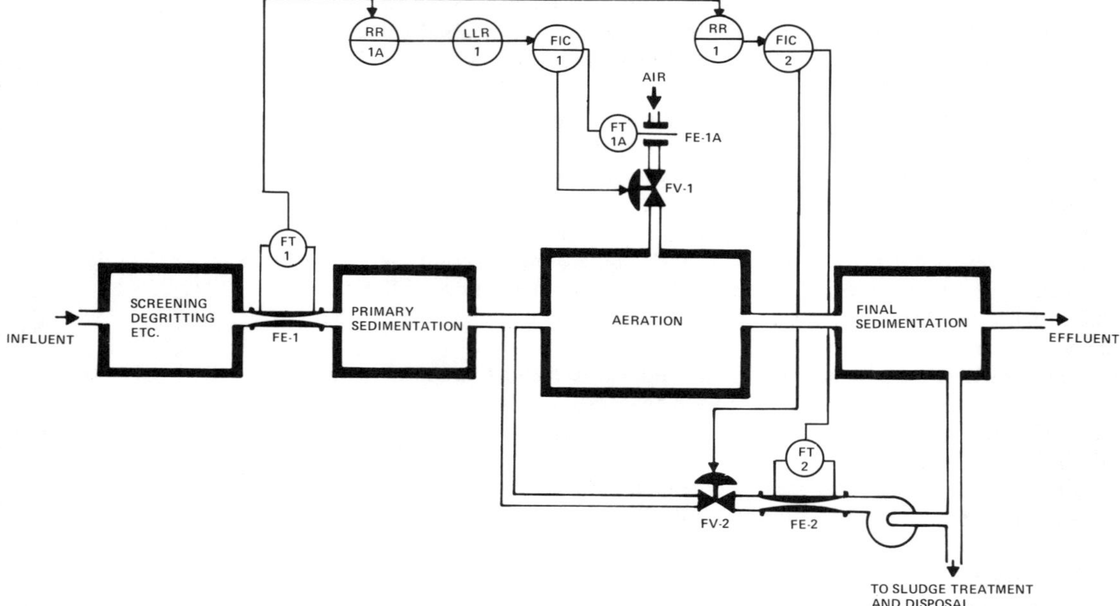

Figure 13-1. Basic activated sludge process.

previously developed and enters the aeration tank. Aeration is necessary because the bacteria and other microorganisms that make up the activated sludge must have oxygen for their life processes. The aerated mixture then passes to the final settling tank in which the activated sludge is separated, leaving a clear liquid to be discharged to the receiving watercourse.

Instrumentation

The control of the activated sludge process revolves primarily around control of the return activated sludge flow and control of oxygen content in the aeration tank. The most common technique involves the simple ratioing of both return activated sludge flow and air to influent flow. These flows are measured by elements FE-1, FE-1A, and FE-2. A transmitted signal from the raw waste flow transmitter, FT-1, goes through an adjustable ratio relay, RR-1, which then sets the index of flow controller FIC-2 on the return activated line. The ratio of raw flow to return flow can now be adjusted by supervisory personnel, based on plant laboratory tests. Control of airflow can be accomplished in a similar manner by the use of FE-1A, FT-1A, RR-1A, and FIC-1, with low limit relay LLR-1 incorporated to ensure that the airflow does not fall below a desired minimum.

With the advent of reliable, continuous, analytical-type elements, other measurements such as ORP, turbidity, pH, and residual chlorine are coming into common use for process control and operator guidance in more sophisticated waste treatment systems.

ADVANCED PROCESS

Figure 13-2 is a diagram of a more extensive waste treatment system which includes the basic system shown in Figure 13-1 in addition to some of the more recently developed analytical measurements.

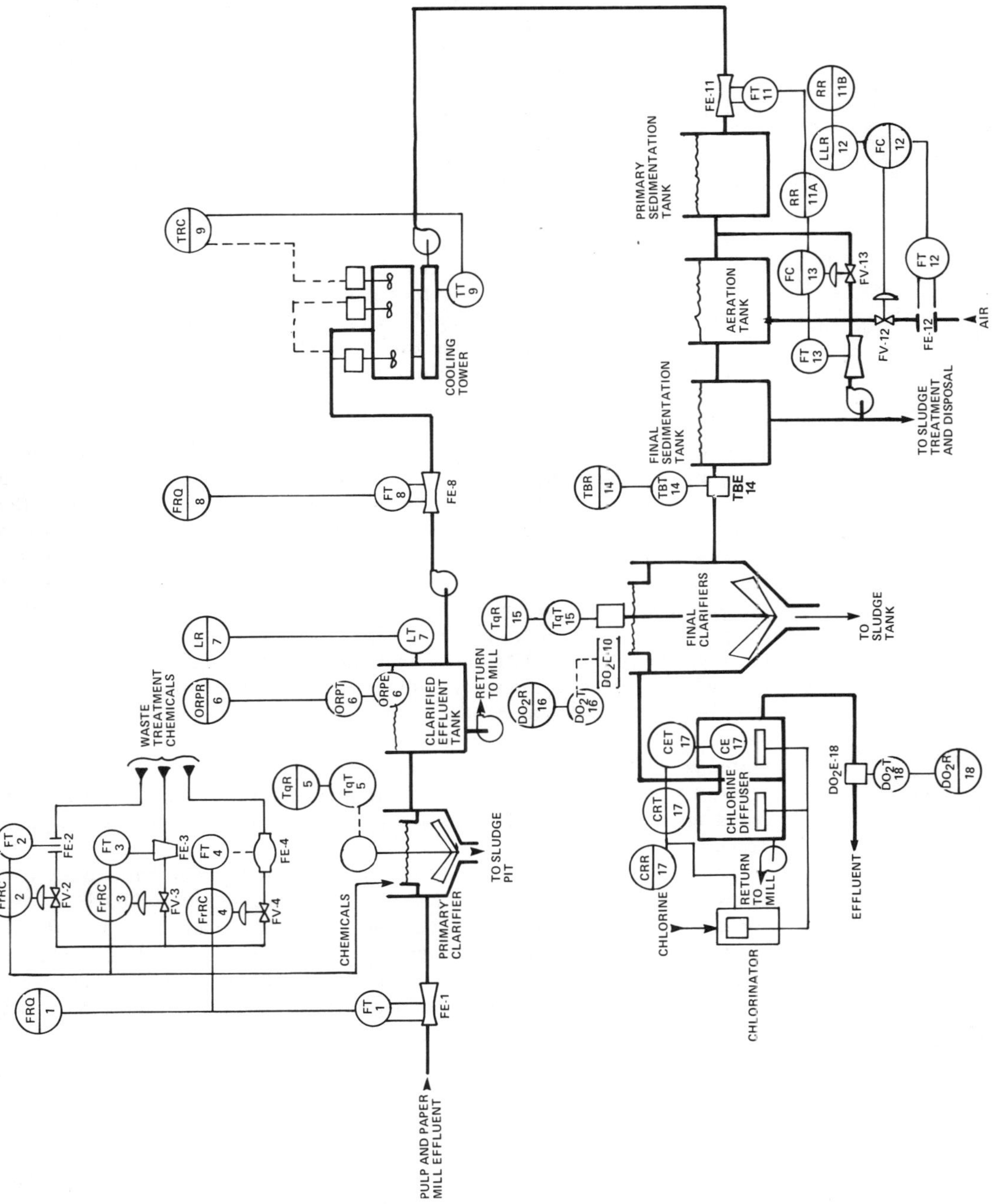

Figure 13-2. Advanced activated sludge process.

Instrumentation

The combined mill effluent from various sections of the mill is metered to the primary clarifier by a venturi-type flow element, FE-1, and is recorded and totalized on FRQ-1. The various waste treatment chemicals added to the primary clarifier are ratioed to the effluent flow by recording ratio controllers FrRC-2, FrRC-3, and FrRC-4. The type of primary element used for chemical flow measurement, be it an orifice plate, rotameter, electromagnetic, metering pump, or dry feeder type, depends on the characteristics of the added chemical.

A wattmeter- or ammeter-type measurement is made by TqT-5 to determine the loads on the electrical motor's primary clarifier agitator drive and recorded as torque on TqR-5. Oxidation-reduction potential in the clarified effluent tank is measured by either a submersion-type electrode assembly, ORPE-6, or by continuously moving a sample and running it through a sample flow chamber electrode assembly. The ORP measurement is transmitted by ORPT-6 and recorded on ORPR-6. It is used primarily for operator guidance in detecting possible sludge overflow from the primary cleaners.

Level recorder LR-7 records the level in the clarified effluent tank, while FRQ-8 records and totalizes the flow from the clarified effluent tank to the cooling tower. Temperature in the cooling tower is recorded and controlled by TRC-9 through the regulation of variable-pitch fans.

Flow rate from the cooling tower to the primary sedimentation tank is measured by FE-11 and transmitted by FT-11, whose output is fed to ratio relay RR-11A. This, in turn, remotely adjusts the set point of recirculation rate flow controller FC-13. This arrangement maintains the recirculated flow rate from the final sedimentation tank at a ratio of influent flow to the primary sedimentation tank. This ratio can also be adjusted by the plant operator. The amount of air to the aeration tank can be controlled in a similar manner by having the flow rate to the primary sedimentation tank adjust the set point of airflow controller FC-12 through ratio relay RR-11B and low limit relay LLR-12, which assures that the airflow will not fall below a desired minimum. A turbidity measurement is made on the effluent from the final sedimentation tank by turbidimeter TBE-14 in order to detect any undesirable increase in suspended solids. The measurement is transmitted by TBT-14 to turbidity recorder TBR-14.

A wattmeter- or ammeter-type measurement is made by TqT-15 to determine the electrical motor's final clarifier agitator drive and is recorded as torque on TqR-15.

The dissolved oxygen level in the final clarifiers can be continuously measured by a galvanic-type probe, DO_2E-16, and transmitted by DO_2T-16 to recorder DO_2R-16. Dissolved oxygen can also be measured and recorded after the addition of chlorine in the same manner, with DO_2E-18, DO_2ET-18, and DO_2R-18.

In order to meet certain stream standards, the effluent must be chlorinated before discharge. This is sometimes done by havng a chlorine residual measuring element, CE-17, located in the chlorine diffuser or on a sample from the diffuser, whose signal is converted and transmitted by CET-17 to a liquid feed chlorinator feeding the chlorine diffuser. The residual chlorine is also recorded on CRR-17.

EFFLUENT CONTROL

A number of schemes have been used to control the amount of waste water discharged into streams from pulp and paper mills, including the use of pH, conductivity measurements, and ratio control. Figure 13-3 shows a simplified method of a version using ratio control in which the effluent from the mill is admitted to a river through a submerged, motorized, automatically controlled gate.

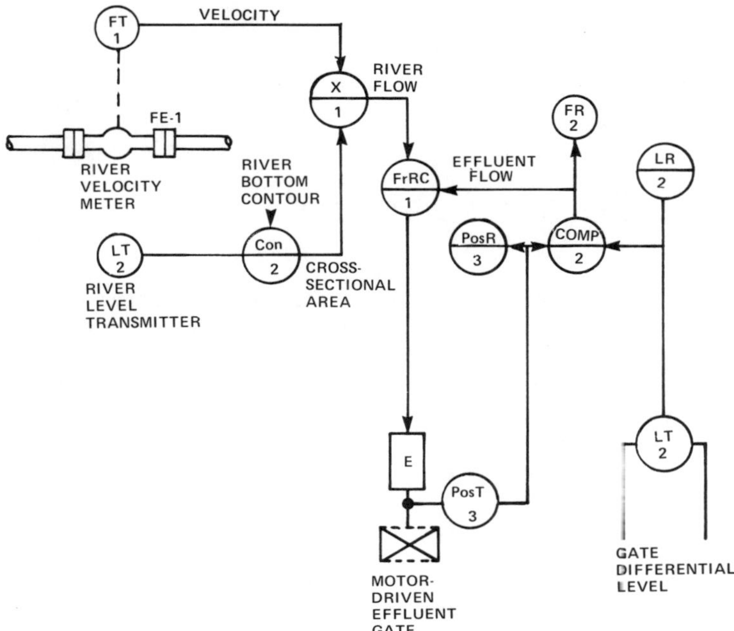

Figure 13-3. Mill effluent control system.

The effluent flow is ratioed to the river flow in order to discharge a predetermined optimum amount of effluent at all times.

The river flow is determined by making a river velocity measurement using a submerged electromagnetic-type flowmeter, FE-1, with pipe sections installed on the inlet and outlet of the flow tube. This velocity measurement is multiplied by multiplying relay X-1, with a signal representing the river cross-sectional area obtained by making a level measurement with LT-2 at a point where the contour of the river bottom can be measured. The relation between river cross-sectional area and river level can be determined and the measurement of the river level converted into cross-sectional area by converter Con-2.

The multiplied signal, representing river flow, is transmitted to and becomes the secondary or "wild" flow of ratio recording controller FrRC-1.

The primary or "controlled" flow is the effluent flow which is determined by measuring the differential head across the effluent reservoir's discharge gate with differential head-type level transmitter LT-2, and by measuring the opening of the discharge gate with position transmitter PosT-3. These signals are combined in computing relay COMP-2 which calculates the effluent discharge flow rate. This flow rate is recorded on FR-2.

The output from flow ratio controller FrRC-1 is converted into a pulse duration signal which operates electric contacts. These, in turn, cause the operation of a reversible-motor electric drive on the effluent discharge gate to automatically maintain a desired ratio of mill effluent to river flow.

14 Computers in General

The character of process control in the pulp and paper industry has undergone a drastic change, particularly since the development of applying computers to process control. Today, almost every company, large or small, in the paper industry has either investigated or has installed and is using a process control computer somewhere in its mill.

DIGITAL COMPUTER USES

Business and Scientific Computers

As in other industries, the first uses for the digital computer in the paper industry were in business-oriented operations. Consequently, use of digital computers by large pulp and paper organizations became commonplace in such operations as making inventory reports, maintenance schedules, payrolls, production schedules; processing orders and invoices; sales forecasting; determining costs and profits, etc. Eventually, the digital computer was pressed into service by engineering personnel to perform more technical functions. In order to solve problems that require more complex mathematical computations involving simultaneous equations, linear programming, differential equations, and other interrelated mathematical functions, scientists have made extensive use of the digital computer. These computations are quite often done by simulation. A mathematical model of the operation to be explored is constructed and a random selection of probabilities is used in the model to determine their effect and establish those that will produce the optimum solution to the problem. Computers used for these purposes are generally referred to as *electronic data processing* (EDP) computers.

Process Computers

Naturally, the next step in the ever-expanding use of digital computers in industry has evolved in their application to the process itself. When the computer is used in this manner it is considered to be a *process computer* and usually operates in *real time*, as contrasted to the business and scientific EDP computers which do not. The term *real time* refers to the actual time during which the process is going on with input information and process variables, such as flow, pressure, temperature, etc., being continuously monitored and fed to the computer. The EDP computer mode of operation is based principally on input data being held in the form of punched tape and cards with subsequent handling at the computer's convenience. This mode of computer operation is also known as *off-line*. Process computers may also operate off-line when there is no physical contact between the process and the computer. Communication between the two is provided by the operator who selects process variables from operating logs, feeds them into the computer, and takes action based on its readout. On the other hand, when input data enter the computer directly from the process, it is considered to be operating *on-line.*

Priority interrupt, in which variables being monitored are given an order of importance with the most significant acted upon first, therefore, becomes a characteristic feature with process computers operating on-line and in real time. If two variables change at the same time, the one with the higher priority is attended to first. In cases of a particular urgency, the computer immediately interrupts its operations in progress to act upon the process upset detected with high priority assigned to it.

The process computer can be used in several ways: (1) as a data processing device, which receives signals from process instruments and presents computed values for use by an operator; (2) as a sequence controller, which obtains most of its data from an operator and then exercises direct control over the process; or (3) as a full-fledged process controller, which notes process conditions and takes action without operator intervention. These uses are usually associated with the analog instrumentation generally employed for process regulation. The block diagram in Figure 14-1 shows this instrumentation arranged in the basic loop configuration. The three process computer schemes just mentioned involve this loop or variations of it.

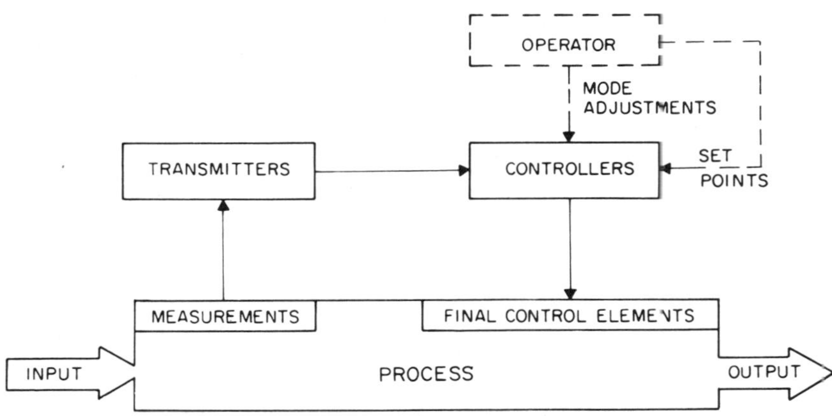

Figure 14-1. Basic process instrumentation loop.

Data Logging. The first attempts in using the digital computer for process-related regulation purposes were as data loggers to automatically collect process operation data. In essence, data logging has been practiced in the paper industry for many years, using a conventional graphic circular chart and strip chart recorders. However, digital data loggers convert analog process measurements into more concisely recorded digital data and present this information in the form of a list (or log) printed out on a typewriter at regular intervals. A typical log would contain the time, point, location, and value—expressed as a decimal number in engineering units—of the particular measurements being periodically scanned by the computer. A log, consisting of a single point or a complete list, can be obtained on demand at any time at the plant operator's request. Limited "after-the-fact" logs can also be made to study history of process operation data by calling up information previously stored in memory. Data trends and average logs can also be produced automatically or on demand. In addition, data loggers can perform other functions such as flow integration, alarm scanning, and simple calculations such as mass flow.

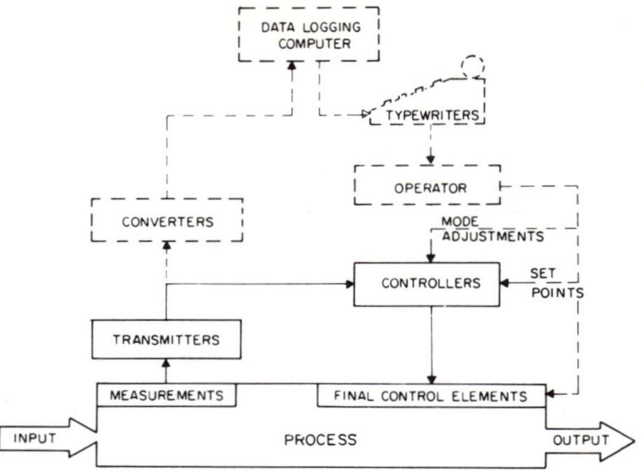

Figure 14-2. Data logger used as operator's guide.

Although initially, data loggers were primarily used for the acquisition of data for management information purposes, they constituted an important step toward the use of digital computers for direct-process control in the paper industry by introducing data gathering and handling techniques to the field of process control. These techniques eventually resulted in the use of this information by plant operators as guides. In this case, the computer not only gathers information, but also performs the mathematics and logic necessary to determine the changes that should be made in plant operations in order to advise the operator what should be done. The operator then makes the necessary changes to the process equipment and/or set point and mode adjustments to the controller in process control loops, as instructed by the computer through messages automatically written out on a typewriter. A horn or light is usually provided to advise the operator that there is a message for him on the typewriter. This is sometimes referred to as using the process computer in an *open loop* control fashion, which implies that the operator is required to complete the cycle from the process to the computer and back to the process as shown in Figure 14-2.

Sequence Controlling. When the computer is considered as a sequence controller, the operator feeds into the computer information which he obtains by observing the process operation, raw material characteristics, product specifications, and equipment conditions. Then the computer performs any mathematics or logic required in the data and adjusts process conditions by direct manipulation of analog control instrumentation or process equipment in accordance with the results obtained from this information of preestablished programmed basis. Figure 14-3 illustrates this concept, with the operator being interfaced between the process and the computer. In this respect, the approach can also be considered as open loop control. This approach can be used to start up a machine or system on a precalculated schedule. It is a function similar to those performed by cam set controllers, such as conventionally used for batch digester cooking control, but has the means provided by which information on wood species, chip/liquor ratios, time-temperature-pressure programs, and the operation of the steam flow and relief valves for optimum loading conditions can be communicated through the computer.

Supervisory Process Computer Control. An extension of the sequence computer control concept is commonly known as *supervisory process computer control.* Direct communication between the computer and the process is established

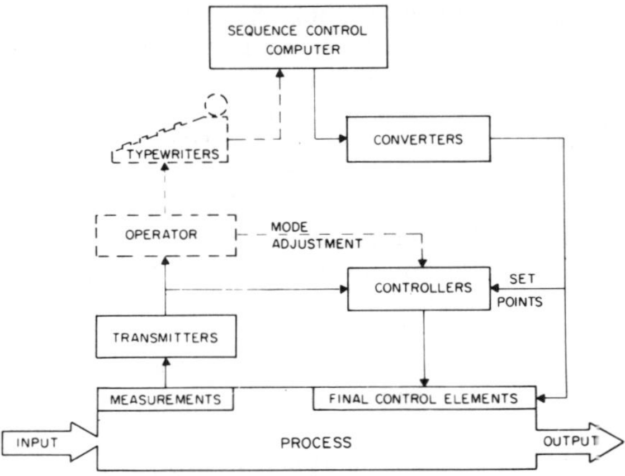

Figure 14-3. Process computer used as a sequence controller.

without direct operator intervention. As illustrated by the block diagram in Figure 14-4, the operator is not included in the operation of the control cycle, so this concept can be classified as a *closed loop* type of operation. With this type of system, the computer receives process measurement signals directly, calculates the best operating conditions, and automatically adjusts the analog controller set point. In essence, the computer performs the same function as the operator in setting the set points of the analog controllers in a conventional process control loop. However, the controllers' mode settings are still made by the operator. Although it is possible to design a system to permit the computer to automatically adjust the controller modes as well (called *adaptive control action*), this is not as practical with this type of computer system as in others.

Other provisions (typically typewriters) are made available for direct operator-to-computer communications. This configuration of process computer control has also been designated as *set point control* and *digital-directed analog control.* In case of computer outages due to failures and maintenance, backup process

Figure 14-4. Process computer used for supervisory control.

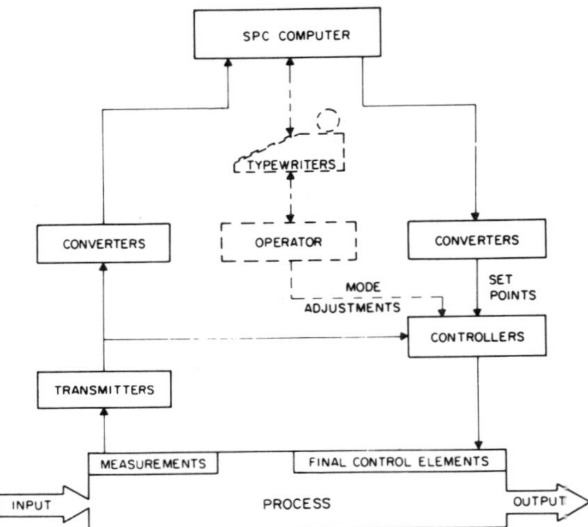

control operation is provided by designing the system so that it will operate on local conventional control loops with set point adjustments being made by the operator. Automatic/manual stations are provided in conjunction with the analog controller, or alone, to allow manual control of the process when the computer is out of service.

Direct Digital Control. As previously mentioned, the controller in a conventional control loop is basically an analog computer. It actually computes equations which relate the various controller settings of proportional band, integral, and derivative. It accepts a measurement signal, compares it with a set point, and changes the controller's output to correct for any differences. Therefore, the controller's function is merely calculation. A digital computer is ideally suited for making calculations. This leads to the next logical consideration: further extending the computer's use for process control. That is, not only can it replace the operator's function in the conventional control loop, but it can also replace and perform the function of the analog controller and send its output directly to the valve and/or other final control elements, as well as perform other needed or desired functions. This then becomes another type of closed loop computer control concept called *direct digital control* (DDC), as depicted in Figure 14-5. Although the illustration shows only one complete loop, in actual operation the computer is time-shared among many control loops. It samples a measurement, performs the necessary calculations based on the set point and other control loop data stored in memory, and makes the necessary changes to the output signal going to the corresponding final control element. It then samples the next measurement and performs the calculations for that loop, and so on, repeating the cycle for the remaining loops. The computer performs this operation at such extremely fast speeds that it appears as if continuous analog control is being accomplished on each loop. With this scheme, the proportional band, integral, and derivative responses of this control loop are part of the computer program, making automatic adjustments of these modes by the computer a practical approach to achieving automatic tuning and adaptive control functions.

DDC saves a good part of the cost of individual controllers. However, the

Figure 14-5. Process computer used for direct digital control.

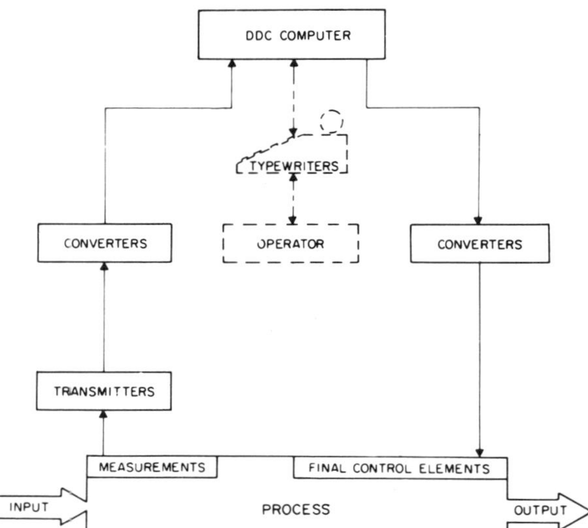

process control is more completely dependent on the computer. Therefore, provisions for backup process control become a very important consideration when the computer is out of service for any reason at all. There are several methods being used today to ensure continuity of process operation during computer outage. The first is to pass all computer output signals through manual stations so that the process may be operated by manual control until the computer is brought back on-line. The second method is to pass all vital output signals through backup analog controllers so that the process may be run by conventional automatic control when the computer is down. A third method is to employ two digital computers, each with the facility for operating the process should a failure occur in the other. The backup system must always be ready to take over control; therefore, it must contain self-checking procedures which ensure that the system is available for backup. Provisions must also be made for continuously updating the backup system with the latest process information, such as set points and tuning parameters, in order to prevent upsets during the transfer of control from one computer to the other. The other method of providing backup protection consists of any and all combinations of the first three methods.

Special-Purpose Computers. Most of the computer techniques mentioned so far fall into the class of general-purpose computers. Another approach is offered by a number of so-called special-purpose computers which are designed on the same lines and employ much of the technique of the general-purpose computers, but they are generally smaller and dedicated to specific purposes. Due to their smaller size, they have sometimes been referred to as *minicomputers*. Several such special-purpose computers would appear to be a much more economical proposition than a full-sized general-purpose computer. However, if the number of such computers on any one application continues to increase, a stage will soon be reached when a large-sized, general-purpose computer would be technically and economically more desirable. This concept has gained increasingly wide acceptance on and around the paper machine, with some units also being used in other areas of the pulp and paper making operations.

Computer Systems Configuration. A typical configuration of a computer system used to control a single-unit process such as a digester, a bleach, or a paper machine is shown in Figure 14-6. Computer control systems have also been de-

Figure 14-6. Computer control system configuration—single-unit process control.

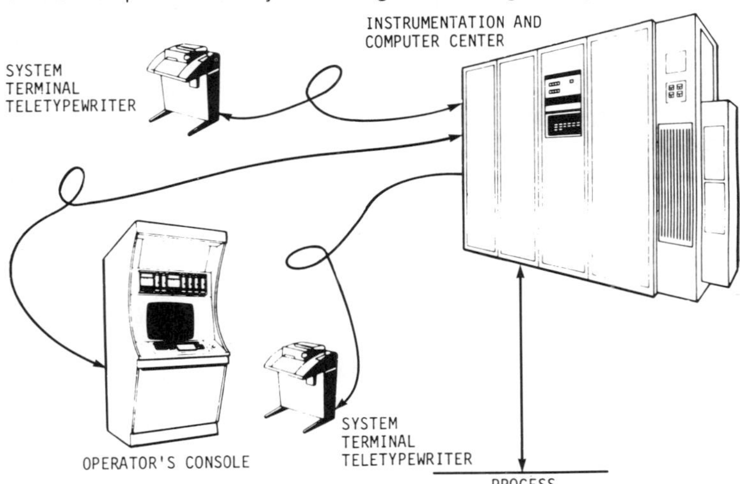

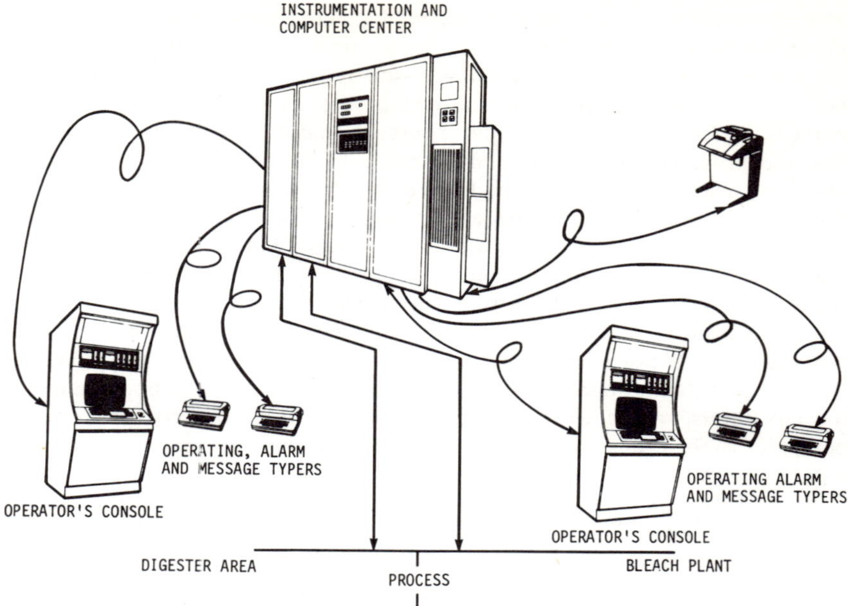

Figure 14-7. Computer control system configuration—multiple-unit process control.

signed and are being used to control a number of similar unit processes or a combination of different unit processes. For instance, one system can be individually used to control a number of digesters or bleach plants, respectively. The system can also be used to control a combination of digesters and bleach plants. This configuration is shown in Figure 14-7.

PROCESS CONTROL COMPUTER APPLICATIONS

Among the primary incentives for installing a digital process computer control system are manufacturing cost reduction, improved product efficiency, improved product quality, process control enforcement, and safer process operating conditions. Numerous applications exist for the installation of process control computers in the paper industry which can be classified under three general categories: (1) those found on chemical type processes, primarily in the pulp mill; (2) those concerning the physical operations of stock preparation and sheet manufacturing in the paper mill; and (3) those operations involving utilities and by-products processing, such as water and waste treatment, power boilers, and tall oil plants.

These processes, which require such inputs as steam, chemicals, air, water, electricity, and wood, need interrelated close contact in order to achieve the most efficient utilization in their processing and final conversion to paper. Figure 14-8 illustrates a feasible approach to a control system to accomplish this task. The entire system is depicted as consisting of a number of computers organized much like a corporation, with individual units assigned to do specific operations in the local mill areas under the direction of a centralized management-control type computer. The next three chapters are concerned with some of the specific tasks that can be and are being done by the process control computers located in the various pulp and paper mill areas.

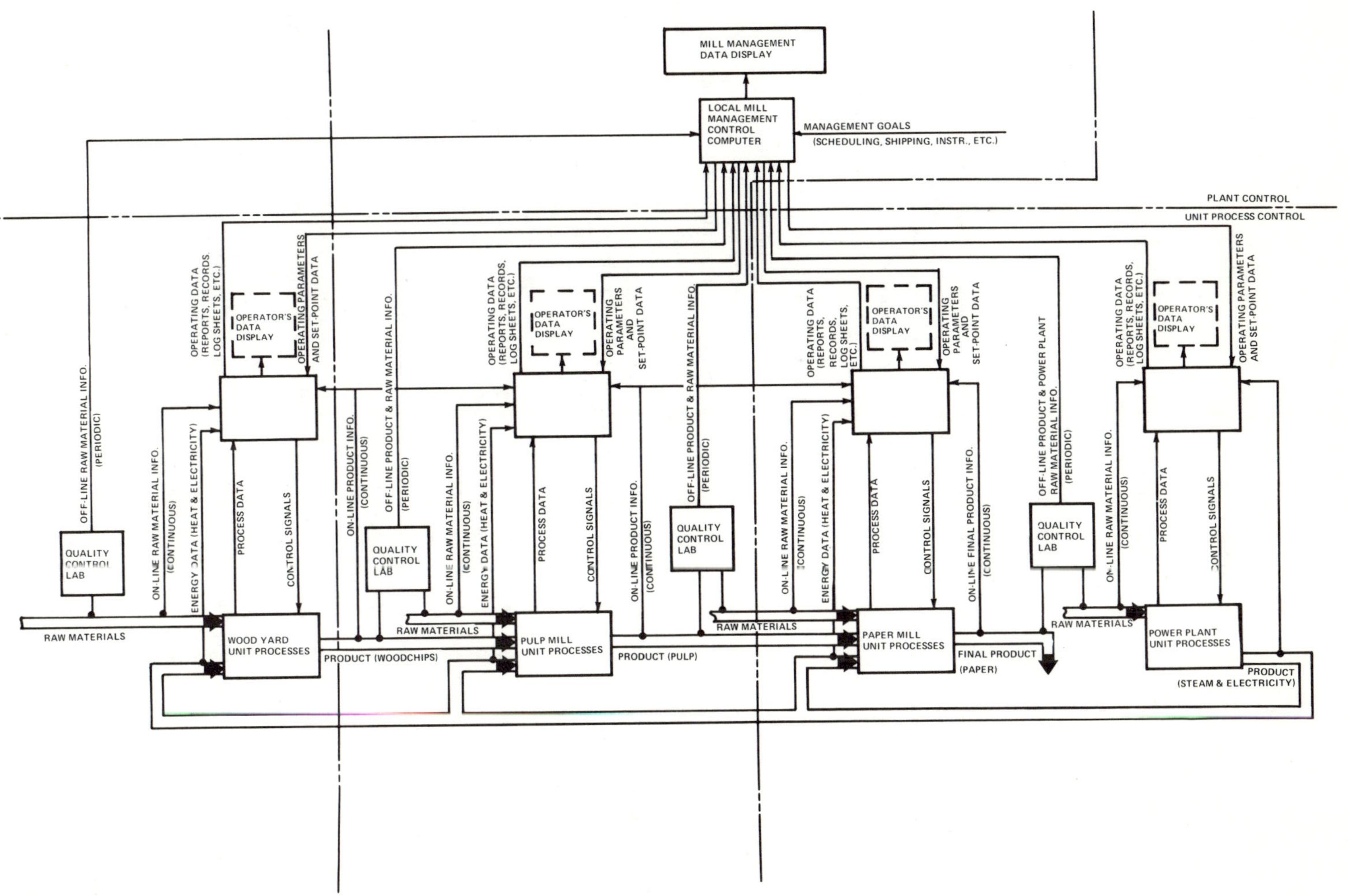

Figure 14-8. Process control system. A centralized computer directs individual units in local mill areas.

15 Computers in the Pulp Mill

Applications in the pulp mill, although not as numerous as those found in the paper mill, also possess the potential of realizing improvement of process operations by the use of computer control techniques. Figure 15-1 shows the unit processes in the pulp mill to which a computer control system can be applied to perform such major functions as: (1) scheduling of various units to provide efficient utilization of existing equipment, (2) prediction of process loading to avoid upset conditions, (3) tracing various types and grades of pulp through several units to provide profiles on processing conditions and pulp quality, and (4) providing "operator guide" instructions which will result in more uniform operations.

BATCH DIGESTERS

In sulfate chemical pulping operations using batch-type digesters, computers are capable of production scheduling to ensure that digester discharge blow tanks will not run dry or overflow, and will still maintain the required number of batches per day in the most efficient manner through the operator guide concept. Computers can be used to automatically calculate the chip rate required for maximum digester loading based on such input information as size of digester, type of wood, wood moisture, and density of chip packing. In the same manner, the computer can calculate the correct digester liquor charge based on type of wood, chip moisture, final weight chip load, type of cook, size of digester, white liquor strength, total liquor volume requirements, and corrections required due to K or Kappa number test on pulp from the previous cook. The computer can also provide for liquor fill, steaming, and control of functions. Additionally, the relative cooking rate could be calculated and integrated against time in order to obtain a factor to determine when the cook is finished, sometimes referred to as the *H* factor. The computer could have control of the digester from the time the cook capped the digester until it alerted the operator that it was time for the digester to be blown. Monitoring of the cooking cycle and calculation of the proper additions of white and black cooking liquors would provide a more uniform pulp with less screenings and greater yields, more constant K numbers, and increased turpentine yields.

Several basic concepts have been used in the application of computer control systems to batch-type digesters. A common one is to have the computer calculate the cooking temperature and the time required to reach a given K or Kappa number by using an H factor model of the cooking process. Figure 15-2 is a block diagram of such a system. The structure is shown as consisting of four general sections:

1. Basic control functions
2. Coordinated control
3. Optimization
4. Information system

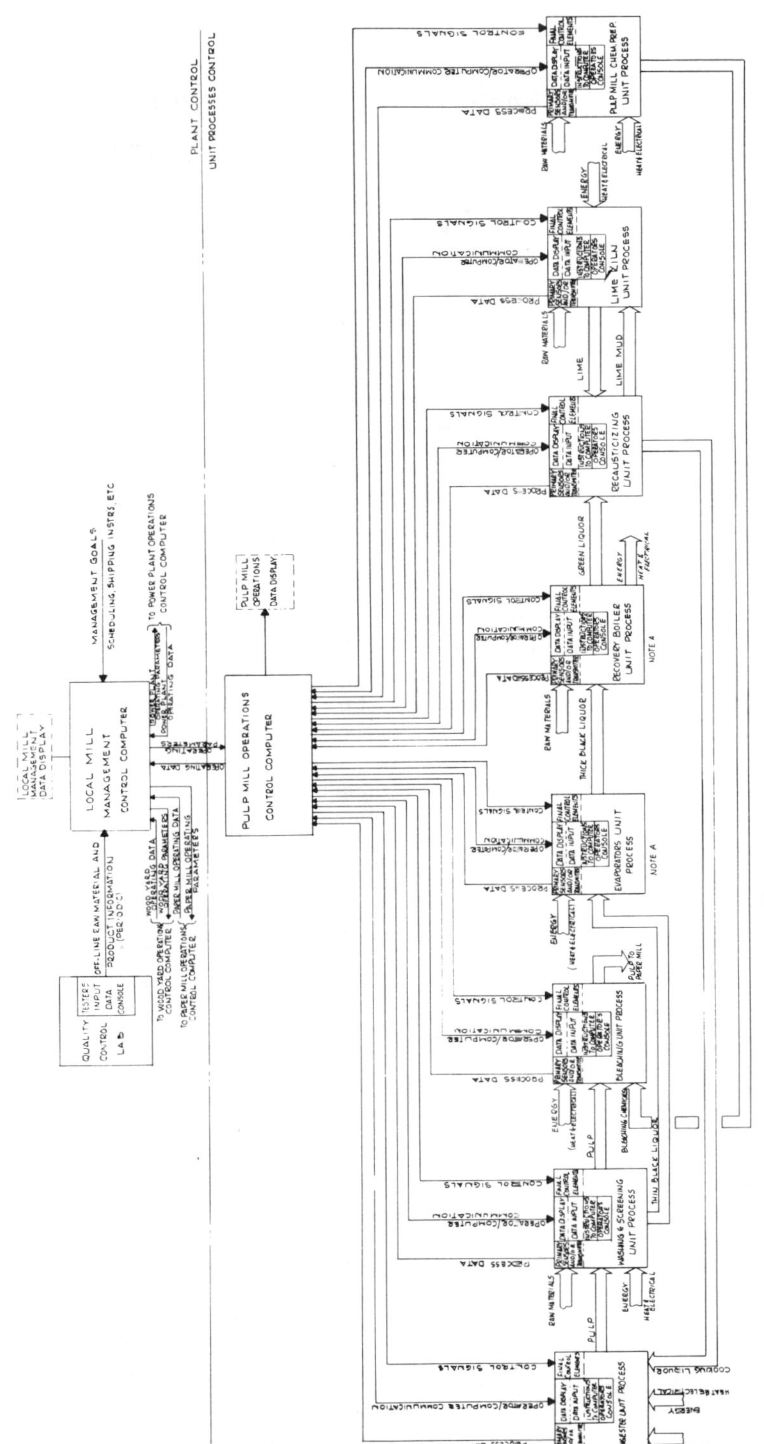

Figure 15-1. Application of computer control in a pulp mill.

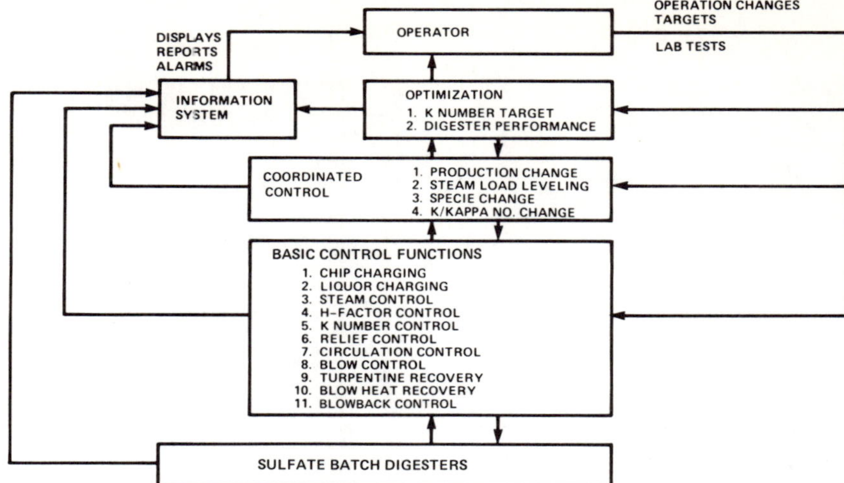

Figure 15-2. A typical control system structure—sulfate batch digester computer control system.

This system concept uses specified target set points, measurements, and laboratory test results to compute the appropriate amount of cooking liquors (chemical), heat, and it sets the blow times to achieve the required quality of pulp at the desired production rate. Production changes and K/Kappa number changes are coordinated so that they can be accomplished with minimum upsets to the process. Steam load leveling is used to coordinate the digesters and minimize steam load upsets. Target optimization is done on the K/Kappa number control in order to reduce K number variations. The operator interacts with the system via an operator's console using interactive displays. These displays pro-

Figure 15-3. Basic control functions—batch digester computer control system.

OPERATOR'S
MAN MACHINE INTERFACE

PRODUCTION SCHEDULING & CONTROL
KAPPA OR K NO. CONTROL
STEAM LOAD LEVELING

CHIP CHARGE CONTROL
LIQUOR CHARGING
WHITE CONTROL
BLACK CONTROL
CAP AND UNCAP
RELIEF AND AUTOMATIC BLOWBACK
STEAM CONTROL
FLOW/PRESS. TEMPERATURE
STEAM
DIGESTER
DIFF. TEMP. & CIRCULATION CONTROL
M
CHIP SILO
WHITE LIQUOR
BLACK LIQUOR
BLOW CONTROL
CYCLONE & BLOW TANK
HEAT RECOVERY

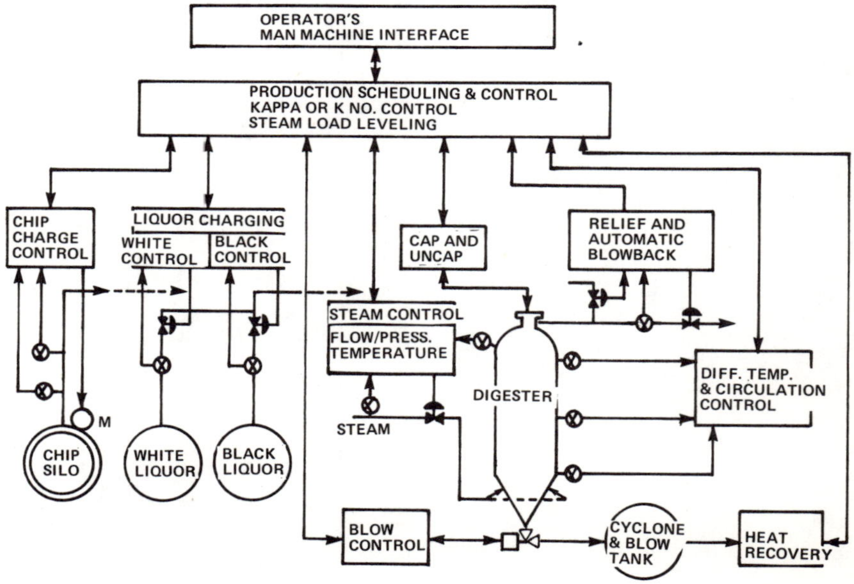

vide for initiating production changes and steam load leveling, for changing target set points, and for entering laboratory test results. A demand digester graphic display with periodically updated variable values is provided for each digester. The system also normally includes the provision for producing hard copy management reports. Typical basic control functions performed by such a batch digester computer control system (shown in Figure 15-3) are:

1. Chip charging
2. Liquor charging
3. Steam control using H factor correction
4. K/Kappa number control
5. Relief control
6. Circulation control
7. Blow control
8. Automatic blowback
9. Production change coordination
10. Steam load leveling
11. Blow heat recovery
12. Turpentine recovery

A brief functional description of each of the above control functions follows.

Chip Charging

The computer takes complete control of charging each digester after the operator has initiated the signal to begin the charge. As shown in Figure 15-4, this scheme controls the operation of chip bins, conveyors, weightometer (and/or level gauges), chip diverter, chip chute, and capping valves.

The computer starts the conveyor, filling the digester to a given level with chips, based on a gamma gauge level signal. The actual chip weight is then read from the weightometer and this is used in the liquor charging program to achieve a liquor/wood and alkali/wood target. The program can be provided for digesters with manual capping operation where the operator has to cap and uncap

Figure 15-4. Batch digester chip charging.

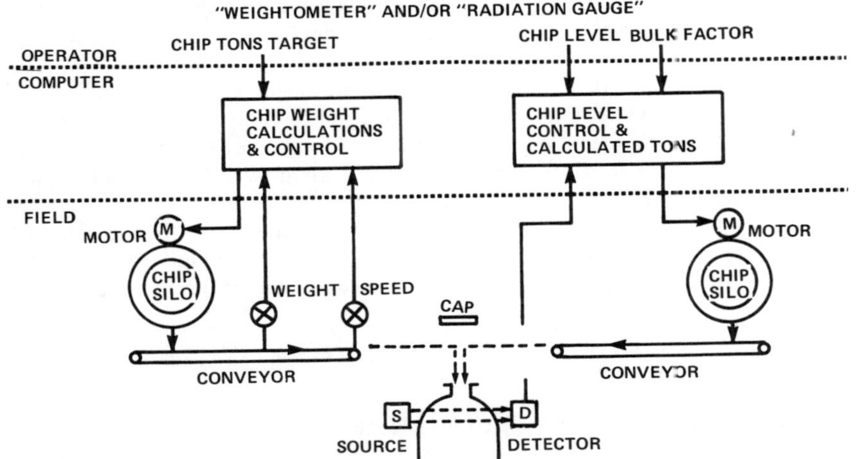

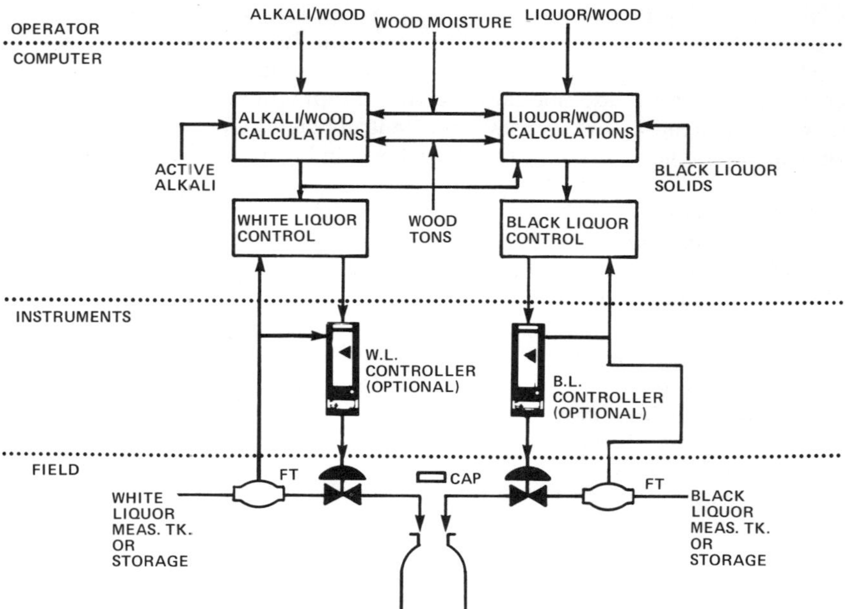

Figure 15-5. Batch digester liquor control.

digesters, as well as position the chip chute and start-stop the conveyor belt, or for automatic chip charging digesters with automatic capping operation.

Liquor Charging

There are several ways to load a digester with chips and liquor. The most common is to add liquor while the chips are being loaded. This provides for chip compaction during the charge.

With the scheme shown in Figure 15-5, the correct liquor charge is determined after all the chips have been added because liquor targets, alkali/wood ratio, and liquor/wood ratio depend on the chip tonnage.

The liquor is added in two parts. Part one is the base load of liquor added during the chip charges. Part two is the final "trim," calculated after the chip charge is complete.

Presteaming can be done at the appropriate tonnage, if desired.

The amount of white and black liquor charged to each digester is based on achieving target ratios. These ratios are adjusted by feedforward-feedback control algorithms. The white liquor target ratio is alkali/wood; the black liquor target ratio is liquor/wood.

The prediction correction scheme uses operator-entered test results to produce the target adjustments. The alkali/wood target is adapted by either K number or effective alkali residual. The liquor/wood target may be adapted by black liquor solids content.

Steam Control with *H* Factor

With the scheme shown in Figure 15-6, the steam flow trajectory is specified to achieve a target H factor. This is translated by an energy balance to a specified

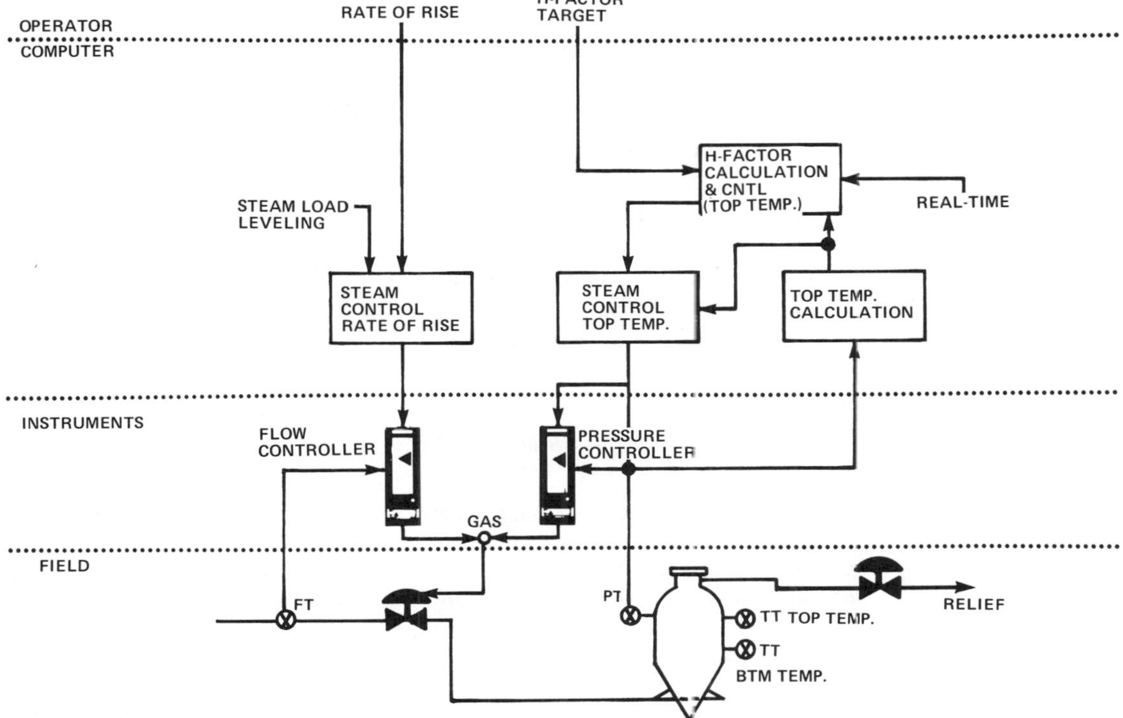

Figure 15-6. Batch digester steam control.

top temperature in the maintenance phase. The temperature rise may be an operator-entered value, or the result of "steam load balancing."

If fluctuations in top temperature occur due to reduced steam availability, the top temperature is adjusted to a level that will allow the digester to be blown at the scheduled time. Both the temperature rate of rise and top temperature have limits that are preset and cannot be exceeded.

The target H factor is dynamically determined, based on changes in sulfidity, predicted K number, and effective alkali residual. This gives better control than using H factor alone, where only the time and temperature are taken into account.

Instruments are arranged so that steaming is achieved by autoselecting steam flow or pressure rise control through a low-pressure autoselecting relay when in backup status and when the system is not on computer control.

K/Kappa Number Control

The diagram in Figure 15-7 shows a K/Kappa number control concept which uses a predictive and dynamic model technique called model reference control to adjust the H factor target. The control scheme uses a steady-state prediction model that takes into account the alkali residual, the sulfidity, the alkali/wood ratio, and the shive and knot content. The output of this model is a predictive K/Kappa number that is the input to the K/Kappa number controller.

The prediction model and dynamic model are corrected at periodic intervals, based on the K/Kappa number lab test. This entered lab test is compared with the

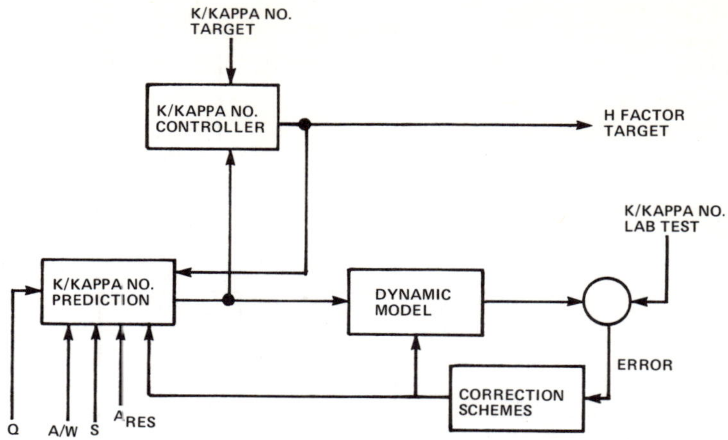

Figure 15-7. Batch digester K/Kappa number control.

output of the dynamic model and the error is passed to a corrective scheme for both models. A model of a blow system is provided for K/Kappa number samples taken downstream of the blow tank. This allows for the proper time-frame referencing of samples taken in the wash and screen rooms.

Relief Control

Control of digester relief is available in any of three options. Each option is designed for specific use in achieving a production goal, such as minimizing steam use, maximizing turpentine recovery, etc.

The system shown in Figure 15-8, the first option, uses a measurement of relief flow made with either a target flowmeter or a differential pressure transmitter.

Figure 15-8. Batch digester relief control.

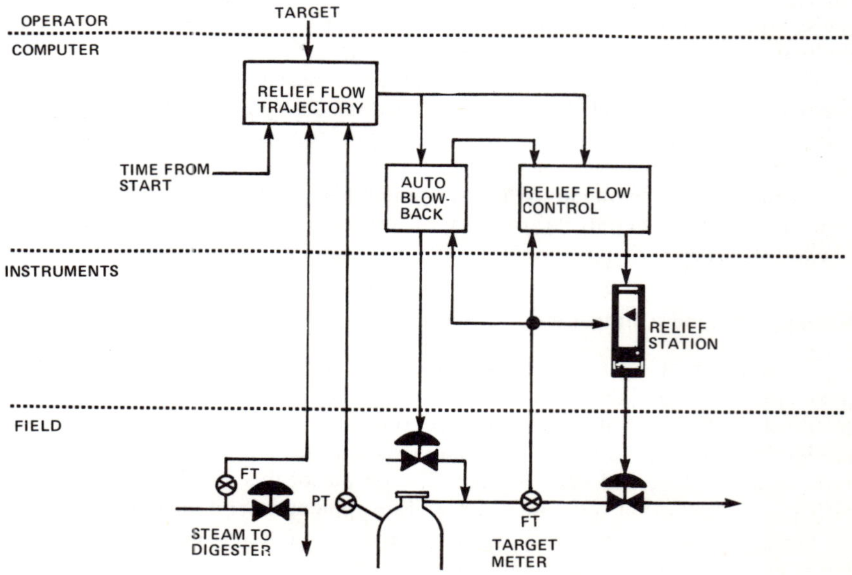

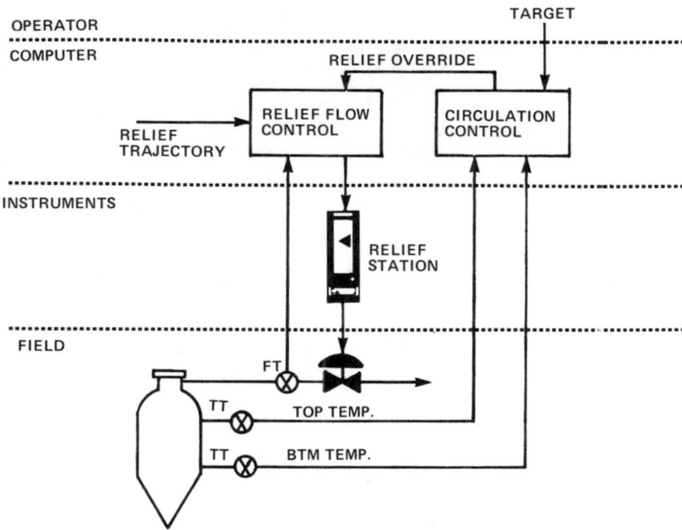

Figure 15-9. Batch digester circulation control.

The relief system is then controlled to maintain a specified flow rate trajectory of relief.

In the second option, if a relief flow measurement is not available, the valve opening can be controlled to a specified trajectory.

The third option is to control the false pressure in the digester to a given trajectory. This gives good circulation in the digester.

The relief flow trajectory is set to maximize turpentine recovery.

Circulation Control

Digester circulation control, shown in Figure 15-9, uses the top and bottom digester differential temperature to override the relief flow control.

If the differential temperature on the digester is excessive, the override increases the relief flow target. This, in turn, increases the natural digester circulation by decreasing digester pressure, thus lowering the boiling point.

Blow Control

Based on H factor calculations and production coordination, the blow times are accurately calculated. The operator is notified before the predicted blow. He then has the option to blow the digester himself or let the computer do it.

Automatic reblow is performed by an operator-entry pushbutton if it is determined that pulp still remains in the digester after the blow.

The level in the blow tank is checked before notifying the operator that the digester is ready to blow, and an alarm is given if the digester level is too high.

Automatic Blowback

Automatic blowback of the relief system is controlled by the computer. The blowback is based on a scheduled time basis, but optionally it can be based on controlled relief flow rate or on pressure difference in the relief system.

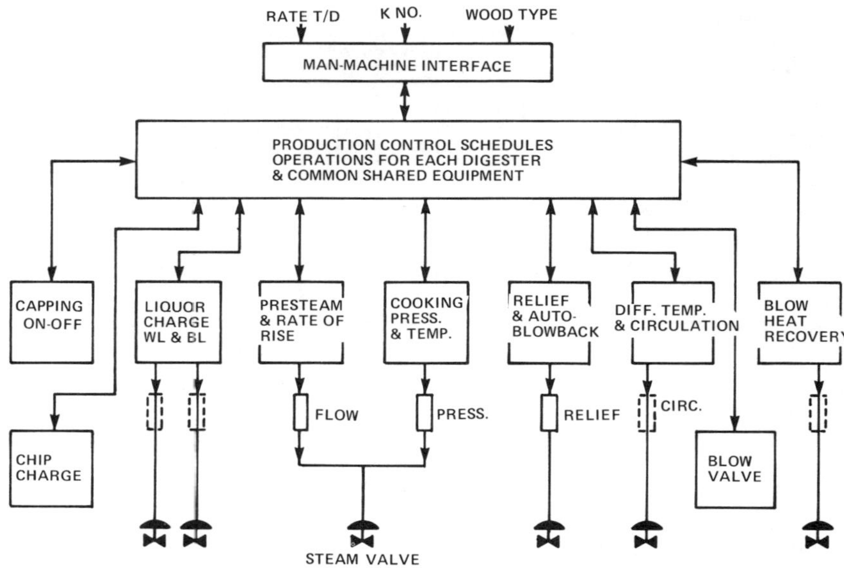

Figure 15-10. Batch digester production control.

Production Change Coordination

With the concept depicted in Figure 15-10, the operator can change the production rate by entering either the new spacing or the new production rate. A feasibility determination is then made, based on a computer upper limit H factor. If the new production rate is infeasible from a quality (K number target) constraint, the operator is warned. He then has the choice of changing blow time (production) or getting off-spec pulp. Upper limit H factor and feasibility determination are as follows:

Upper limit H factor calculation:

$$H_{UL} + F\ (\text{TOP TEMP}_{\text{MAX}}, \text{SPACING}, \#\text{of Digesters}, \text{Max Rate of Temp Rise})$$

Feasibility determination:

$H_{TARG} > H_{UL}$ Alarm Operator

$H_{TARG} \leqslant H_{UL}$ Insert New Blow Times for Digester

This program also coordinates production changes to the steam load balancing program.

Steam Load Leveling

Typically, a batch digester uses a large amount of steam during a cook. In addition, the steam flow varies during the bringing-up phase. During the maintenance phase, the steam phase will drop. During the blow, fill, and capping of each digester, the steam flow is zero.

As a result of this, the steam demand in a digester house fluctuates. This changing demand puts a severe strain on the powerhouse. There are times when not enough steam is available for cooking and when more steam is produced than the digester can use. During these times, either production is lost or energy is wasted.

The steam load leveling program evens out the digester steam demand and forces the digesters to use less steam than normal. By maintaining production rate and K number quality during these intervals, energy is conserved without sacrificing production or quality. There are two options for steam load leveling:

Option A. Each digester's steam flow is computed, based on the available steam flow and the operator-entered maximum. The steam flow reduction is prorated for each digester.

Option B. The program calculates a new temperature rise for each digester in the bringing-up phase. The calculation is based on an energy balance and the current steam flow available. The priority protocol of steam allotment is based on top temperature proximity.

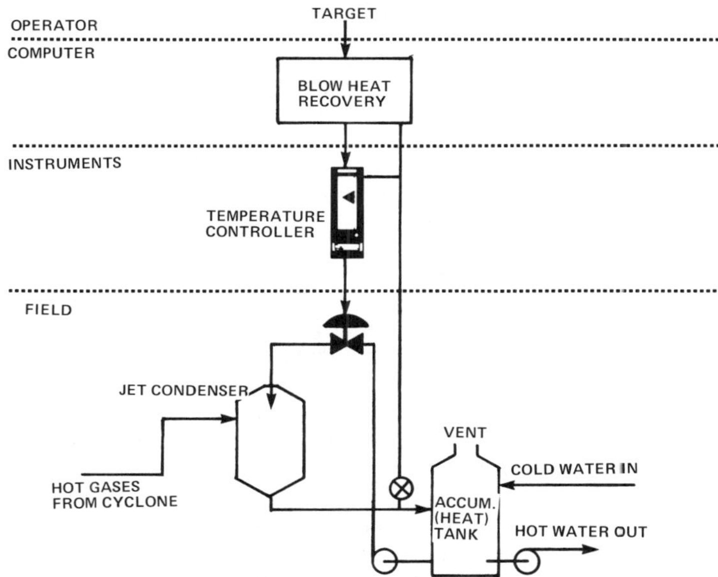

Figure 15-11. Batch digester blow heat recovery.

Blow Heat Recovery

In coordination with each blow, the heat recovery exchange is regulated to recover the maximum amount of heat from the blow, as shown in Figure 15-11.

The water flow to the jet condenser is controlled when the digester starts to blow to transfer heat to the hot water accumulator tank. The input to the controller is from the temperature going to the accumulator.

To prevent the condenser from drawing a vacuum when the shower water is turned on before the blow gases come in, consideration should be given to adding a pressure controller on the jet condenser that will apply steam into the condenser. To prevent too much gas from leaving the accumulator, the gas temperature can be controlled by adjusting the hot water flow out of the accumulator.

Turpentine Recovery

The turpentine recovery system, shown in Figure 15-12, is applicable for softwood cooking. The relief gases from each digester are fed to a cyclone separator

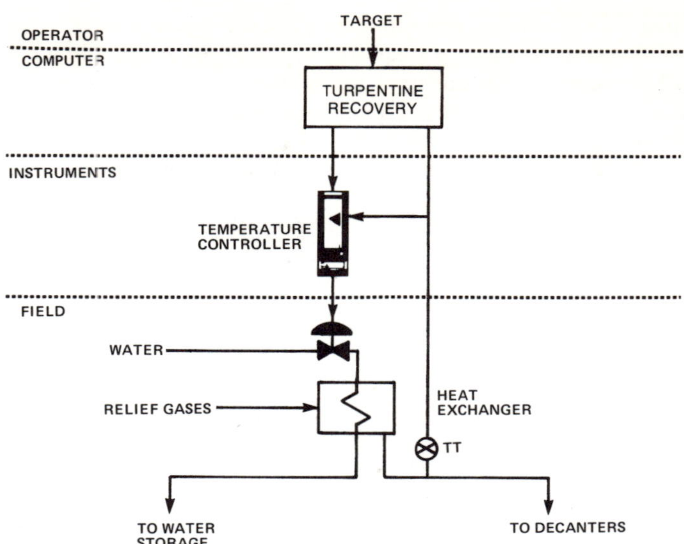

Figure 15-12. Batch digester turpentine recovery.

and then on to a heat exchanger or condenser. The heat exchanger causes the steam and condensable gases to condense. Basically, the turpentine recovery control scheme operates to maintain a predetermined temperature on condensed effluent from the condenser. The effluent continues on to decanters for further processing.

The relief flow trajectory is specified to maximize the turpentine recovery. Normally, the relief valve will be open during the rate of rise phase to the recovery system. When the top temperature is reached, the relief flow is diverted to the blow heat recovery system.

Another concept of a computer control system for a sulfate batch digester is designed to provide real time calculations which are specific for each cook, with the computer selecting the best set of cooking conditions to reach the targeted K number. A simplified schematic of such a system is shown in Figure 15-13. Control over an individual cook is achieved by measuring the effective alkali concentration during the early portion of the cook. The rate of delignification (pulping) is primarily controlled by three cooking factors—time, temperature, and alkali concentration. Based on the determination of the effective alkali concentration, the time-temperature cycle of the remaining part of the cook is adjusted to achieve the H factor required to reach the targeted K number for that specific cook.

Effective alkali can be determined by a condometric titration with standardized reagent on intermittent samples taken from the digester's liquor circulating lines. Another method of determining effective alkali along with sulfidity, which is another important cooking variable, is by the use of a multiple ion-selective electrode measuring sensor on a continuous sample stream from the digester during the cooking process.

CONTINUOUS DIGESTERS—VERTICAL TYPE

When the sulfate chemical pulping is done in a continuous-type digester, computers can be used for many of the same functions. Heat and material balances

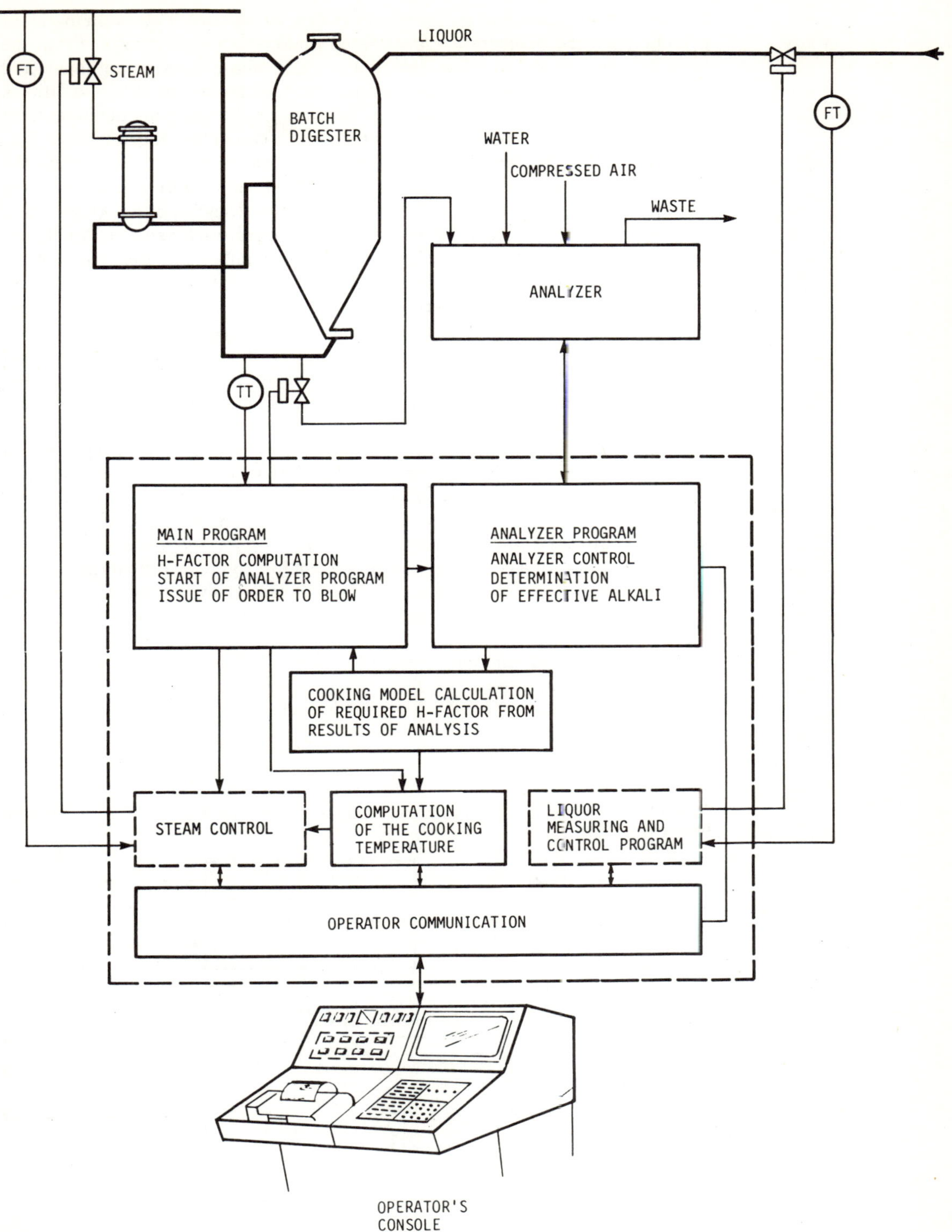

Figure 15-13. Schematic of a typical batch digester computer control system using effective alkali analysis.

are important for successful control of this type of digester. The mathematics of calculating these balances are performed on-line by the computer. From the input information on chip feed, black and white liquor flows, steam flows, and temperatures, it is possible for the computer to regulate the alkali/wood ratio, K number of the pulp, and chip load to the digester. To bring about the correct alkali/wood ratio, the operator enters the desired target into the computer. The computer makes the necessary corrections in the white liquor flow to achieve the target band on species, production (chip flow), and white liquor (cooking chemical) strength. The desired K/Kappa number (degree of cooking) of the pulp is maintained through the regulation of temperatures. Samples of pulp are taken periodically, tested in the laboratory for K number, and the results of the test are entered into the computer. The computer then makes the necessary corrections to cooking temperatures to bring K/Kappa numbers into the target range in accordance with a program for that particular species. Difficult measurements to make, such as cooking zone and chip moisture, are determined by having the computer make heat and mass balances automatically. The computer can also be programmed to take the time delay through the digester (residence time) into account so that it can anticipate changes in material input to the digester. Experience has shown that major savings from computer control of a continuous digester have come from: (1) higher yields, (2) reduced chemical usage, (3) increased throughput from existing facilities, and (4) a more uniform product. All of these are directly related to closer, faster, more accurate control, preventing up and down excursions in quality.

The structure for a computer control system for a typical vertical-type sulfate continuous digester is shown in Figure 15-14. It can be divided into four general sections, similar to the batch digester's computer control system:

1. Basic control function
2. Coordinated control
3. Optimization
4. Information system

Figure 15-14. A typical control system structure—continuous digester computer control system.

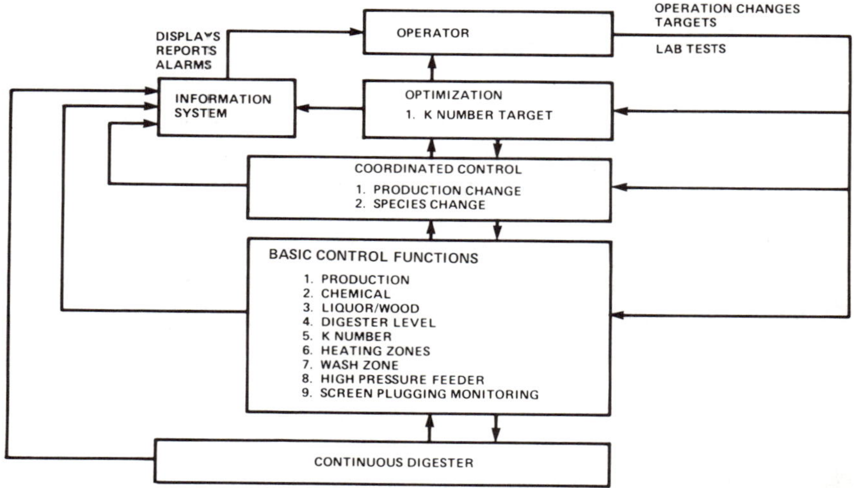

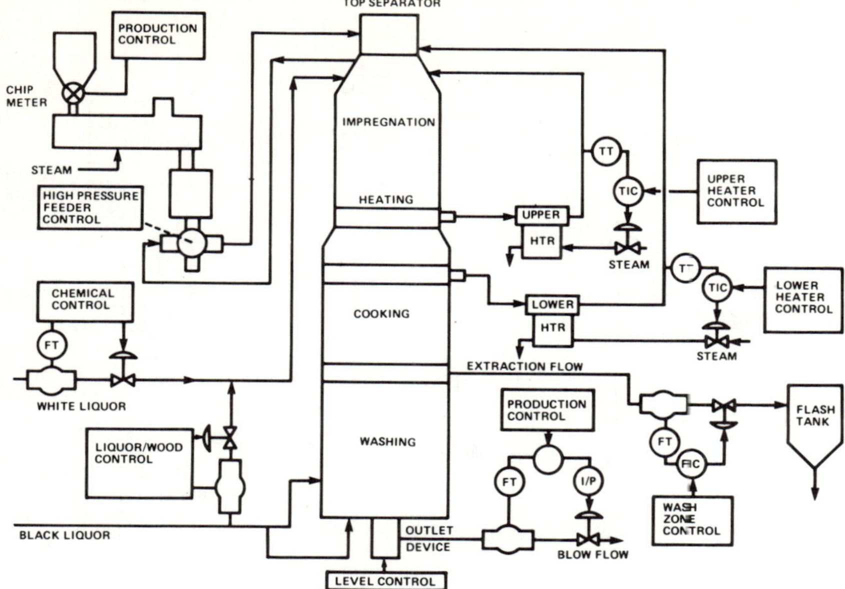

Figure 15-15. Basic control functions—sulphate vertical type continuous digester computer control system.

Figure 15-15 is an overview of the basic control functions found in such a system, listed as follows:

1. Production control
2. Chemical control
3. Liquor/wood control
4. Digester level control
5. K/Kappa number control
6. Upper and lower heater temperature control
7. Wash zone control
8. High-pressure feeder control

A brief description of each of the above control functions follows.

Production Control

This control strategy provides for production control in order to maintain the balance for wood-in and pulp-out, and to control production-out at the desired set point. Production control, as shown in Figure 15-16, involves both the wood input to and the pulp output of the digester. The wood input is calculated by using a speed-dependent pocket-fill factor, the chip bulk density, and the chipmeter speed. Yield is then used to calculate production (which is controlled to an operator-set target) by manipulating the chipmeter speed control. Production from the digester is set to the input production by manipulating the blow flow. By maintaining a constant blow flow/production ratio, a steady average blow consistency is assured, and upsets to digester operation are avoided.

The wood input computation assumes constant chip packing in the chipmeter pockets at varying speeds. Pocket-fill factor is the tuning parameter that

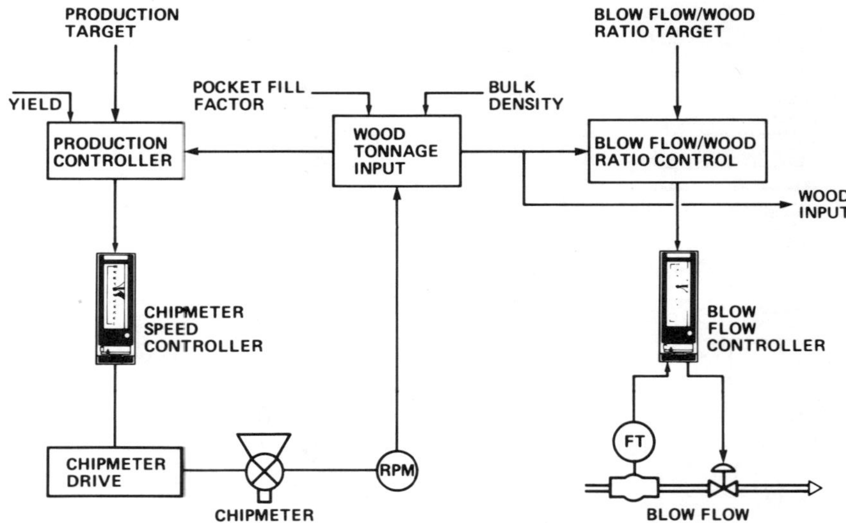

Figure 15-16. Continuous digester production control.

accounts for nonlinear filling of the chipmeter pockets at varying speeds. Yield is the ratio of pulp out/wood in and is used to compute the actual production from the wood input.

Chemical Control

The appropriate amount of chemical is provided by the system which maintains a desired alkali/wood ratio by maintaining the white liquor flow to the digester.

The purpose of controlling the alkali/wood ratio is to maintain a constant chemical application to the bone-dry wood entering the digester. Figure 15-17

Figure 15-17. Continuous digester chemical control.

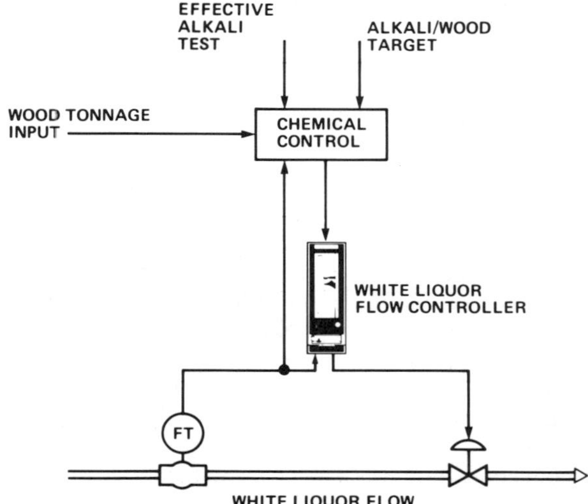

shows that the alkali/wood ratio is calculated by using the wood into the digester, the white liquor flow, and the operator-entered effective alkali laboratory test. The alkali/wood ratio is controlled to the operator-entered target by manipulating the white liquor flow.

A dynamic model is used to provide rapid adjustment in the ratio when a target change is made and to feed forward the alkali/wood ratio to the K/Kappa number prediction model for improved K/Kappa number control. This procedure is needed to overcome the high capacity and long time constant in the top recirculation of the digester.

Liquor/Wood Control

Liquor-to-wood control provides for maintaining a desired liquor/wood ratio. The objective is to maintain a constant ratio between the total liquid entering the digester and the bone-dry wood. As shown in Figure 15-18, the liquor/wood ratio is calculated from the total liquor and the wood going to the top of the digester. The total liquor is determined by adding together the white and black liquor flows, the water in the chips as moisture, and the water condensed from steam in the steaming vessel. The black liquor flow is then manipulated to maintain the operator-entered target liquor/wood ratio.

The use of an operator-entered value of chip moisture and an assumed amount of condensate steam per ton of chips will generally be sufficient to give good control. The direct measurement of one or both of these can be used optionally if large variations are experienced.

Digester Level Control

The level control is used to trim the production control for any variation in the balance of wood in and pulp out that cannot be eliminated by production control alone.

Stable digester level control is required to ensure a constant chip flow through the digester. The result is a constant residence time for chips in the cook zone.

Figure 15-18. Continuous digester liquor/wood control.

CHIP MOISTURE
LAB OR MEASURE
LIQUOR/WOOD
TARGET
WOOD INPUT
WHITE LIQUOR
STEAM FLOWS
LIQUOR/WOOD
RATIO CONTROL
BLACK LIQUOR
FLOW CONTROLLER
FT
FROM
STORAGE
TANKS
BLACK LIQUOR FLOW
TO
DIGESTER

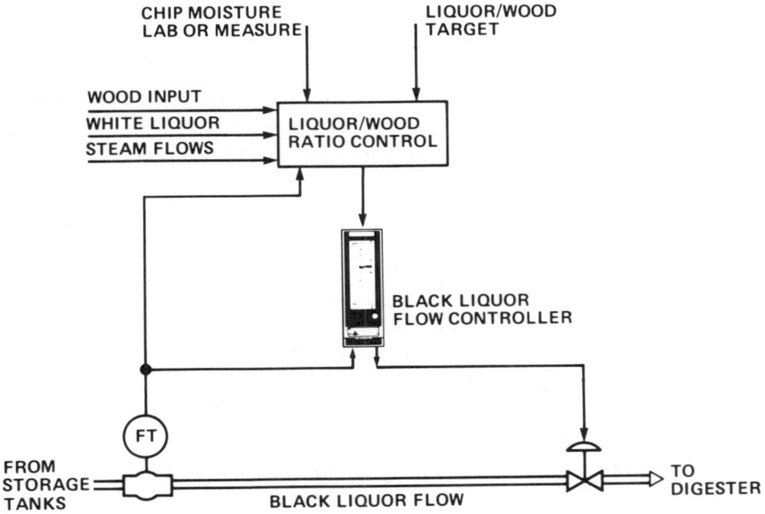

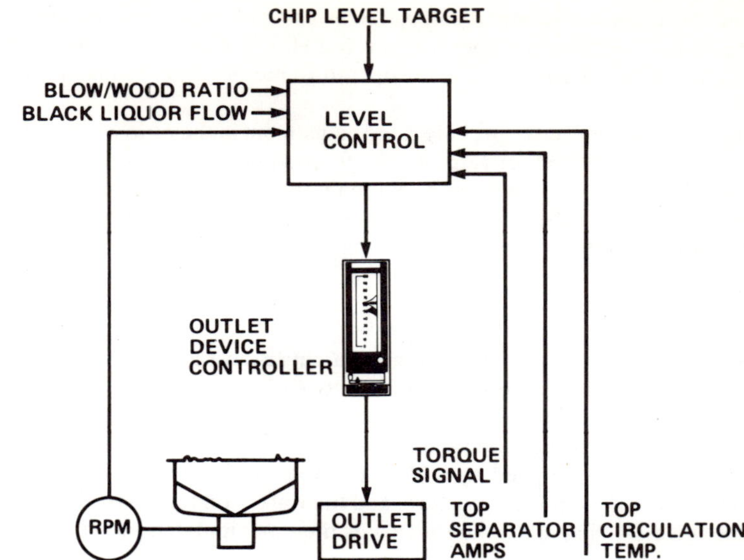

Figure 15-19. Continuous digester level control.

Figure 15-19 illustrates how the production control task prevents most upsets to the chip level in the digester by matching the blow production to the feed rate. When level disturbances do occur, the outlet device speed is manipulated to maintain a constant level. If the outlet device speed goes outside the set limits or if a level upset persists, then the blow/wood ratio is adjusted until the level is brought under control and the outlet device is back in range.

A special-purpose nonlinear control algorithm is used for chip level control. Other primary and/or backup control actions besides outlet device speed and blow/wood ratio may be specified. An emergency backup action adds black liquor to the top of the digester when the top separator amperage gets too high.

K/Kappa Number Control

K or Kappa number control varies the temperature set point, based on current alkali/wood ratio, residence time, and temperature, in order to minimize the K/Kappa number variation and control K/Kappa number at a desired target set point.

The primary goal of continuous digester control is the production of pulp at a target K/Kappa number with minimum variations. The reduction in K/Kappa number variations allows the mill to increase the target and, thus, to increase yield without producing off-spec pulp.

In Figure 15-20, the K/Kappa number of the pulp is predicted from the lower heater's cold side temperature target (the major manipulated variable), the alkali concentration, and the anticipated cook zone residence time. This predicted K/Kappa number is used as the input measurement to the controller, which adjusts the lower heater's cold side temperature target.

A dynamic model is used to track the pulp through the digester and blow tank for later comparison with the laboratory tests of K/Kappa number at the blow line or washers. The error calculation resulting from this comparison is used to correct the models.

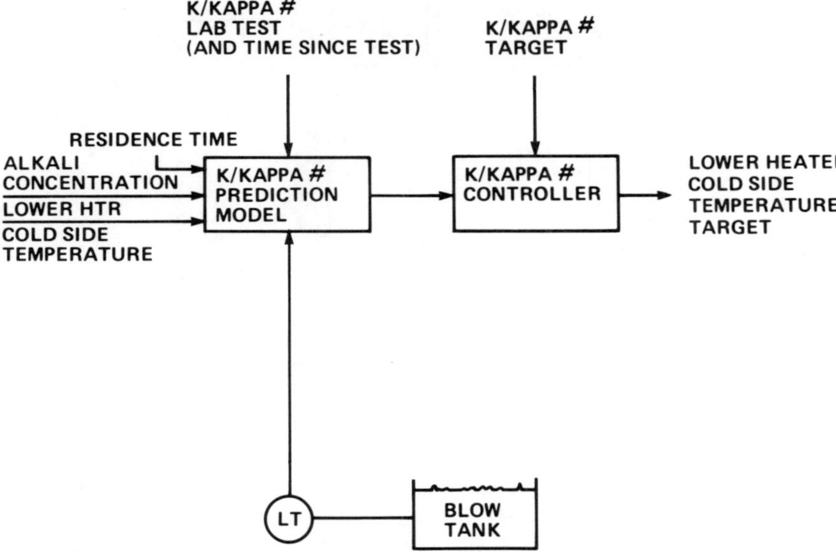

Figure 15-20. Continuous digester K/Kappa number control.

Upper and Lower Heater Temperature Control

The control of the temperature of the upper and lower heaters provides for maintaining the temperature at the top of the cook zone at a desired set point.

The temperature of the chips and liquor in the cooking zone has considerable effect on the delignification reaction and final pulp quality. Therefore, accurate and stable control of this temperature, which is accomplished through feedback and feedforward techniques, is extremely important.

The purpose of the upper and lower heating zone control, shown in Figures 15-21 and 15-22, is to control the temperature of the chip and liquor mass coming out

Figure 15-21. Continuous digester upper heater control.

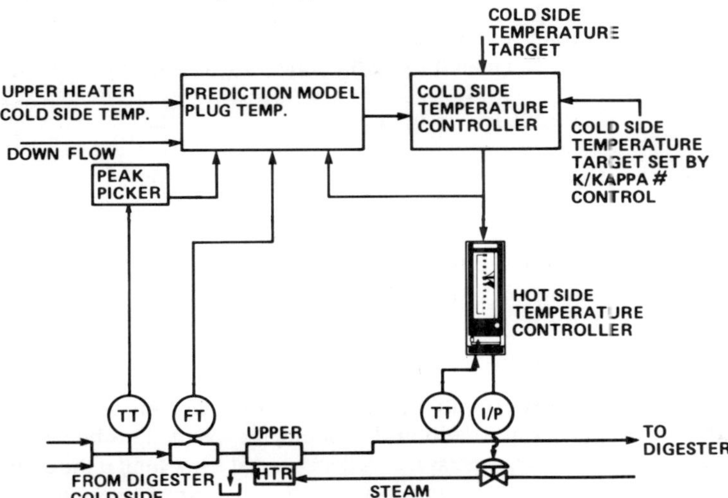

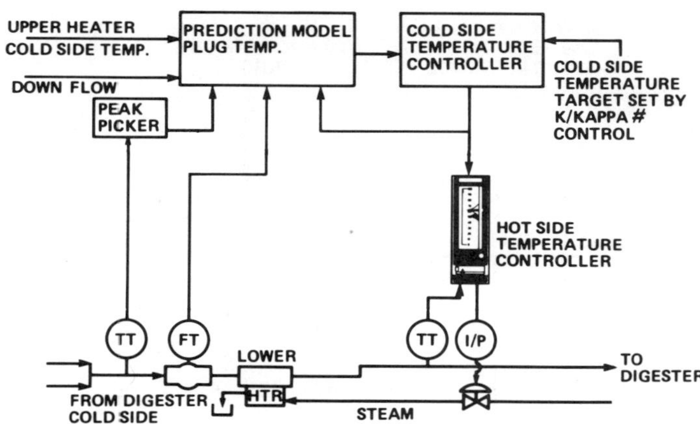

Figure 15-22. Continuous digester lower heater control.

of the respective heating zone at its target temperature. This is achieved by a predictive model of the plug temperature and a controller that sets the hot side set point on the respective heater's steam controller. A dynamic model, representing transport delay and thermal diffusion, is also used to simulate each heating zone's dynamic response.

Feedback corrections to the models come from the high peak temperature of the heating zone screens (cold side temperature), as this temperature most nearly represents the actual plug temperature. If the steam valve position of the lower heater exceeds a high limit, the upper heater's temperature target is changed.

Wash Zone Control

Wash zone control achieves a liquor balance in the wash zone and maintains the desired wash upflow target set point.

Figure 15-23 shows a concept which maintains a constant washing of the pulp. An inferred wash upflow is controlled to an operator-entered target. A flow-balance calculation is used to infer the countercurrent wash flow. The extraction flow is manipulated to maintain countercurrent wash flow at the operator-entered target. Changes in extraction flow are made in a manner that will not upset other digester operating conditions.

The wash upflow is calculated using the following flow-balance equation:

$$W_u = F_1 - F_2 + C \times P$$

Where: W_u = wash upflow
F_1 = cold blow flow
F_2 = blow flow
C = blow flow consistency
P = production target

High-Pressure Feeder Control

The high-pressure feeder control shown in Figure 15-24 maintains the speed of the high-pressure feeder at a ratio of the chipmeter speed. This helps maintain chip-chute level and the high-pressure feeder speed at a low limit if the chipmeter is stopped. It also enables the high-pressure feeder to run at an optimum speed.

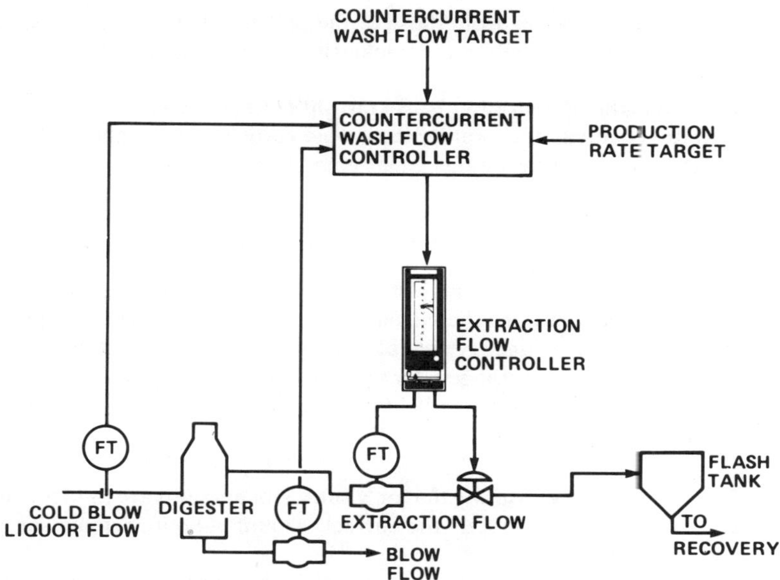

Figure 15-23. Continuous digester wash zone control.

Product Change

Product change coordination is accomplished by making appropriate changes to temperature and chemical targets at appropriate times and ratios, as determined by the computer, to minimize upsets during species changes—for example, hardwood to softwood, and vice versa. This is done by changing the

Figure 15-24. Continuous digester high-pressure feeder control.

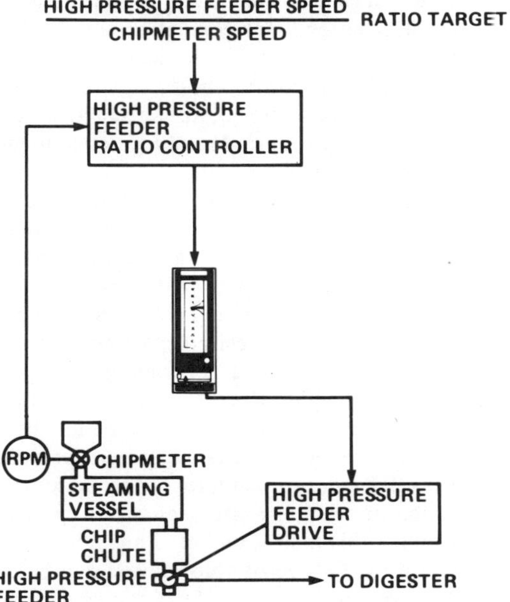

appropriate target set points from those of the old species to those of the new species at various points in time, as the tracked interface of the two species moves through the digester zones.

K/Kappa number optimization is used to shift the K/Kappa number target based on K/Kappa number laboratory tests, which provide for a higher K/Kappa number based on smaller deviations under computer control.

BLEACH PLANT

Like the continuous digester, multistage bleach plants, with their long time delays for pulp to pass through the system, are good candidates for the application of computer control. Since typical bleaching operations consist of a combination of various stages arranged in sequence—for example, chlorination, caustic extraction, hypochlorite, and chlorine dioxide—the primary reasons for using computer control are to:

1. Save chemicals by ensuring that only sufficient amounts are used for each stage of bleaching to obtain the correct amount of liquor removal or oxidation required.
2. Obtain higher yields with the computer system maintaining operating conditions in each stage to minimize degradation of the pulp by decreased yield and strength due to action of the chemicals on cellulose.
3. Improve quality with computer control maintaining the output brightness of pulp at the required level under both steady-state and dynamic operating conditions.
4. Increase throughput.
5. Decrease steam usage.
6. Minimize the total volume of effluent from the plant.

The computer system achieves these objectives by employing a control scheme that optimizes each stage by using quality indicators as input information. This information can be used as feedback correction of previous stages and feedforward correction in succeeding stages. The logical strategy on which these corrections are based can be stored in the computer. Such indicators as brightness, viscosity, and K number measurements could be used as inputs, with the computer calculating and making the required readjustments in chemical flows, retention times, temperatures, pH, etc.

A possible computer control scheme is to have the computer predict the degree of bleachability (K number) of the incoming stock on a fixed-rate scan. The operator then has immediate access to the instantaneous K number entering the bleaching system. When the tonnage rate changes, the operator enters new target tonnages into the computer. The computer immediately gives him the correct pulp flow to the bleach plant. Then the desired chlorine flow is calculated on the basis of predicted K number, the new target tonnage, temperature, retention time, or other variables which might affect it. The chlorine flow is then set and the computer returns to its normal cycle, allowing time for the chlorine to react and the ORP measurement to reflect the change. After a precalculated time, the computer returns to the ORP measurement, reads the value, and establishes this control set point until another tonnage rate change. An optical-type measuring device is also used to make this measurement. Thus, chlorination control and caustic addition based on the degree of chlorination can be calculated by the computer. A constant desired brightness can be obtained with the computer con-

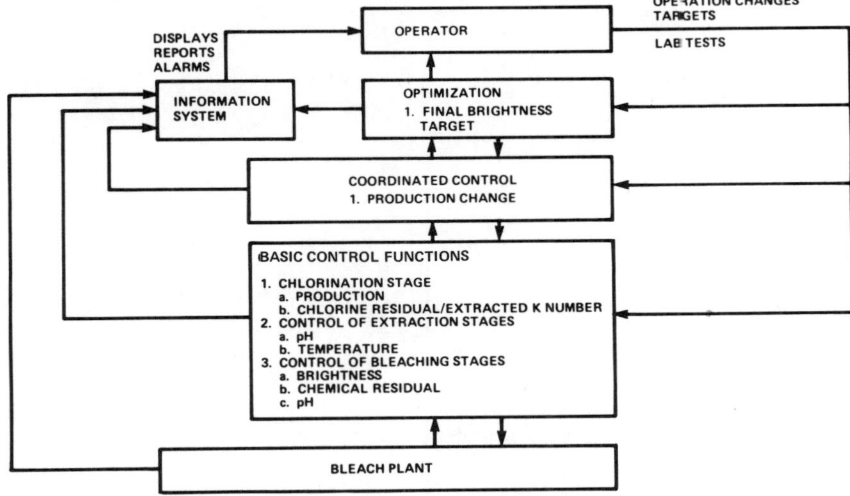

Figure 15-25. A typical control system—bleach plant computer control system.

trolling the hypochlorite stage by predicting the chemical addition value (based on the K number obtained after the extraction stage), retention time, and temperature at a constant pH. The same computer control strategy can be used on the chlorine dioxide stage with brightness being a function of hypochlorite brightness, temperature, tonnage rate, and retention time. This brightness can also be measured with on-line optical-type measuring systems.

As in the digester computer control systems, the structure of a computer control system for a typical bleach plant (shown in Figure 15-25) can be considered as consisting of four similar units. They are:

1. Basic control functions
2. Coordinated control
3. Optimization
4. Information systems

The basic control functions performed by a computer control system for a typical four-stage bleach plant are shown in Figure 15-26 and consist of:

1. Chlorination-stage brown stock control and chlorine flow control.
2. Extraction-stage caustic flow control and temperature control.
3. Bleaching-stage hypochlorite flow control, chlorine dioxide flow control, buffer flow control, and temperature control.

A bleach plant's computer control system can be designed for any number and combination of bleaching sequences of chlorination, extraction, hypochlorite, or chlorine dioxide stages. In the chlorination stage, a target chemical residual or extracted K number is maintained. Extraction stages are controlled to a target pH and temperature. Bleaching stages, such as hypochlorite and chlorine dioxide, are controlled to a target brightness, a target residual and a target temperature. The control objective is usually to maintain pulp quality while minimizing the usage of chemicals and energy.

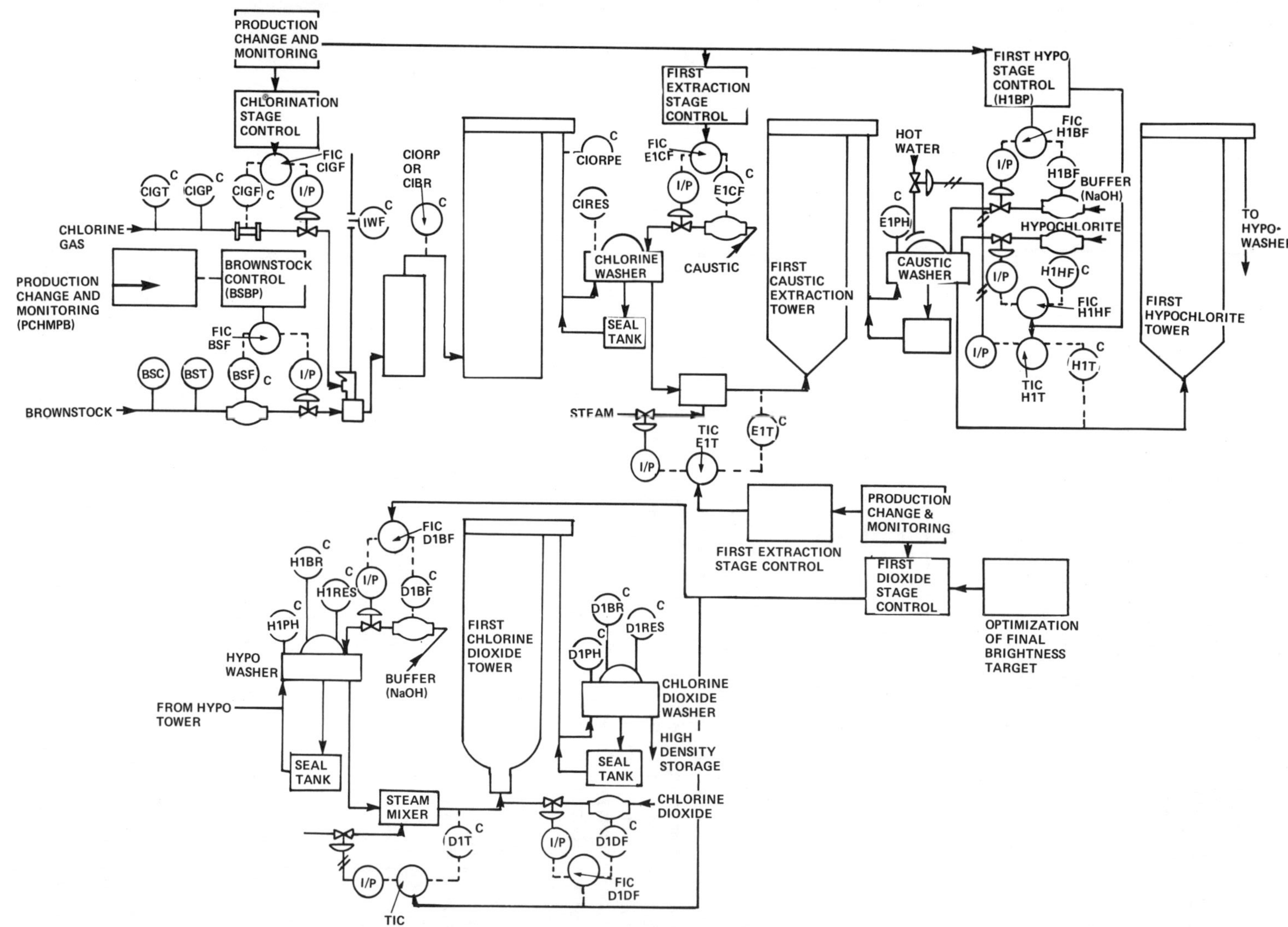

Figure 15-26. Basic control functions—bleach plant computer control system.

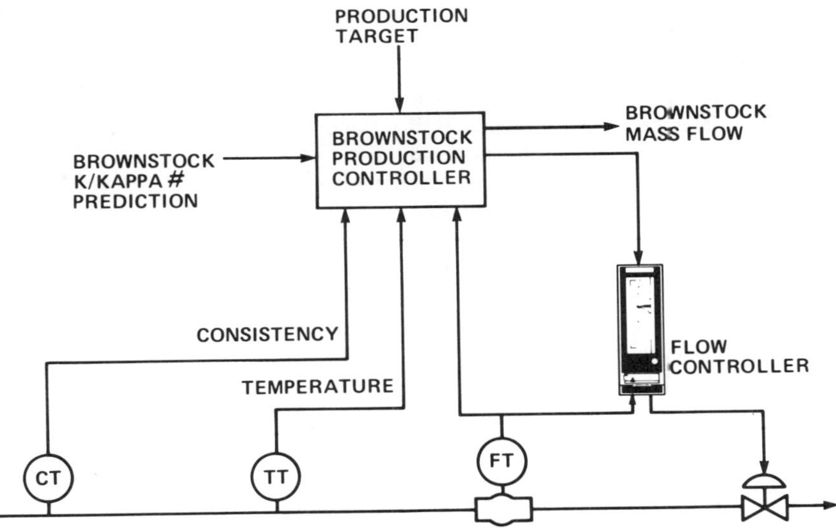

Figure 15-27. Bleach plant brownstock flow control.

Brownstock Control

The brownstock control scheme (shown in Figure 15-27) provides for maintaining a desired production set point in the chlorination stage by controlling the brownstock flow. It compensates the consistency measurement for brownstock K number, brownstock temperature, and brownstock flow variations. The compensated consistency is then used with the production set point to compute the required brownstock flow. This function calculates the compensated consistency and brownstock flow set point. This control of brownstock flow maintains a constant dry fiber flow to the chlorination tower and, in addition, establishes chlorination tower retention time.

Chlorination-Stage Control

A bleach plant's control system can operate the bleach plant under constant residual, constant extracted K number, or a combination of both. The first method is shown in Figure 15-28.

In chlorination-stage control, the proper amount of chlorine is applied to the incoming pulp by controlling the chlorine flow to maintain a target chlorine residual at the stage exit. To achieve this, the required measurement compensation is performed, and chlorine residual and brownstock K number prediction models are used to calculate a predicted chlorine residual out of the stage. Moreover, a chlorine residual dynamic model compensates for the mixing effect and the time delay of the stage.

The chlorine residual laboratory test is used to correct both the predictive and the dynamic models. As an option, the model correction may be accomplished with a secondary ORP measurement, properly compensated to provide a continuous chlorine residual measurement.

In addition, optional compensation can be made to the operator-entered chlorine residual target for brownstock K number, stage-temperature, and stage-residence-time variations. It also computes the extracted K number. Chlorination-stage control supports either a primary ORP or an optical brightness measurement.

This function:

- Calculates percent chlorine.
- Calculates chlorine residual measurement.
- Calculates predicted brownstock K number.
- Calculates predicted chlorine residual.
- Calculates compensated primary ORP measurement.
- Corrects predictive and dynamic models from measurements and/or laboratory tests.
- Calculates compensated chlorine residual target.
- Calculates predicted extracted K number.
- Calculates required chlorine flow set point.

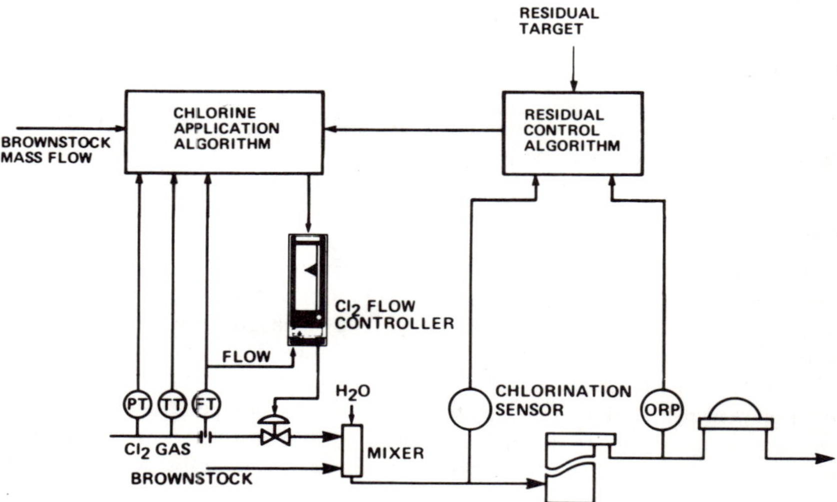

Figure 15-28. Bleach plant chlorine-stage control.

Extraction-Stage Control

The primary function of the extraction stage is to remove alkali-soluble compounds such as chlorinated lignins. The major controlled variables are pH and temperature, as shown in Figure 15-29.

The purpose of extraction-stage control is to maintain a target pH for the stage and to control the stage temperature. The controlled variables are the caustic flow and the stage temperature. Caustic flow control is achieved by using a prediction model for pH for feedforward control and a dynamic model to compensate for the mixing effect and time delay of the stage. Either the laboratory test or a continuous measurement of vat pH is used to provide feedback correction to the models. The computation of the caustic/chlorine ratio is made by using percent chlorine delayed in order that the chlorination-stage time delay may be taken into account.

Temperature control is achieved by increasing the temperature set point from a reference value for increases in production rate and caustic demand. This function calculates percent NaOH, NaOH/chlorine ratio, predicted pH, required NaOH flow, and required stage temperature.

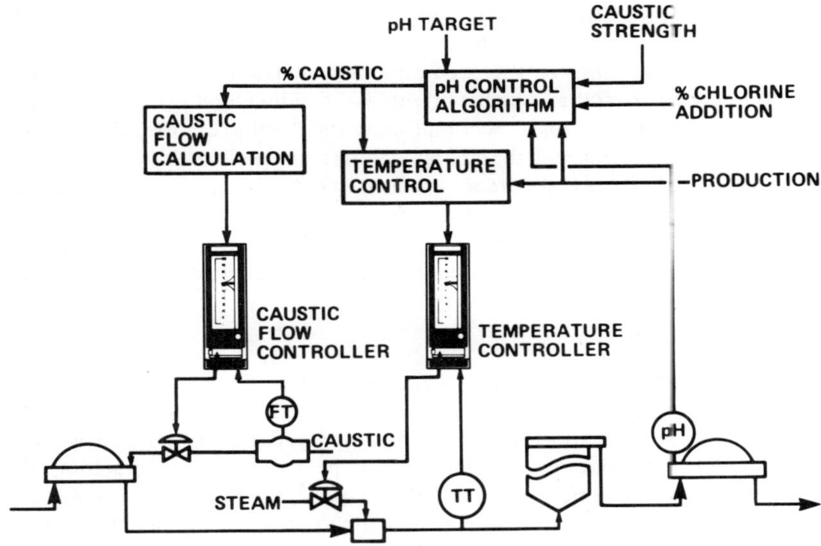

Figure 15-29. Bleach plant extraction-stage control.

Hypo-Stage Control

The purpose of hypo control is to maintain a target brightness, a target hypo residual, and a target pH for the stage. To achieve this, the necessary measurement compensation must be performed and the required variables computed. Moreover, predicted models are used to compute a predicted brightness, hypo residual, and pH, which are used for the feedforward control of the stage. These models also use quality variables from the previous stage, such as brightness and pH. Dynamic models are also used to compensate for the mixing effect and the time delay of the hypo stage.

Figure 15-30 shows that the hypo flow is controlled to maintain the brightness

Figure 15-30. Bleach plant hypo-stage control.

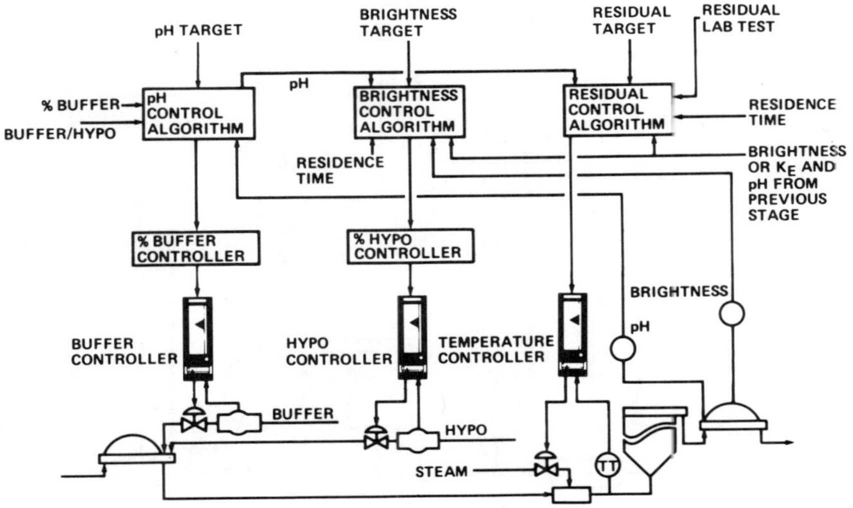

target; temperature is controlled to maintain the hypo residual target; and buffer flow (normally caustic) is controlled to maintain the pH target.

Feedback corrections to the predictive and dynamic models are made by using the laboratory test results for brightness, residual, and pH. For brightness and pH, there is the option of accomplishing the feedback correction by using continuous measurement, with the laboratory tests being used for drift compensation. This function calculates compensated measurements, required variables, percent hypo, percent buffer, buffer/hypo ratio, predicted brightness, hypo residual, and pH. It corrects predictive and dynamic models from laboratory tests and optional measurements; calculates required hypo flow, buffer flow, and temperature.

Dioxide-Stage Control

Dioxide-stage control is similar to hypo-stage control in that they both involve models and strategies for the prediction of residual, brightness, and pH.

The purpose of the chlorine dioxide control is to maintain a target brightness, a target chlorine dioxide residual, and a target pH for the stage. To achieve this, the necessary measurement compensation is performed and the required variables are computed. Moreover, predictive models compute a predicted brightness, chlorine dioxide residual, and pH, which are used for the feedforward control of the stage. These models also use quality variables from the previous stage such as brightness and pH. Dynamic models also compensate for the mixing effect and the time delay of the chlorine dioxide stage.

Chlorine dioxide flow is controlled to maintain the brightness target; tem-

Figure 15-31. Bleach plant dioxide-stage control.

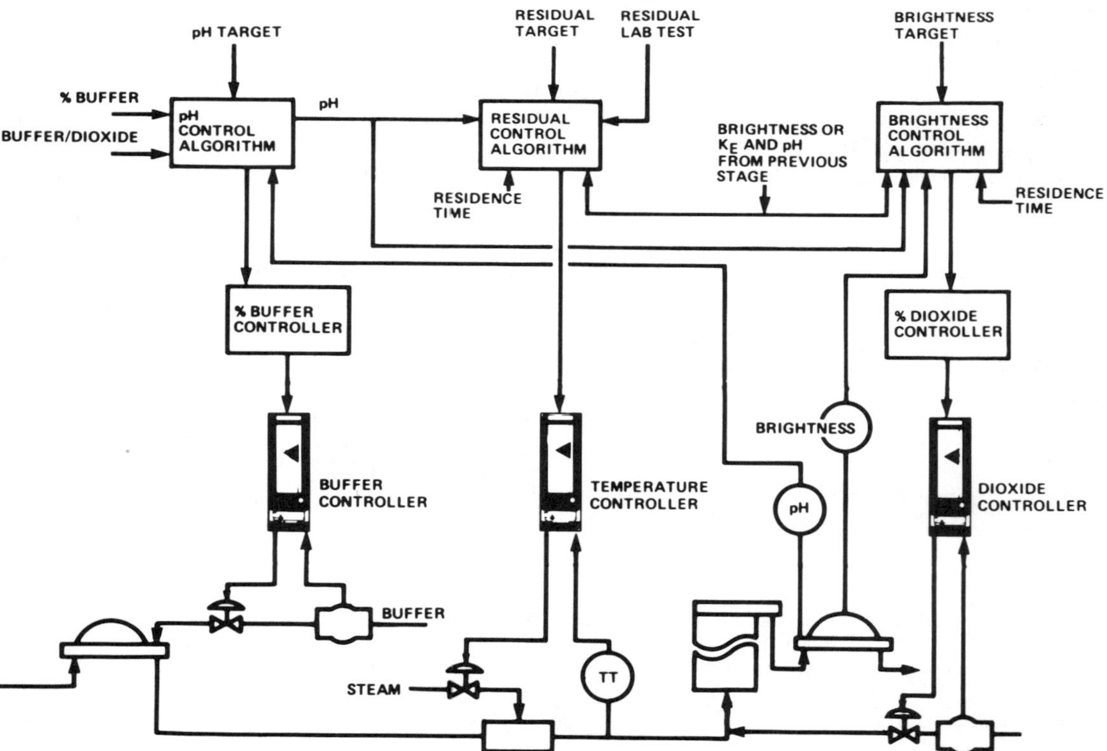

perature is controlled to maintain the chlorine dioxide residual target; buffer flow, normally caustic, is controlled to maintain the pH target (see Figure 15.31).

Feedback corrections to the predictive and dynamic models are made by using the laboratory test results for brightness, residual, and pH. For brightness and pH, there is the option of accomplishing the feedback correction by using continuous measurements, with the laboratory tests being used for drift compensation. This function calculates compensated measurements, required variables, percent chlorine dioxide, percent buffer, buffer/dioxide ratio, predicted brightness, chlorine dioxide residual, and pH. It also corrects predictive and dynamic models from laboratory tests and optional measurements, and calculates required chlorine dioxide flow, buffer flow, and temperature.

CONTINUOUS DIGESTER—INCLINED TYPE

A possible concept of a computer control system for the inclined-type continuous digester is shown in Figure 15-32. The system provides the most critical functions required for the control of its operation. A brief functional description of these control functions follows.

Wood Input Control

The wood feed into the digester is estimated by using chip moisture and weight measurements, and is controlled by manipulating the speed of the auger in the small chip bin. The level of this chip bin will be controlled by adjusting the speed of the drag chain under the main chip storage bin.

Alkali/Wood Ratio Control

The alkali/wood ratio is estimated by using the drywood flow, white liquor flow, and the effective alkali of the white liquor entered by the operator, based on laboratory test results. The ratio is controlled to either an operator-entered set point or a set point supervised by K/Kappa number control. White liquor flow is manipulated to maintain the alkali/wood ratio.

Liquor/Wood Ratio Control

The total estimated liquor entering the digester is ratioed to the estimated drywood flow. The entering liquor flow is estimated from white and black liquor flow with provisions to include steam flow and chip moisture. The ratio is maintained by adjusting total liquor flow.

Temperature Control

The temperature of the digester is controlled by manipulating the steam flow to the digester. The temperature of the liquor entering the digester is controlled to a biased digester set point by manipulating steam flow to the heat exchanger. Feedforward control is used to reduce temperature upsets caused by changes in the incoming liquor flow. Also provision is made to allow changing the manipulated variable used by Kappa control from alkali/wood ratio to temperature.

Reactor Level Control

The close regulation of wood and liquor to the digester prevents most upsets to reactor level. The blow flow is manipulated to maintain digester liquid level. In

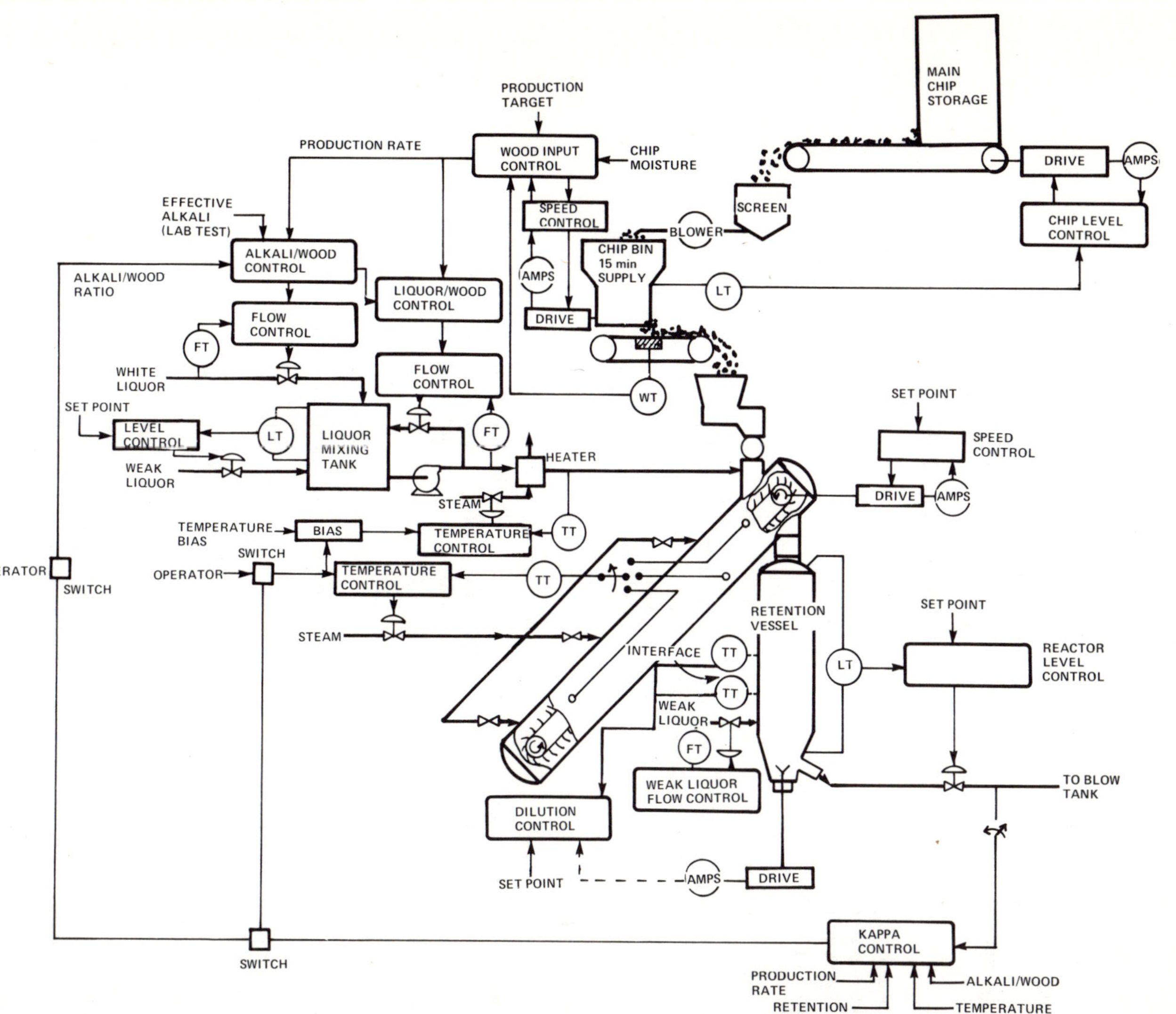

Figure 15-32. Basic control functions—a computer control system for an inclined continuous digester.

addition, the operator has the option of using dilution flow instead of blow flow to control level.

K Number Control

There are two basic types of models used in the K number control scheme. A steady-state or feedforward model is used to adjust the alkali/wood ratio set point based on K number target, alkali/wood ratio, digester temperature, and anticipated residence time.

Dynamic state variable models are used to track the pulp through the digester. The dynamic model keeps track of past control actions, which are used for model corrections.

The alkali/wood ratio set point is automatically adjusted by the K number control scheme in order to drive the K number toward the desired target.

Production Change Coordination

A planned production change includes precompensation of heating zone temperature and white liquor flow to account for the anticipated production change. The goal of the planned production change is to minimize upsets to K number.

Dilution Control

It is recommended that the interface between the cook zone and dilution zone in the digester reactor be stabilized. This will reduce the variation in cook time caused by changing interface levels, thus reducing Kappa variations.

Control of the dilution zone is achieved by the location of the interface as determined by the temperature differences of probes placed in the bottom of the reactor. Two or three temperature probes should be sufficient for this purpose. The black liquor flow to the bottom of the reactor is manipulated to keep the interface between the temperature probes.

Plug Detection and Alarming

The load on the stirrer, located in the bottom of the reactor, is monitored and compared with limits to determine if the vessel has plugged. The operator is notified when a plug starts developing.

OTHER PULP MILL COMPUTER SYSTEMS

Although the application of computer control techniques has not been as extensive in other pulp mill areas as in the digesters and bleach plants, some work has been done or contemplated in applying computers to: (1) maintain proper stock consistencies and soda carry-over from the washers, and evaluate evaporator heat transfer characteristics to define optimum washing schedules; (2) aid recovery furnace and associated steam and electric generating units to meet surges in demand due to startup and shutdown of process equipment, as well as possible reduction of auxiliary fuel and energy requirements; (3) control the density of green liquor going to the recausticizing plant; and (4) improve the throughput of the tall oil plant by computing and specifying such variables as the correct reflux ratio to use to obtain a satisfactory overhead fraction in refining crude tall oil by successive distillation.

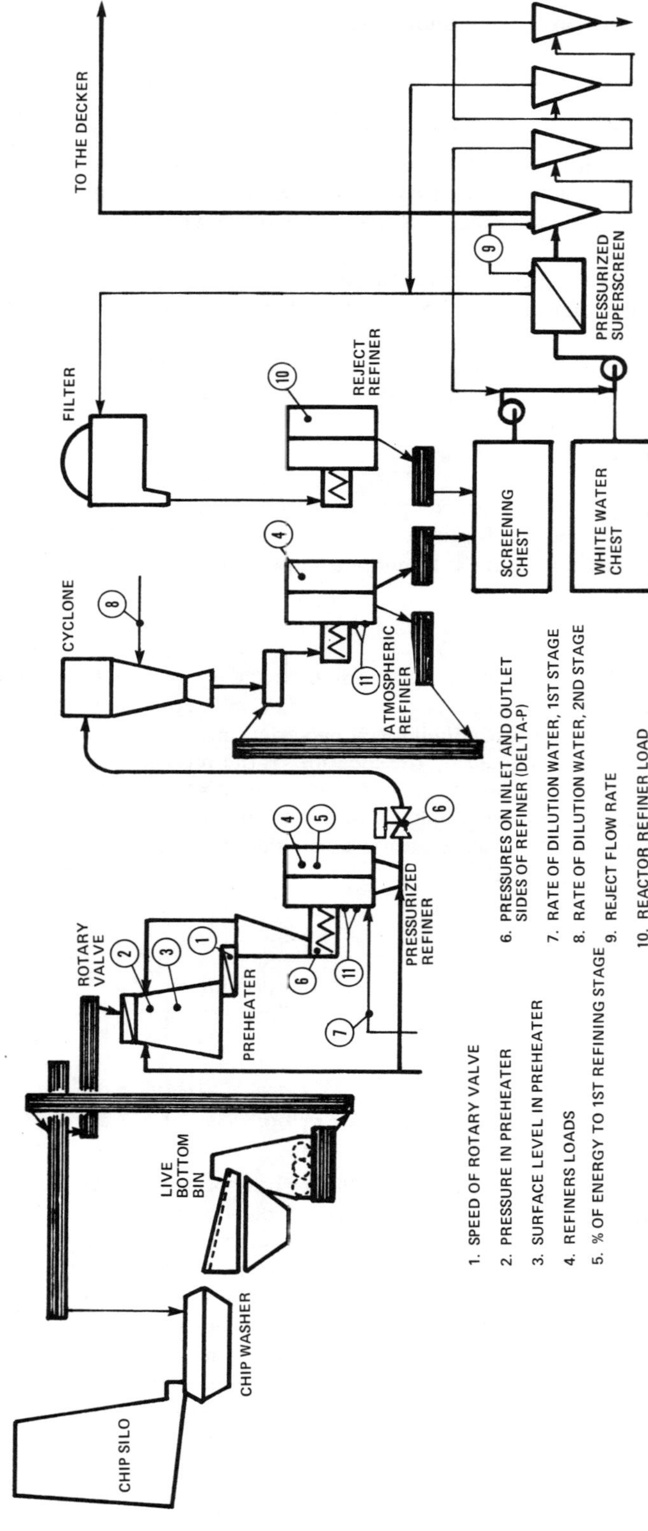

Figure 15-33. Controlling points in a thermomechanical pulping process.

Groundwood Mill

A limited amount of work has been done in computerizing groundwood mill control with emphasis on the grinding operation. The essential feature of one control philosophy is to deal with the groundwood as a single unit. The grinder room with its multiple grinders is controlled in such a way that the quality of the compound pulp from all grinders remains constant at a desired level. This is accomplished by continually monitoring the state of the process to ensure control of the individual grinders in accordance with a predetermined control strategy.

A control strategy can include keeping the freeness of the pulp constant at a target level and maximizing production of the groundwood mill. If target freeness cannot be reached within the energy limits of a grinder, then the pulpstone must be sharpened, with sharpening action effect assessed by the computer.

Log loading and feeding failures can be a significant cause of quality variation and production losses in a groundwood mill. One way of reducing loading failures is by the use of a computerized monitoring system.

Thermomechanical Pulping

Although increased interest in the thermomechanical pulping process is relatively recent, work has already been done in applying computer control systems to its production. A computer system can be used to control such variables as production, preheating temperature and time, energy consumption and distribution, pressure difference across refiners, consistency of the refining stages, reject ratio, and the alignment of discs. As shown in Figure 15-33, a typical two-refining stage thermomechanical process, controlling these variables, can be accomplished by controlling the speed of the rotary valve, pressure in the preheater, refiner power, percent energy to the refiner stage, pressure on the inlet and outlet sides of the refiner, dilution water, screens, reject refiner power, and control pins.

16 Computers in the Paper Mill

Papermaking is primarily a complex mechanical process involving diverse types of equipment. The operational procedures required are difficult to coordinate manually since the performance of any single piece of equipment affects the performance of other areas and of associated equipment.

The control problems caused by complex interaction of variables traditionally subjected to independent control action throughout a paper mill have led to the use of digital computer process control systems designed to take such interaction into account. Figure 16-1 illustrates a possible configuration of a computer scheme that is arranged to accomplish this. The drawing shows the four major unit process areas. However, the majority of the computers installed in paper mills have been concerned with the stock preparation and paper machine areas.

STOCK PREPARATION

Uniform paper manufacturing is easier when using stock whose composition and physical properties, such as freeness, are not subject to rapid variations. Computer control in the stock preparation during the refining and blending operation minimizes these variations in the stock.

Refiners

A computer scheme that can be used for refiner control to maintain constant freeness is shown in Figure 16-2, which employs both feedback and feedforward control concepts. Continuous drainage measurements and periodic freeness laboratory tests provide feedback information on actual freenesses being obtained. When the actual values deviate from the desired values, the computer corrects the error by calculating and executing a change in a computed intermediate variable, which can be horsepower days per ton, kilowatts per ton, or temperature rise of the refiner. This system will also provide feedforward control over disturbances resulting from changes in stock consistency, composition, and flow. The periodic refined stock freeness tests are used by the computer to update the relationship of horsepower days per ton, kilowatts per ton, or ΔT to freeness change. As the varying characteristics of the unrefined fibers are reflected in the refined stock freeness, the intermediate variable of the refiner is adjusted by the computer to return the actual freeness to the desired value.

Stock Proportioning

A computer scheme to proportion dry weights of pulps, chemicals, and additives is shown in Figure 16-3. The weight of these materials is calculated from on-line measurements of all flows, consistency measurements, and periodic off-line measurements of the solid content of additives. The system functions like a flow-ratio controller, whose set points are adjusted by the computer to compensate for

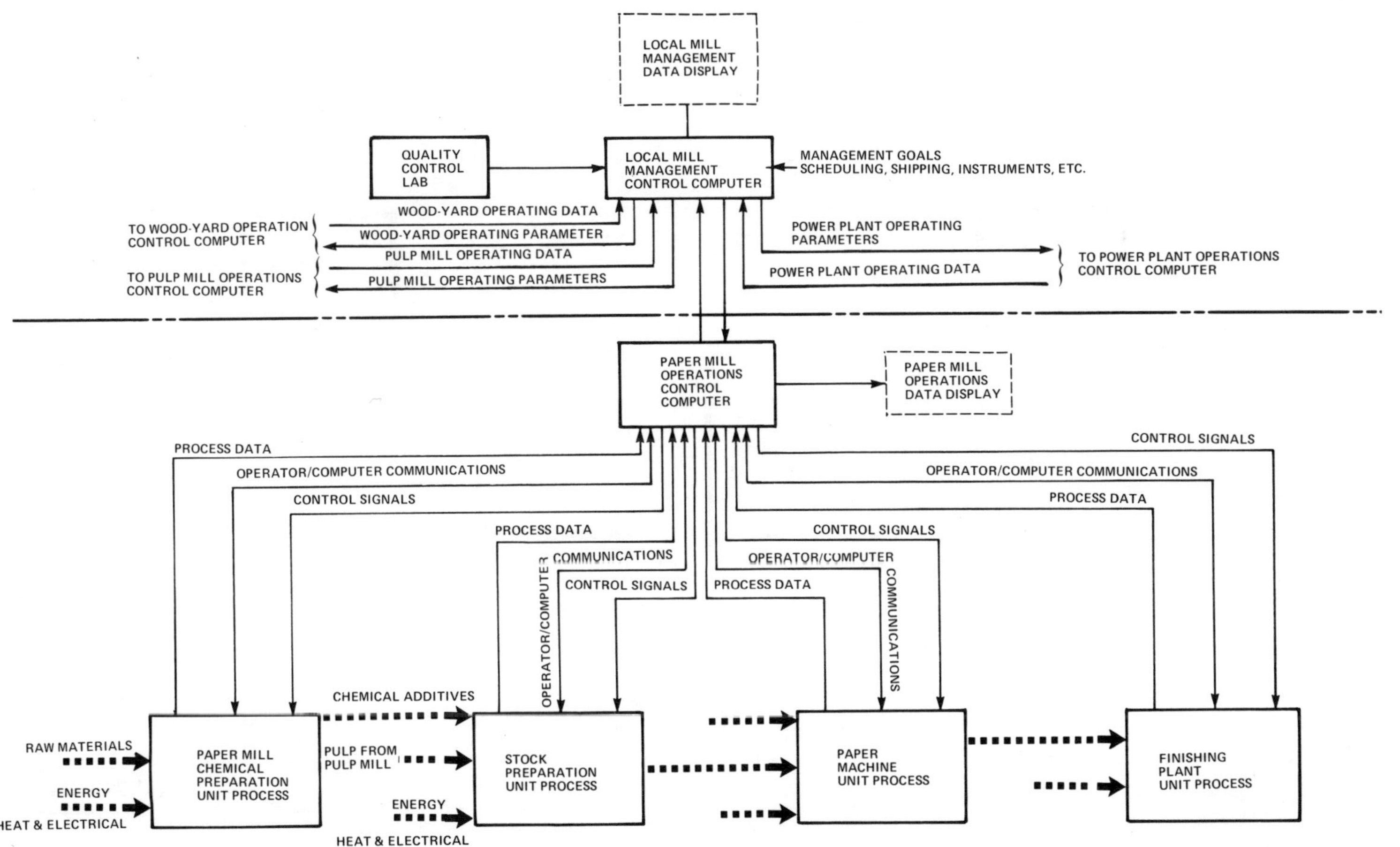

Figure 16-1. A control system for paper mill supervision and operations.

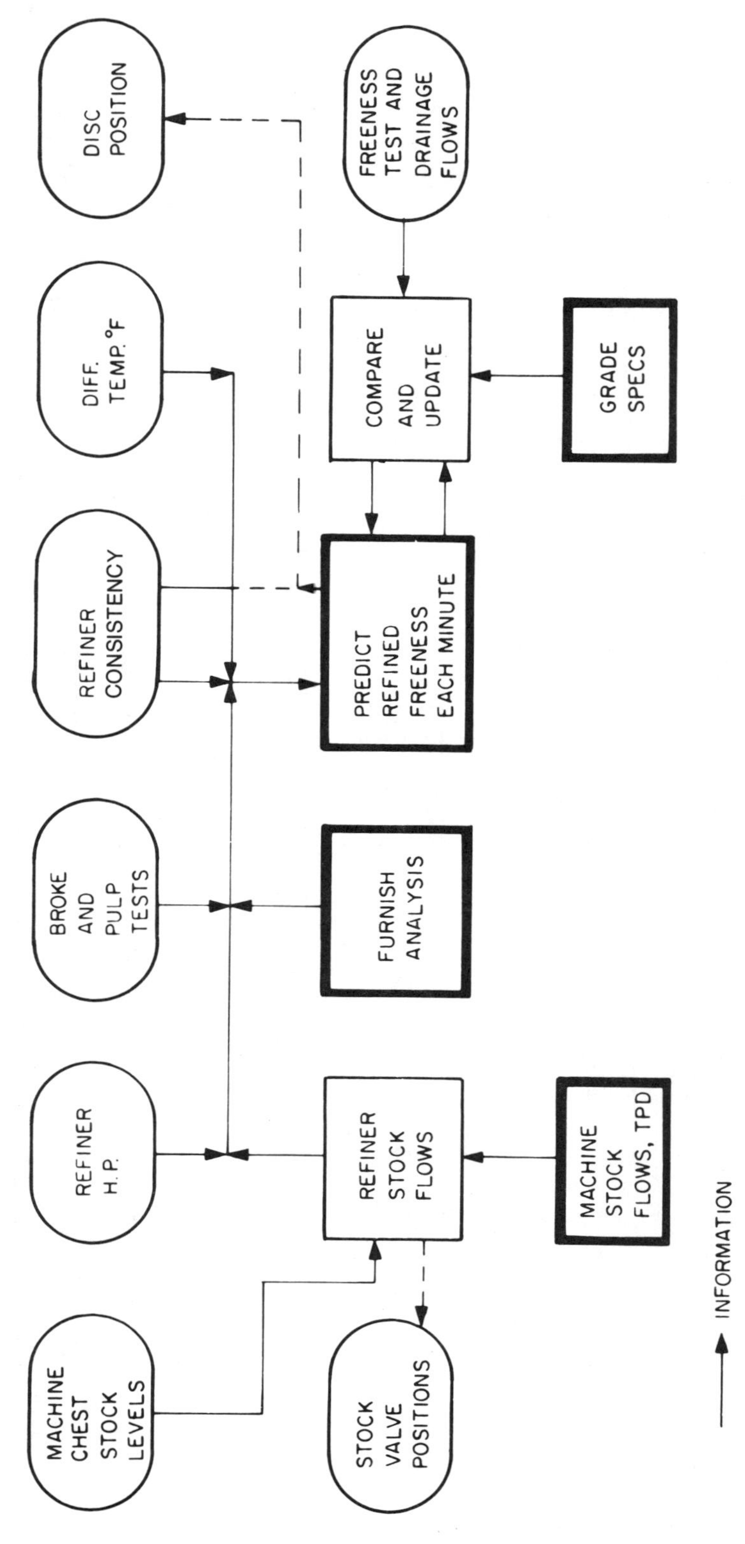

Figure 16-2. Computer control of refiner freeness.

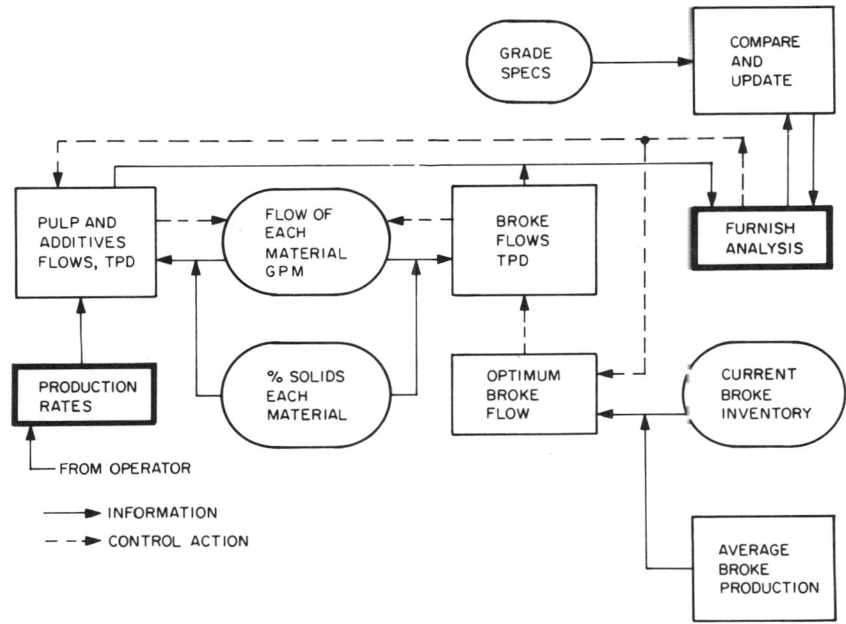

Figure 16-3. Computer control of stock proportioning.

variations in percent solids and consistencies of materials being metered. The computer can also provide the blend of all stock components on a bone-dry basis. Consistency measurements can be compensated if the influence of other variables, such as degree of refining, temperature, and flow, are known. Using compensated consistency and flow, the bone-dry tons per hour flow of each component can be determined and controlled to produce the desired bone-dry fraction being blended in accordance with mill demand requirements.

Totalizing and Logging. During the stock proportioning phase of the stock preparation process, the usual procedure is to integrate and totalize most flows on electromechanical counters for inventory control and cost accounting purposes. Usually included are the flow of stock to individual paper machines and the total flow, in addition to chemical additive and dyes.

Conventional operating practices involve an operator who reads the counters periodically, usually once a day. These readings, together with other pertinent information relative to grades of paper run on each machine, are logged and reported. The data are used in the computation of inventory and costing information. In simple installations (with few machines, furnishes, and infrequent grade changes), this operation can be performed with sufficient speed and accuracy. The complexity increases, however, with an increase in the number of paper machines involved. Data compilation in large, multimachine paper mills requires more people; the possibility of logging inaccuracy is thereby increased. Also, much more time is required to manually collect the data.

A typical computer system, configured as shown in Figure 16-4, is used to automatically perform integrating, logging, and other data acquisition and/or reporting functions on a number of stock proportioning operations.

Correction factors for stock consistency and engineering units used in the integration of flows are continuously calculated and applied by the computer. The data acquisition includes these factors which are derived and applied to

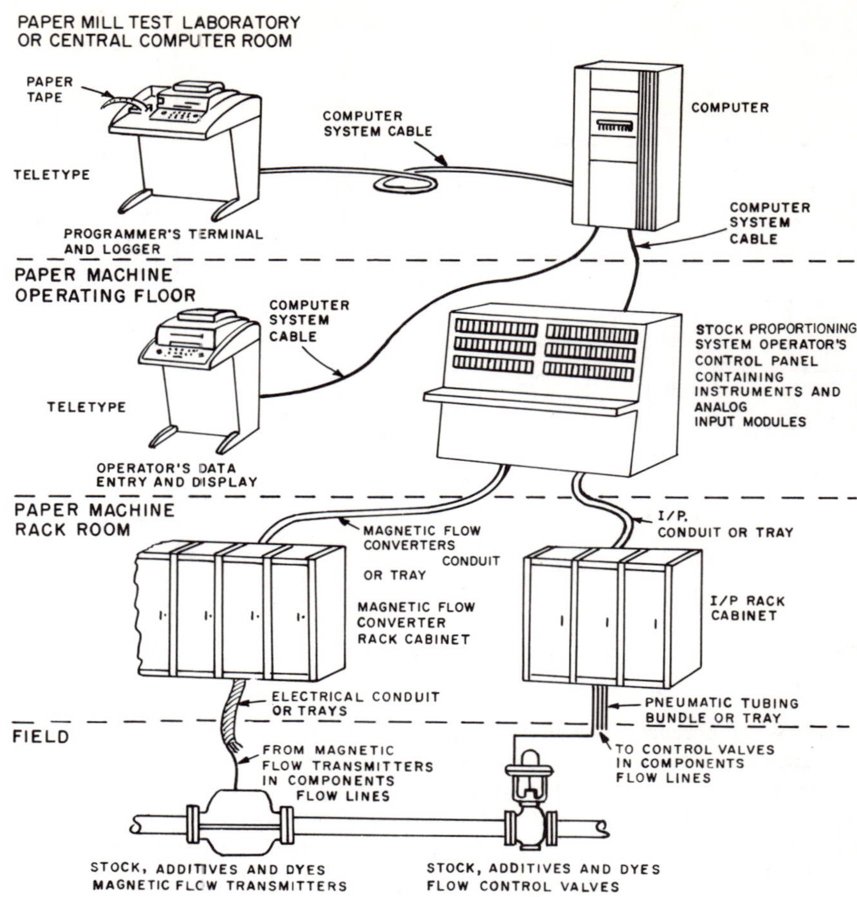

Figure 16-4. A typical computer system applied to stock proportioning.

standardize totalization units and to make adjustments for consistency variations.

Grade reports are also automatically generated, as are compilations of historical record files. These are periodically logged and summarized in both engineering and cost figures, which are used to quickly supply both financial and operating information on individual grades of paper.

PAPER MACHINE

The first applications for computer control in the pulp and paper industry were on the paper machine. This is because, initially, the wet end of the fourdrinier machine was considered as the last operation in the manufacture of paper where the greatest influence could be applied to affect the properties of the final paper product. Logically, this concept extended itself to include the dry end of the paper machine. Paper machine configurations vary from mill to mill, and range of grades manufactured also differs, requiring different operating economics. Utilization of the computer concept can only be determined for a specific machine under consideration. However, there are some general control functions which can be applied to practically all paper machines.

A computer can be used to compute headbox consistency around the wet end of the paper machine, for which there is presently no practical measurement device. It does this by continually computing material balance equations using measurements of the thick stock flow to the paper machine, thin stock flow, additive flow, and thick stock consistency. Freeness measurements can also be made using the same material balance strategy around the wet end.

Rush/Drag

The computer can change sheet formation by adjusting the ratio between wire speed and the speed at which the thin stock is discharged from the slice (called *rush/drag* or *efflux ratio*) and headbox consistency. This is done by having the computer compute the correct headbox pressure value from the wire speed and by controlling the rush/drag at a ratio to maintain the proper relationship required. If a change in consistency is also required, the computer is instructed accordingly and it adjusts the various flows, slice, and headbox pressure in correct time relationship without disturbing the basis weight.

One reported method to accomplish rush/drag control with a computer is schematically depicted in Figure 16-5. The target rush/drag ratio is selected by the operator. A tachometer measures wire speed and the required headbox total head is continuously calculated. Actual head is maintained at the target value by adjustment of a steam flow valve. Other methods accomplish this by manipulating fan pump speed or a streamflow bypass valve. Headbox level is shown being controlled by a local analog controller.

In this case, headbox consistency is controlled in a feedforward manner. A calculated consistency is displayed, based on dry weight at the reel, slice area, jet velocity, and an assumed value for retention.

The capability of incorporating a dry-line adjustment effect is shown as optional in Figure 16.5.

Basis Weight and Moisture

Basis weight of the paper can be controlled by the computer by first setting up standard running conditions for the specific grade being made and then trimming these by feedback from an on-line basis weight measuring gauge. If an error is observed by the computer, the measurement is used to predict what the basis weight would be at a suitable given future moment. A control adjustment is made from this to compensate for the error by taking into account the process dynamics. Values of constants used in the calculations are continually updated. Paper moisture can be controlled by a similar method and interactive effects can be compensated for.

Figure 16-6 is a simplified schematic of a computer concept of controlling paper basis weight and moisture on the paper machine by the use of feedback and feedforward control strategies. Basis weight control regulates conditioned weight or total basis weight by feedback manipulation of the stock flow. Decoupled moisture control regulates moisture by feedback manipulation of the dryer steam pressure.

Speed Change

Changes in machine speed can also be made in a calculated, coordinated manner by a digital computer. When a speed change is required, it is ramped from one level to another, simultaneously ramping and controlling all other interrelated variables; and the computer will adjust the master speed accordingly. The

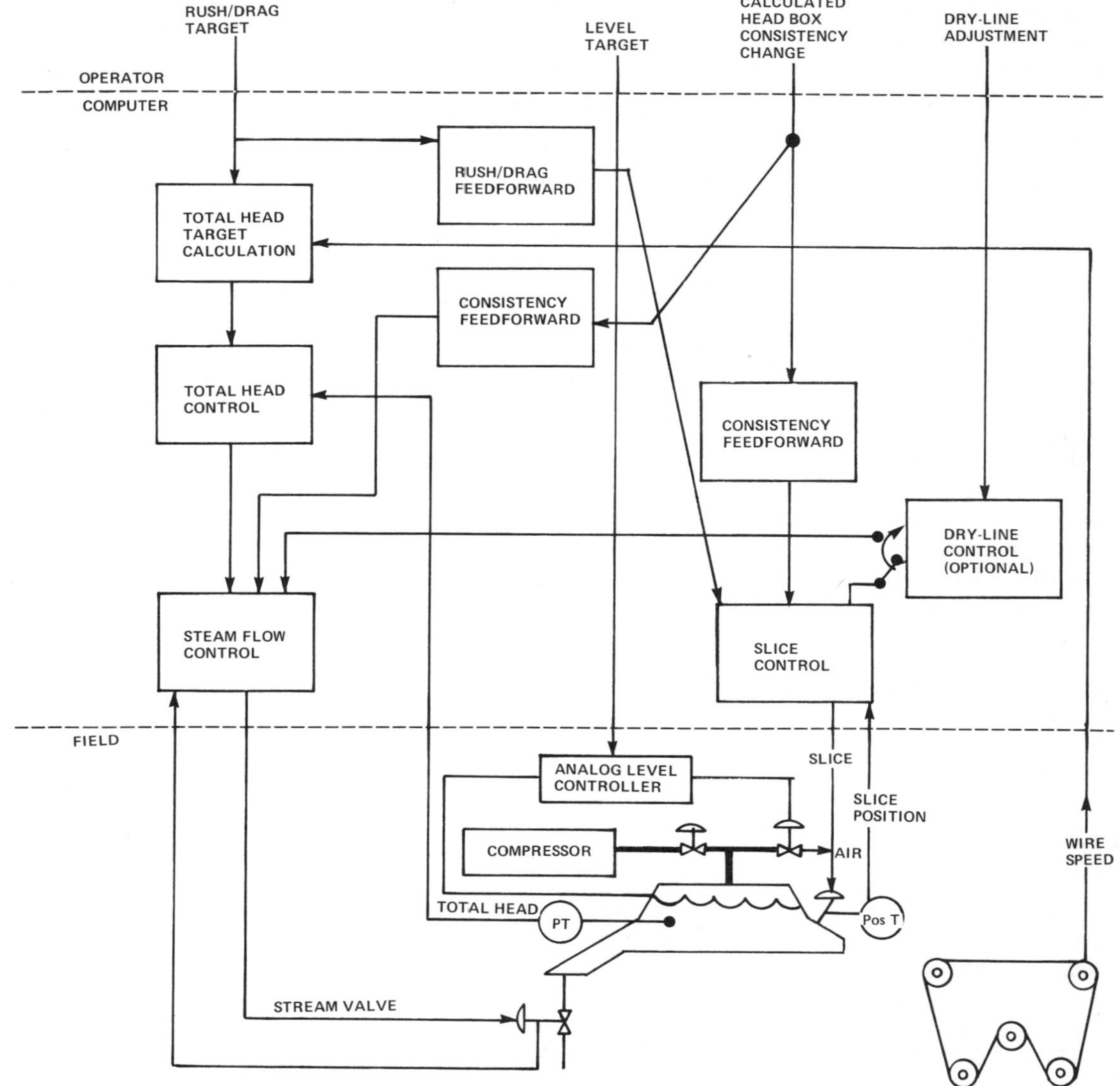

Figure 16-5. Computer control of rush/drag relationship.

computer can also be used to monitor the various speeds of the machine dryer section and calculate the percentage draws along the machine with a digital display to the operating personnel.

The computer control scheme shown in Figure 16-7 permits the paper machine operator to change machine speed while maintaining basis weight control.

When the desired speed change is requested, the computer calculates the change in stock flow required to cancel out exactly the effect of the speed change on weight. The stock flow and speed changes are then made automatically on the machine in small steps. The speed changes are delayed behind the stock flow changes to allow the stock changes to reach the wire.

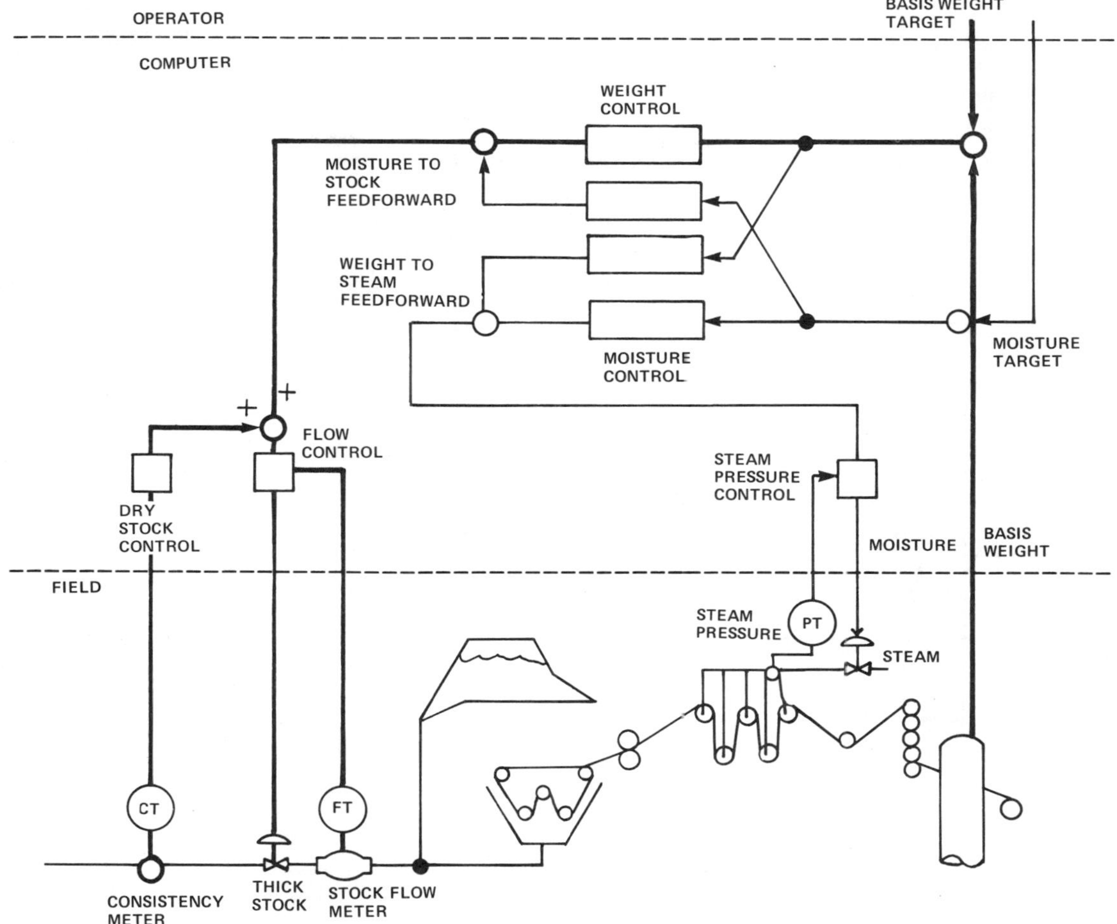

Figure 16-6. Computer control of basis weight and moisture.

Dryer Control

In order to ensure consistent operation of the dry end of the paper machine with respect to other process variables, the steam and dryer drainage system can also be brought under computer control. This will ensure that the dryer pressures are compatible with speed and basis weight under steady-state and transient conditions. To accomplish this, the computer control system must measure the dryer pressures and differentials; measure and calculate bone-dry heavy stock flow; and control the dryer section pressure and/or differential. By the use of a feedforward control scheme, an increase in dryer steam pressures is caused by an increase in heavy stock flow at a rate consistent with the transport lag between the heavy stock valve and dryer sections. Differentials are automatically set in response to changes in machine speeds, ensuring adequate differentials to evacuate dryers while limiting excessive pressure drops.

If the paper machine is dryer-limited, a computer system will allow the paper machine to operate at the maximum practical speed. The computer system will regulate the weight and moisture content by manipulation of the thick stock flow

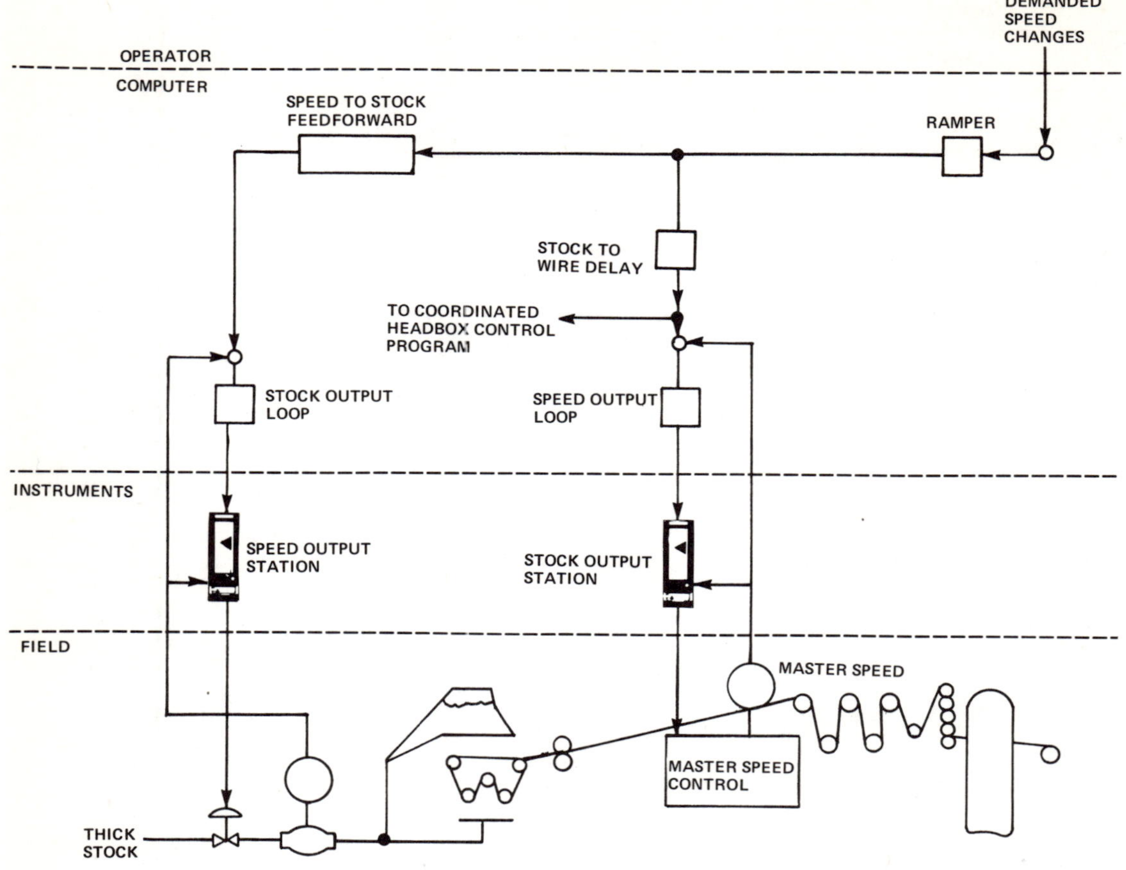

Figure 16-7. Computer control of speed change.

and machine speed, while the steam supply to the dryer is held constant at maximum level.

Coordinated Control

The coordinated computer control concept applied to the overall paper machine operation allows large and small production changes to be made in a smooth and rapid manner, with stable operation at the product target values. This concept is illustrated in Figure 16-8. It offers certain advantages over conventional manual and set point control techniques. Changes in production are conventionally done by adjusting set points or manual controls slowly to avoid upsets; or the adjustments are made rapidly and production is lost until the upsets settle. Coordinated control provides an improved way to make grade and speed changes, and avoid some of the undesirable features of conventional methods. It also offers improvement for control of basis weight and for wet end adjustments during normal production, such as adjustments in the wet end consistency (water in the sheet).

Production changes on a paper machine are typically involved operations consisting of many separate adjustments. The calculation of the correct control actions is greatly aided by the use of a process model. The success of the coordinated control scheme, in terms of both technical achievements and reasonable implementation cost, depends on the approach taken toward process modeling. One practical approach takes advantage of the "building block" or modular organization of the system.

An on-line simulation model is usually used in coordinated paper machine computer control. The simulation model is based on physical material balance equations, mixing, and transport delay. The basic building blocks in the model are made up of sets of algorithms. The parameters of the simulation model can be changed from an operator's panel, with correction of the model being done automatically as machine speeds and wet end flows change. Some of the model parameters (such as retention values) and some of the model inputs (such as dye concentrations) are set to approximate values. This, together with the other approximations in the model, gives an overall approximation of the actual process. The result is an overall control system which operates to minimize disturbances.

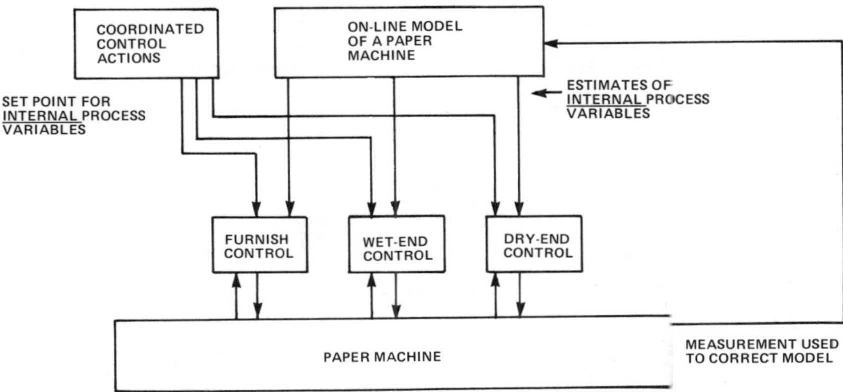

Figure 16-8. Computer-coordinated control of a paper machine.

The quality of the control will depend on the accuracy of the model. The computer provides coordination of the process inputs by providing control based on the estimated values of internal process variables. The major part of the control action is based on the on-line simulation model of the process. Feedback control is used to correct the residual errors due to model inaccuracy and unmeasured disturbances. The paper machine operators also provide some of the "feedback" in terms of adjusting the process inputs.

Grade Changing. Grade changing is a coordination control function which alters weight and moisture targets in coordination with other paper machine variables, such as speed, stock flow, steam pressure, vacuum target, and rush/drag. A typical scheme to accomplish this is shown in Figure 16-9.

In operation, a new grade code is entered into the computer and starts the grade change. The grade change control program ramps machine speed, stock flow, steam pressure, rush/drag ratio, vacuum, and other variables in a coordinated manner, providing a smooth weight and moisture transition from one grade to another.

The computer can also handle grade changes by referring to stored standard running conditions for all probable grades and then setting up the desired operating conditions automatically.

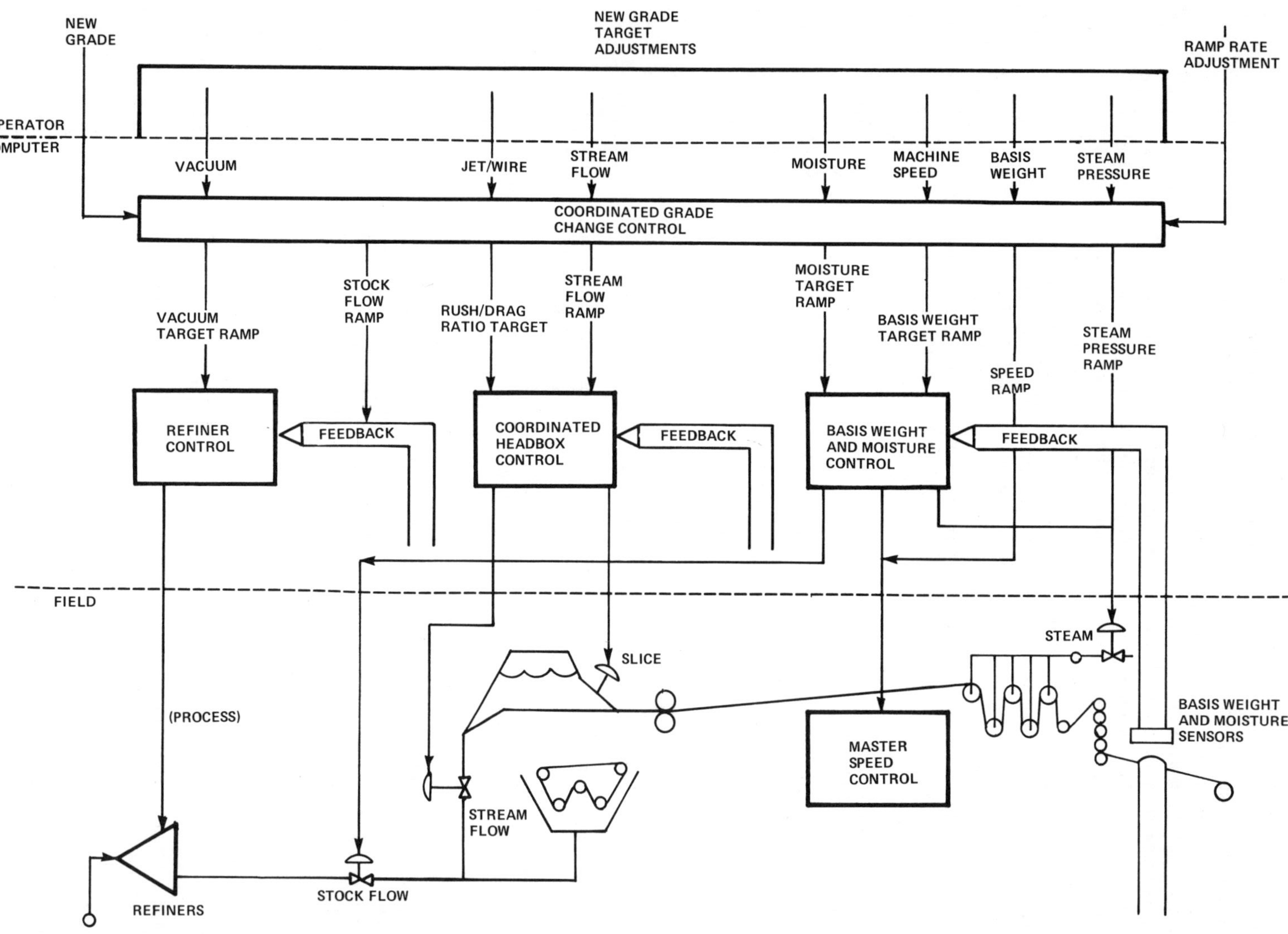

Figure 16-9. Computer control of grade change.

Other Applications

Some work has also been done in the area of using the computer in color control of the paper sheet. The measurement of color has been standardized so that a color can be specified by three factors. Thus, a computer program can be written in terms of the effects of specific dyes on different grades of paper. It is possible to feed the characteristics of available dyes, the furnish to be colored, and the final shade required into the computer and obtain a recipe for the exact match, or the nearest match with the obtainable dyes, together with an indication of the shade error. The computer can then be programmed to make the necessary adjustments to dye flows automatically or print this information out to be used as an operator's guide.

A computer-based system can be used for trimming paper machines. When applied for this purpose, it combines sophisticated and reiterative mathematical methods and the computer's speed of calculation to match a paper company's order with the machine on which the paper is to be made. This determines the best finishing roll sizes so as to keep the machine winder trim loss at a minimum, consistent with finishing process equipment utilization.

17 Computers in Energy Management

The pulp and paper industry is inherently energy intensive, requiring large amounts of thermal, electrical, and mechanical energy for pulp and paper production. Mills purchase various types of fuels and convert these energy sources to steam by the use of power boilers. Waste fuels such as sawdust and bark, as well as recoverable black and red liquors, are used as fuels to fire boilers for steam generation. Steam is used to supply the thermal requirements of the process, and to drive turbines for machinery and turbogenerators for electrical generation. In some geographic areas, hydroelectric stations are a major source of electrical energy. Electricity is also purchased either to supply a mill's total demands or to supplement internal electrical generation capacity.

Due to the worldwide energy crisis, the pulp and paper industry is faced with an increasing financial burden caused by the increased cost of purchased energy, both fuels and electricity, coupled with an ever decreasing availability of certain types of fuels. This problem is further aggravated by proposed federal government programs to impose restrictions on energy consumption by establishing energy efficiency targets for industries.

Many strategies have been and are now being implemented in the pulp and paper industry in an effort to improve the conservation of energy and reduce the cost of its production. They involve such actions as reducing steam, water, and condensate losses in existing mills; conversion of boilers from one fuel to a more economical and a more available fuel; installation of bark, sawdust, and waste-fired boilers; process modifications which are energy-saving related; and many others. Pulp and paper mill equipment manufacturers are designing new equipment and processes which include energy reduction as one of the primary goals. Mills are also looking toward improving their in-house electrical generation efficiencies and increasing generation capacity.

With respect to satisfying the thermal and electrical needs of the process, the focal point of any pulp and paper mill is the powerhouse, which includes the boilers, turbines used to generate electricity, and the distribution systems for various forms of energy required in the mill. At one time, the cost of powerhouse operation, particularly the cost of purchased fuels and electricity, was put at a priority level well below that of the first priority, that of meeting the instantaneous steam and electrical demands of the process. Presently, the cost of powerhouse operation is moving closer to the same priority as meeting steam and electrical users' requirements. Industrial energy management, computer-based systems have been designed and implemented in the pulp and paper industry in order to provide tools for powerhouse operations that are most cost-efficient and, at the same time, respond to process demands effectively through the control of the generation, distribution, and utilization of steam and electrical energy. Some such areas in which computer systems can be applied are:

1. Steam and electric energy-use accounting.
2. Multiple-boiler efficiency and performance monitoring.

3. Economic boiler dispatching.
4. Steam and electric load balancing.

ENERGY-USE ACCOUNTING

A computer program designed for steam- and electric-use accounting helps identify and reduce energy consumption by accounting for all the different forms of energy consumed by the different users throughout the entire mill. Energy-use accounting would permit each digester, bleach plant, evaporator, refiner, paper machine, etc., to be charged fairly for its consumption; would pinpoint wasteful areas for special energy savings efforts; and would show just how effectively each user has reduced its consumption. It can provide management with valuable data for evaluating new projects, selecting cost reduction improvements with the best return on investment, and making better informed decisions on operating maintenance policy. Current usage data can be used to aid operators in trimming their immediate Btu per ton energy consumption, as shown by the functional diagram in Figure 17-1.

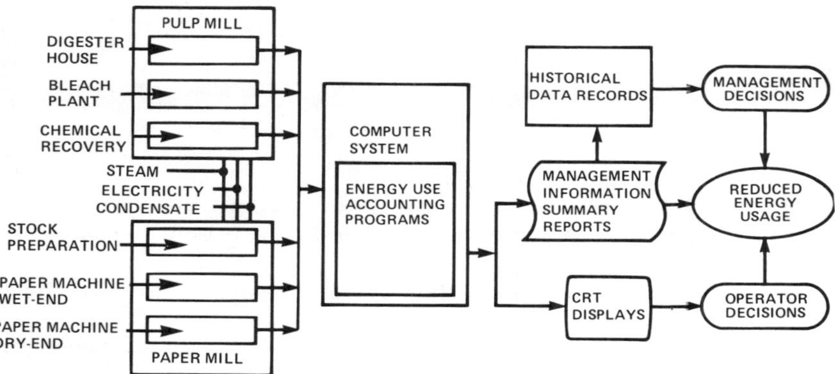

Figure 17-1. Computer system for energy-use accounting.

A typical report (illustrated in Figure 17-2) is shown as displayed on the CRT of the computer system.

MULTIPLE-BOILER EFFICIENCY AND PERFORMANCE MONITORING

A computer program for multiple-boiler efficiency and performance monitoring can optimize boiler operation by monitoring the thermodynamic efficiency and integrated performance of multiple boilers. This procedure improves steam-making efficiency and reduces fuel costs. It is accomplished through immediate notification and identification whenever below-standard conditions occur, through on-line display of current operating conditions, and through printed logs and reports. As shown in Figure 17-3, such a computer system can provide means of obtaining the necessary measurements, performing the required calculations, and producing the necessary communications.

The scheme shown in Figure 17-4 obtains measurements of the major boiler flows, temperatures, pressures, and the oxygen content of the stack gases. These are totaled and averaged periodically for logging and display purposes, and are

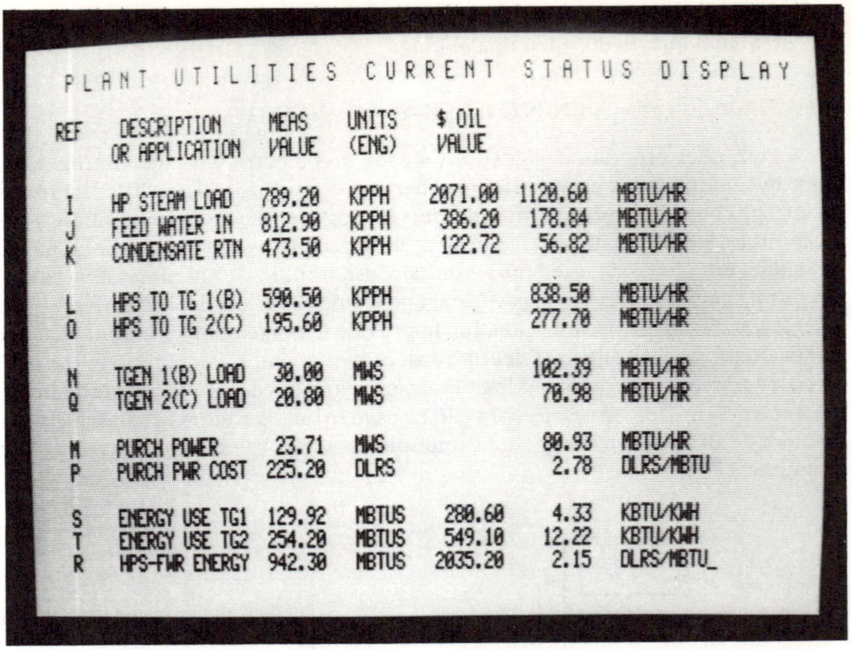

PLANT UTILITIES CURRENT STATUS DISPLAY

REF	DESCRIPTION OR APPLICATION	MEAS VALUE	UNITS (ENG)	$ OIL VALUE		
I	HP STEAM LOAD	789.20	KPPH	2071.00	1120.60	MBTU/HR
J	FEED WATER IN	812.90	KPPH	386.20	178.84	MBTU/HR
K	CONDENSATE RTN	473.50	KPPH	122.72	56.82	MBTU/HR
L	HPS TO TG 1(B)	590.50	KPPH		838.50	MBTU/HR
O	HPS TO TG 2(C)	195.60	KPPH		277.70	MBTU/HR
N	TGEN 1(B) LOAD	30.00	MWS		102.39	MBTU/HR
Q	TGEN 2(C) LOAD	20.80	MWS		70.98	MBTU/HR
M	PURCH POWER	23.71	MWS		80.93	MBTU/HR
P	PURCH PWR COST	225.20	DLRS		2.78	DLRS/MBTU
S	ENERGY USE TG1	129.92	MBTUS	280.60	4.33	KBTU/KWH
T	ENERGY USE TG2	254.20	MBTUS	549.10	12.22	KBTU/KWH
R	HPS-FWR ENERGY	942.30	MBTUS	2035.20	2.15	DLRS/MBTU_

Figure 17-2. Computer CRT display of an energy accounting report.

used to calculate boiler efficiencies and stack losses. Logs can be printed periodically for trend reporting at such intervals as hourly, shift periods, or daily, as well as on demand.

Boiler efficiency, defined as the ratio in Btu per hour of the output to the input, is computed from the enthalpies and flow rates of the steam and feedwater, and from the fuel heating value and flow rates. Stack heat losses can be computed indirectly from fuel and stack gas composition data obtained from laboratory tests, stack gas temperature, and fuel flows. Excess air losses can be computed from the fuel flows, stack gas temperature, and oxygen content.

Current operating data can be displayed at any time on the computer CRT (see Figure 17-5 for a typical display) and data are updated automatically. Special alarm indications and messages can be initiated for any programmed out-of-the-ordinary situation or condition.

Figure 17-3. Computer system for boiler efficiency and performance monitoring.

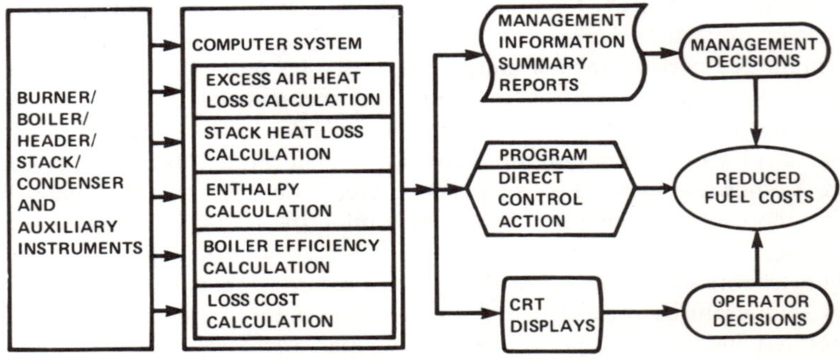

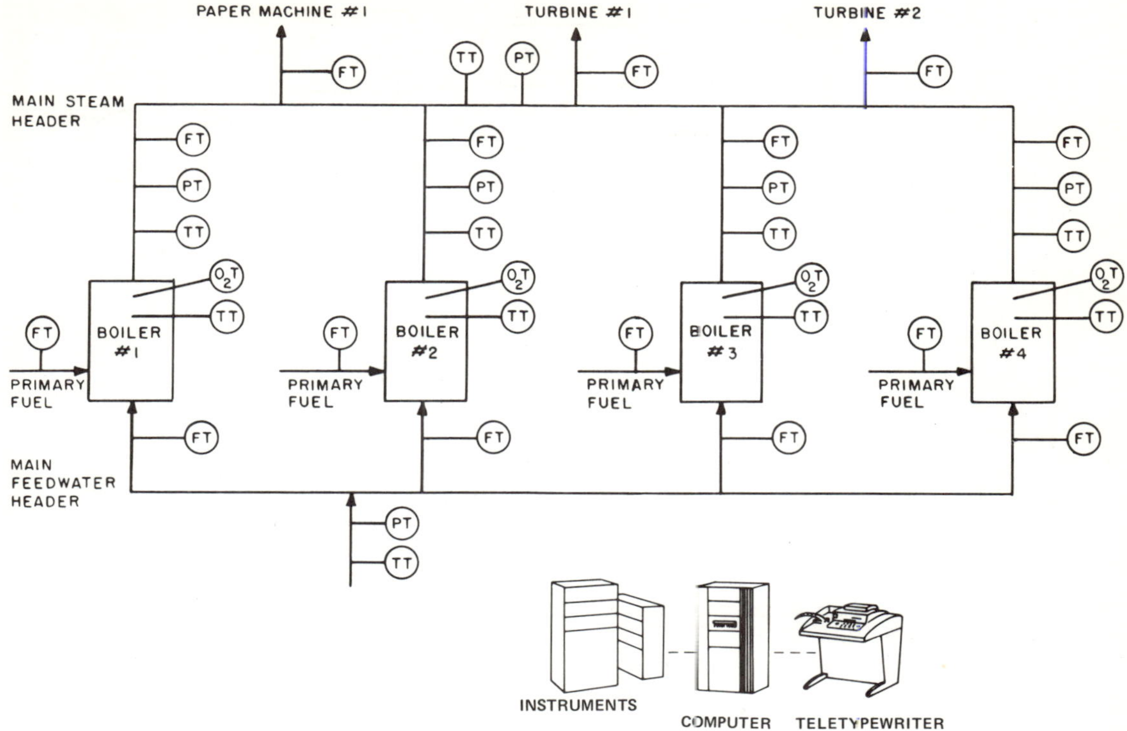

Figure 17-4. Computer applied to multiple-boiler efficiency and performance monitoring.

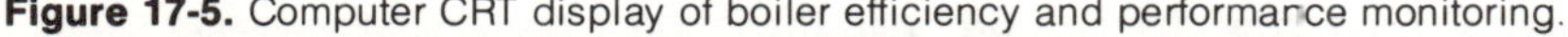

Figure 17-5. Computer CRT display of boiler efficiency and performance monitoring.

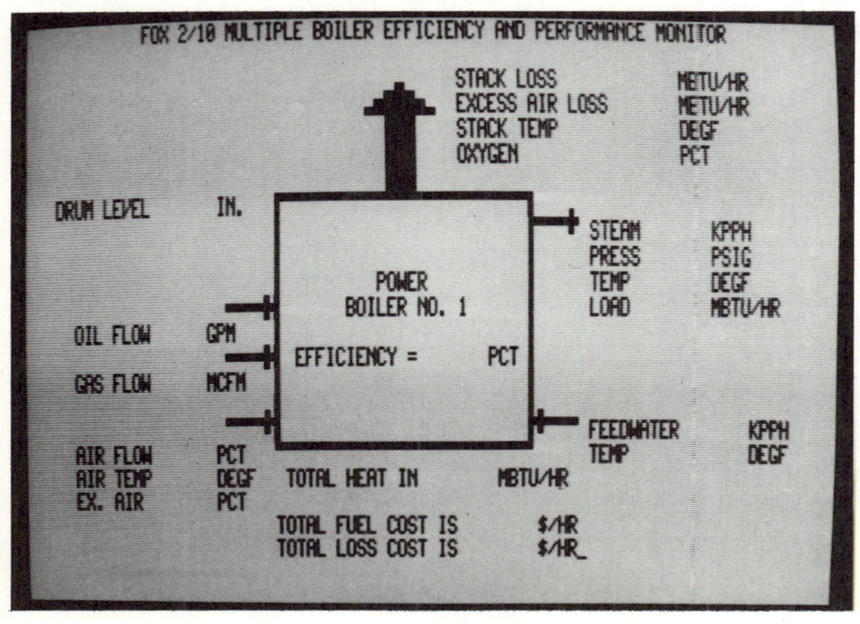

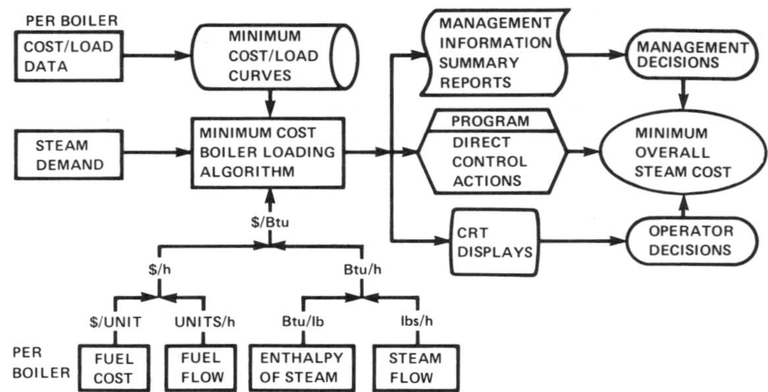

Figure 17-6. Computer system for multiple-boiler economic dispatching.

MULTIPLE-BOILER ECONOMIC DISPATCHING

A computer program for multiple-boiler economic dispatching helps minimize boiler fuel costs in mills which generate moderate to large quantities of steam from multiple-boiler systems. This can be done by a program designed to help the operator monitor and improve the fuel-to-steam energy-conversion efficiency, as shown in Figure 17-6. The heart of the function is a minimum-cost boiler loading algorithm which minimizes boiler loading cost.

Individual boiler cost versus load relationships can be considered and each boiler's economic status communicated through printed management information summary reports and a CRT display. A typical display is illustrated in Figure 17-7. Measurements and calculation values can also be displayed periodically or on demand on an operator's console and can be printed on a logging terminal.

Figure 17-7. A typical computer CRT display of multiple-boiler economic dispatching.

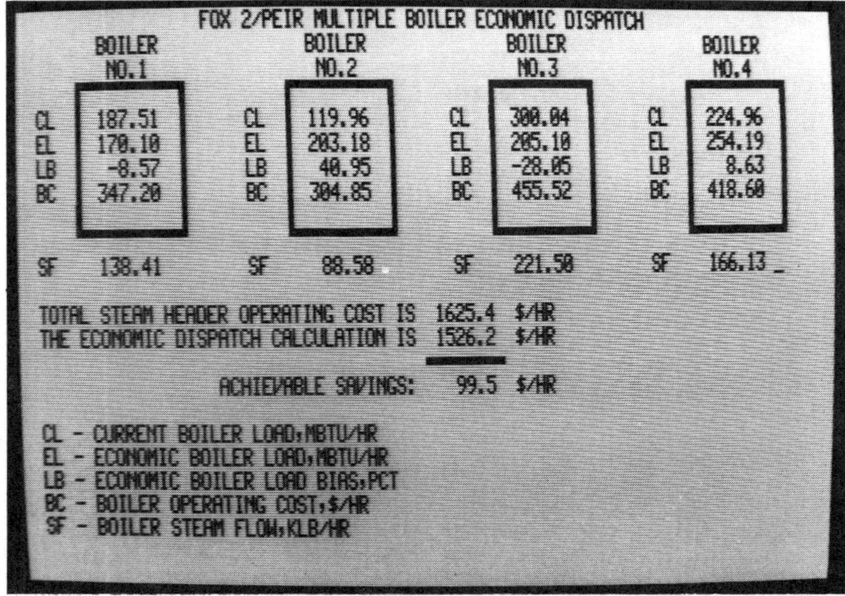

Boiler load in Btu per hour units is calculated from steam output flow, pressure, and temperature. The boiler cost in dollars per hour is derived from fuel flow and fuel cost. These values are measured, and the computer can develop a curve of boiler cost versus load for each boiler. A minimum-cost loading figure calculated for each boiler can guide the operator in adjusting the individual boiler loads. This would result in better fuel utilization, higher boiler efficiency, and lower cost per Btu of steam.

ENERGY LOAD BALANCING

When electrical and process steam demands of the mill are too great for the present electrical purchasing or generating and steam producing capabilities of the powerhouse, load balancing can be performed by a steam-electric load balancing computer program designed to monitor and balance energy usage. Such a program could assist the operator in redistribution of energy (steam and/or electricity) according to a preplanned operating philosophy.

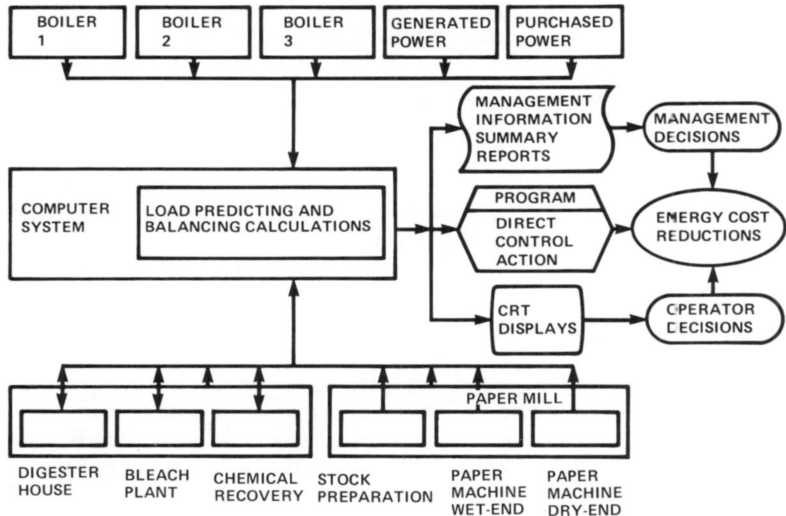

Figure 17-8. Computer system for energy load balancing.

As shown functionally in Figure 17-8, such a system can be designed to monitor the energy flows of specific devices and group them for reporting into categories by user or mill location. Each separate device and each total mill area is measured and checked for both their instantaneous rate of energy use and their predicted total energy draw over a time interval. If an energy excess is measured or predicted, the operator is alerted so that he can take corrective action.

During mill operations, the quantity of energy used is compared to the quantity of energy supplied. If the predicted use is greater than the present supply, a search for any additional energy available from the supplying devices is made and, if none is found, sheddable loads are identified and reported. This preplans suggested operator actions to future load upsets.

System measurements and set point values can be displayed on a CRT operator's console or printed on a logging terminal at any time. The values can also be reported in periodic management information summaries and on-demand logs.

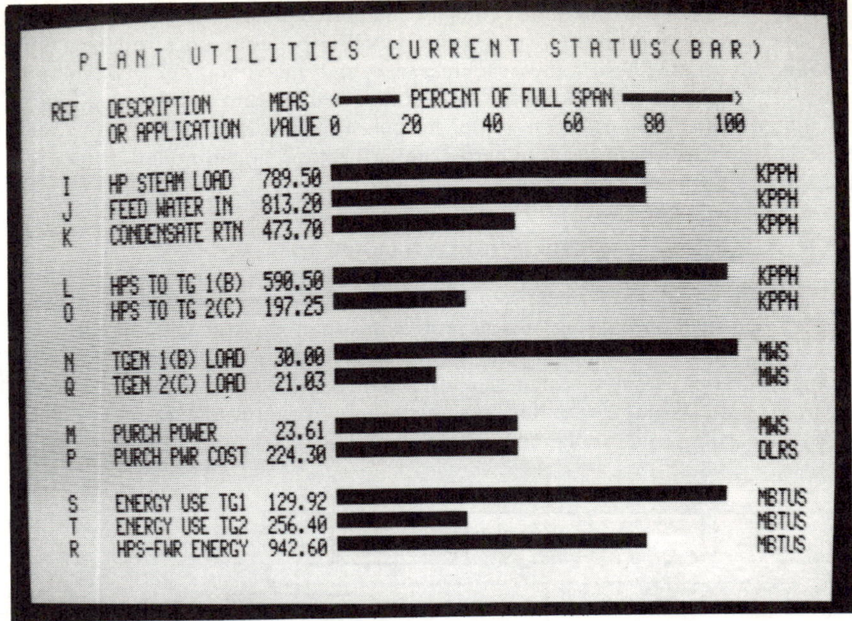

Figure 17-9. A typical computer bar graph display of utilities status.

The operator can also enter or change the equipment shedding priority in order to accommodate changes in mill equipment use.

Figure 17.9 shows a typical CRT display of utilities status in bar graph form.

OTHER ENERGY MANAGEMENT COMPUTER FUNCTIONS

Many additional programs can be designed and developed to allow computer systems to be used in a number of other ways to improve energy management efforts in the pulp and paper industry. For instance, a steam header optimization program can be implemented whose objective would be to regulate header pressures to achieve optimum usage of steam—based on minimized venting and condensing—and, at the same time, to provide user steam requirements within allowable limits.

Most mills use steam extensively to drive machinery, generate electricity, and provide process heat. Efficiency is maximized when all of the steam used for heat has already passed through turbines where work was extracted. In some instances, however, the demand for heat may require less steam than turbines are supplying, and the excess of low-pressure steam may have to be vented. More work may be obtained from the turbine as its back pressure is reduced. A heat-work balance may then be achieved by minimizing the pressure into which the turbines discharge. The pressure set point can then be reduced as much as possible and still satisfy all users.

An automatic extraction-condensing turbine evaluation function can also be incorporated. This feature monitors the performance of the turbine versus its design criteria and indicates the effect of deviation from the best extraction/condensing ratios.

Appendix A
References

Ainsworth, J. H. *Paper—The Fifth Wonder*. Kaukauna, Wis.: Thomas Printing & Publishing Co., Ltd., 1959.

Britt, K. W., ed. *Handbook of Pulp and Paper Technology*. 1st and 2nd eds. New York: Van Nostrand Reinhold Company, 1970.

Calkin, J. B., ed. *Modern Pulp and Paper Making*. 3rd ed. New York: Reinhold Publishing Corporation, 1957.

Cole, E. J., and Todd, M. *Pulp and Paper Mill Instrumentation*. New York: Lockwood Trade Journal Co., Inc., 1957.

Considine, D. M., and Ross, S. D., eds. *Handbook of Applied Instrumentation*. New York: McGraw-Hill Book Company, 1964.

Evans, J. C. W., ed. *Pulp and Paper Mill Process Instrumentation*. New York: Lockwood Publishing Co., Inc., 1969.

Fadum, O. K. *Pulp and Paper Computer-Coordinated Control Systems*. Pub 443. Foxboro, Mass.: The Foxboro Company, 1976.

Grant, Roger L. *Introduction to The Terminology, Equipment and Concepts of Computers*. New York: Lockwood Publishing Co., Inc., 1971.

Lavigne, J. R. *An Introduction to Paper Industry Instrumentation*. San Francisco: Miller Freeman Publications, Inc., 1972, rev. ed. 1977.

———. *Instrumentation for the Pulp and Paper Industry*. Pub 355B. Foxboro, Mass.: The Foxboro Company, 1975.

———. *Power and Recovery Boiler Control Systems for the Pulp and Paper Industry*. Pub. 356B. Foxboro, Mass.: The Foxboro Company, 1976.

———. *Stock Proportioning Systems for the Pulp and Paper Industry*. Pub 133C. Foxboro, Mass.: The Foxboro Company, 1976.

Libby, C. Earl, ed. *Pulp and Paper Science and Technology*, Vols. 1 & 2. New York: McGraw-Hill Book Company, 1962.

Liptak, B. G., ed. *Instrumentation in the Processing Industries*. Philadelphia: Chilton Book Company, 1973.

Lowe, K. E., ed. *Practical Computer Applications for the Pulp and Paper Industry*. San Francisco: Miller Freeman Publications, Inc., 1975.

MacDonald, R. G., ed., and Franklin, J. N., tech. ed. *Pulp and Paper Manufacture*, Vols. 1-3, 2nd ed. New York: McGraw-Hill Book Company, 1970.

Shinskey, Francis G. *Energy Conservation through Control*. New York: Academic Press, 1978.

Appendix B
Glossary of Pulp and Paper Terms

The pulp and paper industry is deeply rooted in prescientific craftsmanship. Therefore, many of the terms used in the language describing the processes are expressive and colorful. The following is not intended to be either a complete or official list of terms and definitions used in the pulp and paper industry. However, it is representative of those terms most commonly used.

ABSORBENCY. That property of a paper which causes it to take up liquids with which it is in contact.

ACTIVE ALKALI. NaOH and Na_2S content in sulfate cooking liquor expressed as Na_2O.

ACTIVITY. The active alkali content divided by the total alkali content of sulfate cooking liquor, expressed in percent:

$$\left(\frac{NaOH + Na_2S}{NaOH + Na_2S + Na_2\,CO_3 + \frac{1}{2}Na_2S}\right)\%$$

AIR BRUSH COATING. A method of coating paper on one side in which the coating mixture is applied by a metal roller and distributed by a thin, flat jet of air coming from a slot in a metal blade which extends across the machine. The air removes excess coating and smooths the surface.

AIR DRIED. The artificial drying of paper by contact with air over skeleton dryers after the paper has been surface sized.

AIR DRY. Refers to moisture-free pulp divided by factor 0.9.

ALPHA PULP. A specially processed chemical wood pulp of high alpha-cellulose content.

ALUM. Aluminum sulfate used for precipitating the rosin size onto the pulp to give water-resistant properties to the paper.

ANILINE DYES. Derivatives of coal tar, classified according to degree of brightness or fastness to light. Basic dyes have extreme brightness but are not fast to light, while acid dyes are less brilliant but are more lightfast.

APPARENT DENSITY. The weight per unit volume of the sheet. It is commonly calculated by dividing the basis weight by the caliper, although it must be recognized that the numerical value thus obtained is dependent on the definition of the ream.

APPEARANCE. The effect on the sense of sight resulting from observation of color, brightness, finish, cleanliness, formation, and other visible characteristics of paper or boards.

ASH. The inorganic residue obtained by igniting a specimen of pulp, paper, or other cellulosic material in such a way that the combustible and volatile compounds are removed. The *ash content* is the percentage of such residue based on the weight of the test specimen.

BARKING DRUM. Large rotating cylinder in which pulpwood sticks are tumbled against one another to remove the bark.

BASIS WEIGHT. The weight in pounds of a ream (either 480 or 500 sheets) of paper cut to a given size. The standard size ream varies with different grades of paper according to trade practices.

BEATER. A large mixer in which the pulp is mixed with the other ingredients of paper.

BED PLATE. The metallic plates in the floor of a beater directly beneath the roll. Pulp fibers are trapped between the bars of the roll and the bars of the bed plate, and are there subjected to a crushing, cutting, brushing action associated with beating.

BETA CELLULOSE. That portion of a pulp or other cellulosic material which dissolves in an alkaline solution.

BLACK LIQUOR. Spent cooking liquor from a sulfate digester. It contains dissolved organic wood materials and residual active alkali compounds from the cook. It is the name applied to liquors recovered from the digesters, up to the point of their incineration in the recovery plant.

BLANKS. Boards made either on a cylinder machine or in multiple ply on a fourdrinier and pasted into several plies. If uncoated, they are called *plain mill blanks;* if coated, they are called *coated blanks.*

BLEACH. A chemical used to purify and whiten paper pulps.

BLEACHING. The process of purifying and whitening pulp by chemically treating it to alter the coloring matter so that the pulp has a higher brightness.

BLISTER. A defect arising in paper when dried too suddenly on the drying cylinder or when the felts are not in good condition, leaving air between the felt and the sheet.

BLOW. Releasing the pressure and discharging the contents of a digester into a blow tank.

BLOW TANK. A tank into which pulp from the digester is blown.

BODY STOCK. The base stock or coating raw stock for plain or decorated coated papers. Also termed *base paper.*

BOGUS. A descriptive term applied to papers and paperboards manufactured principally from old papers, or inferior or low-grade stock in imitation of grades using a higher quality of raw materials.

BOND PAPER. The name *bond* was originally given to a paper which was used for printing bonds and stock certificates. It is now used in referring to paper used for letterheads and many printing purposes. Important characteristics are finish, strength, freeness from fuzz, and rigidity.

BONE DRY. Refers to moisture-free pulp.

BOOK PAPER. A class or group of papers having in common physical characteristics that, in general, are most suitable for the graphic arts, exclusive of newsprint. These physical characteristics are varied to meet the requirements of the type of impress employed and the objective use of the article produced.

BOXBOARD. Paperboard used for fabricating boxes.

BREAK. A complete tear or rupture of the web of paper or paperboard during manufacture or some subsequent operation which utilizes rolls of paper.

BREAST ROLL. A large diameter roll around which the fourdrinier wire passes at the machine headbox just at or ahead of the point where the stock is admitted to the wire by the stock inlet.

BRIGHTNESS. The reflectivity of a sheet of pulp or paper for blue light measured under standard conditions.

BROKE. Paper trimmings or damaged paper as a result of breaks from the paper machine or finishing operations. Usually returned for reprocessing.

BROWN STOCK. Unbleached pulp from a digester. It derives its name from its dark brown appearance.

BRUISING. A refining action which results in internal fibrillation.

BRUSH COATING. The process of applying a semifluid mixture of the pigment and the binder by means of a revolving cylindrical brush and smoothing the coating so applied by means of oscillating flat brushes which contact the coated sheet as it is being drawn tightly over a moving rubber apron or a revolving drum.

BRUSH FINISH. A coated paper run under stiff brushes after coating and before calendering to give a high finish.

BRUSHING. A refining action which results in external fibrillation.

BULK. The thickness of a sheet or relative thickness according to the basis or substance weight of the sheet.

BURST. The pressure in pounds required to rupture one square inch of paper under stipulated testing procedures.

BURSTING STRENGTH. Resistance of paper to rupture under pressure, as indicated in pounds per square inch on a burst or *pop* tester.

CALENDERING. Smoothing the surface of a paper sheet by applying pressure or friction.

CALENDER STACK. Steel rolls at the dry end of a paper machine which smooth and level the sheet of paper.

CALIPER. The thickness of a sheet measured under specified conditions, expressed in thousandths of an inch, millimeters, or points.

CAST COATING. Coating paper by pressing against a solid surface while the coating is in a highly plastic condition.

CAUSTICITY. The NaOH content divided by the active alkali content of sulfate cooking liquor expressed as percent:

$$\left(\frac{NaOH}{NaOH + Na_2S}\right)\%$$

CAUSTICIZING. Converting green liquor to white liquor.

CAUSTICIZING EFFICIENCY. A percentage found by dividing NaOH by NaOH + Na_2CO_3 in the causticizing process, both items being expressed as Na_2O.

CELLULOSE FIBER. The fibrous material remaining after the nonfibrous components of wood have been removed by pulping and bleaching operations.

CHEMICAL CONSUMPTION. The pounds of Na_2SO_4 added per day, divided by the tons of air-dry pulp produced per day.

CHEMICAL LOSS. A percentage found by dividing sodium sulfate (expressed as Na_2O) in new chemical by total alkali to the digesters.

CHEMICAL LOSS IN EVAPORATORS AND FURNACE. A percentage found by dividing the total alkali to the evaporators, plus sodium sulfate (expressed as Na_2O) in chemical, minus total alkali in the green liquor, by the total alkali to the digesters.

CHEMICAL LOSS IN PULP WASHING. A percentage found by dividing the total alkali to the digesters, minus the total alkali to the evaporators, by the total alkali to the digesters.

CHEMICAL LOSS IN RECAUSTICIZING AND MUD WASHING. A percentage found by dividing the total alkali in the green liquor, minus the total alkali in the white liquor, by the total alkali to the digesters.

CHEMICAL PULP. Pulp obtained by cooking wood with solutions of various chemicals.

CHEMICAL RECOVERY. A percentage found by dividing the total alkali to the digesters, minus the sodium sulfate (expressed as Na_2O) in new chemical, by the total alkali to the digesters after correcting for any change in the liquor inventory.

CHEMIGROUNDWOOD. A process in which a whole or full-sized pulpwood log is cooked with chemicals and steam prior to grinding. The resultant products, after grinding this cooked log, are called chemigroundwood.

CHIPBOARD. A paperboard used for many purposes where strength and quality are not needed. It is of relatively low density and made from waste (usually mixed) papers.

CHIPPER. Machine with rotating knives which cuts pulpwood logs to chips about 1-inch square and ⅛-inch thick.

CHLORINE DIOXIDE. A chemical (ClO_2) used in pulp bleaching in a solution of water, usually in one or more of the latter stages of a multistage sequence.

CLASSIFIER. A device for separating pulp fiber in liquid suspension according to fiber length and particle size by depositing the fractions on known mesh wire cloth for final bone-dry weighings.

CLOTHING. A term applied to paper machine felts and fourdrinier wires.

CLOUDY. A term indicating unevenness or irregular formation in look-through of paper.

COCKLING. A rippling effect given to the surface of a sheet of paper which has not been properly dried.

COLOR. Dyes added to wood pulp to give the finished product the desired color.

COMPRESSING. A refining action which results in densification of a fiber.

CONIFEROUS. A tree belonging to the Coniferae species, so-called because the tree bears cones, as in the pines and firs.

CONSISTENCY. The percentage expression of the relative weight of dry pulp in a given weight of pulp and water mixture.

CONTINUOUS DIGESTER. A process vessel for producing chemical pulp continuously.

COOK. The operation of treating any cellulosic raw material with chemicals, usually at elevated pressures and temperatures, for the purpose of removing impurities and producing pulp suitable for papermaking.

CORE. A tube on which paper is wound for shipment.

CORRUGATED BOARD. Board after is has passed through a corrugating machine.

COUCH ROLL. A roll primarily involved in dewatering and picking off, or couching, of newly formed paper web from the wire on which it was formed and partially dewatered, and in the transfer of the web to the wet press felt for further dewatering.

CROWN. The difference in diameter between the ends of a press roll or calender roll necessary to allow for deflection so that the nip pressure will be uniform over the full width of the press or calender. The increase in diameter of the middle over the ends expressed in thousandths of an inch is called the crown of the roll.

CRUSHING. A refining action resulting in permanent fiber deformation.

CURL. Tendency of a sheet of paper to coil or roll up at the edges. Caused by changes in weather or faulty drying on the paper machine.

CURLING. A refining action which permanently reduces the radius of curvature of a fiber.

CUTTER. A machine which cuts rolls of paper into predetermined sheet sizes.

CYLINDER MACHINE. One of the principal types of papermaking machine, characterized by the use of wire-covered cylinders or molds on which a web is formed. These cylinders are partially immersed and are rotated in vats which contain a dilute stock suspension.

DANDY ROLL. A skeleton cylinder covered with a woven wire cloth, or with an arrangement of fine longitudinal wires, crossed at close intervals by heavier circumferential wires. The former structure produces wave and the latter, laid paper. The dandy roll is one method of applying water marks to paper while wet. It is positioned on the top of the wire in front of the couch roll.

DECIDUOUS. A tree of a species which loses its leaves annually.

DECKER. A drum vacuum filter which is used to thicken pulp for storage and produces white water for reuse in the process.

DECKLE. The straps on the wet end of a paper machine which prevent the fibers from overflowing the sides and which determine the width of the web of paper which can be run on any given machine.

DIGESTER. The vessel used to treat pulpwood, straw, rags, or other such cellulosic materials with chemicals to produce pulp.

DIGESTING. Cooking in a digester.

DISSOLVING PULP. Special grade of chemical pulp made from wood or cotton linters for use in the manufacture of regenerated cellulose (viscose rayon and cellophane) or cellulose derivatives such as acetate, nitrate, etc.

DOCTOR. A device for keeping the surface of a roll clean by a scraping action.

DRAW. The tension applied to the paper between sections of a paper machine, such as the press section or dryer section.

DREGS. Settable solids which comprise the underflow from clarifiers in the causticizing process.

DRY END. That part of the paper machine where the paper is dried.

DRYERS. The steam-heated cylinders over which the paper in the web is passed to be dried.

EFFECTIVE ALKALI. $NaOH + \frac{1}{2} Na_2S$ content in sulfate cooking liquor expressed as Na_2O.

ELMENDORF TEST. A test which determines the tearing resistance of paper.

FELT. A woven cloth used to carry the web of paper between the press and dryer rolls on the paper machine.

FELT SIDE. The top side of the sheet when it is formed on the wire.

FESTOON DRYING. Air drying paper by washing in a single continuous web, in short festoons, or loops, on traveling poles or slats moving through a drying chamber in which the temperature and humidity are controlled.

FIBER. A unit of paper pulp which is longer than its diameter.

FIBER AXIS RATIO. Ratio of fiber width to fiber thickness.

FIBER COARSENESS. Weight per unit length of fiber.

FIBER DEBRIS. Pieces of the wall which have been separated from the body of a fiber.

FIBER FLEXIBILITY. The ability of a fiber to bend or deform.

FIBRILLATION. The external disruption of lateral bonds between surface layers of a fiber which results in partial detachment of fibrils or pieces of the outer layers of the fiber and the internal or lateral bonds between adjacent layers within a fiber.

FIBRILS. Threadlike elements of the wall of a native cellulose fiber.

FILLER. A substance added to pulp in paper manufacturing to fill the spaces between wood fibers.

FILM-COATING. Application of light, pigmented coating to provide a more uniform and smoother surface.

FILTER. Any kind of device which can be used to remove or partially remove suspended solids from a liquid.

FILTRATE. Washer and filter effluents.

FINE PAPERS. Writing papers which usually possess good pen and ink writing characteristics, such as bond, mimeo, ledger, duplicator, manifold paper.

FINES. Small particles of pulp which include fiber debris and small particles.

FINISH. A term used to describe the cutting, counting, sorting, trimming, and packing of paper. Also refers to the conditions of its surface. A high finish gives a smooth, hard surface; a low finish gives a relatively rough, toothy surface.

FLOW-ON COATING. Flowing a suspension of pigment and adhesive directly onto the wet web as it is being formed on the papermaking machine.

FOLDING ENDURANCE. A test that measures the number of double folds that can be given to a strip of paper, under specified tension, before it will break.

FORMATION. Arrangement of the fibers in a sheet of paper.

FOURDRINIER. The wire part of a type of papermaking machine where the wet sheet of paper is formed on a wire screen from liquid pulp.

FREE. Refers to pulp or paper stock which parts readily with its water, as contrasted to wet or well-hydrated stock. Also a sheet of paper which contains no groundwood.

FREENESS. The ability of a mixture of pulp and water to give up or retain water. The ease or lack of ease with which the mixture or slurry will drain is referred to as freeness or slowness.

FURNISH. The list of ingredients which make up the blend of pulp fibers and additives in the wet end mixture fed to the paper machine.

FUZZ. Fibrous projections on the surface of a sheet of paper caused by excessive suction, insufficient beating, or lack of surface sizing.

GRADE. A term applied to a paper or pulp which is ranked (or distinguished from other papers or pulps) on the basis of its use, appearance, quality, manufacturing history, raw materials, performance, or a combination of these factors.

GREEN LIQUOR. The name applied to liquor made by dissolving the recovered chemicals (smelt) in water and weak liquor preparatory to causticizing.

GROUNDWOOD PAPERS. A general term applied to a variety of papers made with substantial proportions of mechanical wood pulp together with bleached or unbleached chemical wood pulps (generally sulfite), or a combination of these, and used mainly for printing and converting purposes.

GUILLOTINE. An instrument for trimming paper with a downward cutting action similar in operation to the guillotine used in France.

HARD-SIZED. A type of paper which has been treated with considerable size to resist moisture penetration.

HEADBOX. This is a device for maintaining a constant head to a piece of equipment such as a pump, screen, paper machine, etc. The level is maintained by throttling incoming flow or by continuous overflow from the box as a recirculated item. The first wet end component of a fourdrinier machine. It distributes the furnish over the moving wire screen.

HUMIDITY. The moisture condition of the air. Actual humidity is the number of grains of moisture in the air at any given time. Relative humidity is the percent of moisture relative to the maximum moisture which air at any given temperature can retain without precipitation.

HYDRATION. To swell the fibers and then partially break them down into fibrillae and mucilage or gel; any process for altering cellulose fibers to increase their ability to absorb water.

HYGROEXPANSIVITY. That property of a material which causes it to expand or contract when its moisture content is changed (as when the relative humidity of the surrounding atmosphere is changed).

HYGROSCOPIC. That property of paper or other substance which makes it prone to absorb water.

IMBIBITION. Sorption of water by fiber without increase in volume.

INDENTED. A paper or paperboard with raised knobs developed or formed into the sheet in the primary process.

JOB LOT. Paper produced in excess of an order or small lots of discontinued lines; or paper rejected because of defects or failure to conform to specifications, or paper which, although of standard quality at the time of manufacture, has become nonstandard because of a change in standards subsequent to manufacture.

JORDAN. A type of refiner whose working elements consist of a conical plug rotating in a matching conical shell. The outside of the plug and the inside of the shell are furnished with knives or bars commonly called *tackle*.

JUMBO ROLL. A roll of paper with a large diameter, usually over 9 inches.

KAPPA NUMBER. The permanganate number of a pulp measured under carefully controlled conditions and corrected to be the equivalent of 50 percent consumption of the permanganate solution in contact with the specimen. It gives the degree of delignification of pulps through a wider range than does the permanganate number. *See also* Permanganate number.

KINKING. A refining action producing fibers containing an abrupt change in the radius of curvature.

KNOTTERS. Special pressure screens separating knots and uncooked wood materials from the blow tank brown stock.

KRAFT. The sulfate or alkaline process. It means *strength* in German.

KRAFT PULP. A strong pulp made largely from pine by the sulfate process.

LAP MACHINE. A machine to produce sheets or laps of pulp for storage.

LIGNIN. The interfiber bonding material in wood, and it is the material separated from cellulose fibers in the cooking and bleaching process.

MACHINE DIRECTION. The direction of paper parallel to its forward movement on the paper machine. The direction at right angles to this is called the cross direction.

MACHINE DRIED. The process of drying paper on the paper machine by passing the damp sheet or web over steam-heated cylinders or drums.

MACHINE FINISH. The finish applied on the paper machine, commonly called M/F.

MACHINE GLAZED. The finish produced in glaze on the wire side of a sheet as it is passed in contact over a single, large diameter, steam-heated cylinder on the Yankee machine. The finish is commonly referred to as M/G.

MACHINE WIRE. The continuous, copper meshed wire which is the traveling surface on which the web of paper is formed. It is usually referred to as the *wire*.

MERCAPTAN. Ill-smelling substances formed in cooking wood by the sulfate process (C_2H_5SH).

MOISTURE CONTENT. The percent of moisture found in finished paper. The amount varies according to atmospheric conditions because paper may either absorb or emit moisture.

MULLEN. Mullen or burst is the pounds per square inch of pressure required to rupture the sheet. It is normally reported as point calculated as follows: Bursting ratio = bursting strength x 100 x weight in pounds (500-25x40) ream.

MULLEN TESTER. Equipment which tests burst strength.

NEWSPRINT. A generic term used to describe paper of the type generally used in the publication of newspapers.

NIP. The point of contact between two rolls, as in a calender stack.

OPACIMETER. The instrument with which the degree of opacity may be measured.

OPACITY. That property of a paper which prevents show-through of dark printing, etc., on or in contact with the backside of the sheet.

PAPER. A homogeneous formation of primarily cellulose fibers which are formed in water suspension on the machine wire and bound together by weaving of the fibers and by bonding agents.

PAPERBOARD. A thick, stiff type of paper made of wood fiber, straw, or waste paper, or a combination of these materials. Common kinds are boxes, milk cartons, etc.

PAPER CLAY. A natural white, low-free silica substance with high retention, color, and suspending properties used for filling or coating paper.

PAPER MACHINE. The machine on which the fibers and other components of papers are formed, pressed, dried, calendered, wound on reels, slit into appropriate widths, and wound into rolls or cut into sheets in certain cases.

PERMANGANATE NUMBER (K NUMBER). A measure of the amount of lignin left in chemical pulp after digestion, determined by laboratory tests on a sample. The more lignin there is in the pulp, the larger the K number. *See also* Kappa number.

pH VALUE. The degree of acidity or alkalinity measured on a scale from 0 to 14, with 7 as a neutral point. From 0 to 7 is acid; 7 to 14 is alkaline. Numerous instruments are available for measuring the pH value.

PIGMENT. An insoluble powder material of mineral and inorganic compounds.

PINHOLES. Imperfections in paper which appear as minute holes when looking through the sheet, which originate from foreign particles that are pressed through the sheet.

PITCH. In the paper industry, pitch is the material (largely a mixture of fatty and resin acids and unsaponifiable organic substances) that can be extracted from wood, mechanical, or chemical pulps, by means of organic solvent.

PLY. The separate webs which make up the sheet formed on a multicylinder machine. Each cylinder adds one web or ply which is pressed to the other, the plys adhering firmly upon drying.

POINT. A term used for an impression of thickness of a sheet of paper in one-thousandths of an inch (0.001 in.).

POP TEST. Slang term for bursting test, originating from the popping sound when the paper bursts.

POROSITY. A test which measures the time required for a given amount of air to flow through a sheet of paper. This test is a measure of how closely the fibers are compacted and bonded together. A paper of high porosity is one that is quite dense and relatively nonporous. This test is made with a densometer.

PULP. The crude fibrous material produced either chemically or mechanically from fibrous wood cellulose and used as raw material for papermaking.

PULPING. The manufacturing process for transforming pulpwood, rags, and other raw material into a pulp which, after suitable beating, refining, bleaching, etc., can be made into paper or paperboard.

RATIO OF ACTIVE ALKALI TO WOOD. The ratio of active chemical in the liquor to dry wood charged into the digester. In the sulfate process, the chemical is sodium hydroxide and sodium sulfide.

RATTLE. The crisp, crackling sound produced by shaking or crumpling a sheet of paper to indicate its rigidity or stiffness.

REAM. A number of sheets of paper, either 480 or 500, according to grade.

REAM WEIGHT. The amount which one ream of paper weighs.

REDUCTION. A percentage based on the analysis of green liquor and obtained by dividing Na_2S by Na_2SO_4 + Na_2S, all expressed as Na_2O.

REFINER. A device which separates fibers from one another or cuts fibers in shorter lengths by an action in which the pulp passes between the discs centrifugally to the periphery.

REFINING. The application of energy to papermaking fibers which causes changes in their physical characteristics via mechanical and/or hydraulic means.

REFRACTIVE. Ability to bend a light from a straight course. Materials differ in refractivity, which is measured as refractive index.

REWINDER. A machine which takes rolls from the winder, slits or rewinds them into smaller rolls.

RIGIDITY. Stiffness, resistance to bending, inflexibility.

RIPPLE FINISH. A dimpled or rippling effect given to the surface of paper by means of an embossing roll.

ROLL COATING. A process in which the coating is applied by rolls and subsequently smoothed by means of reverse rolls contacting the freshly coated surface.

ROSIN. A natural resin obtained after distilling the turpentine from the gum of the southern pine. When cooked with caustic soda or soda ash, it forms a soap that is used for the sizing or waterproofing of paper.

SALT CAKE. A makeup chemical added to black liquor. Chemically, it is sodium sulfate (Na_2SO_4).

SAVE-ALL. A device which screens paper fibers from water. Used to save pulp which might otherwise be lost to the sewer.

SCREEN. A device used to separate undesirable objects from pulp.

SCREENING. The operation of passing chips over screens to remove sawdust, slivers, and oversize chips, and passing pulp or paper stock through a screen to reject coarse fibers, slivers, shives, knots, etc.

SCREENINGS. The knots of pulp and foreign matter which are screened from pulp.

SEMICHEMICAL PULP. A pulp produced by a mild chemical treatment of the raw material followed by a mechanical fiberizing operation.

SHAVINGS. The strips of paper cut from the sheet or roll in trimmer, rewinder, or cutter.

SHEARING. A refining action which results in two contiguous parts of the fiber sliding relative to each other, in a direction parallel to their plane of contact.

SHEET ROLL. A roll of paper that is to be finished into sheets on finishing room equipment.

SHIVE. A bundle of incompletely separated fibers which may appear in the finished sheet as an imperfection.

SHIVES. Uncooked wood particles which show up in the finished sheet.

SHORTENING. A refining action causing a physical separation of a fiber across the axis into two or more pieces.

SIDE ROLL. The roll of paper run in conjunction with a regular run in order to complete the fill of a paper machine.

SIZE. A water-resisting material which is added to paper.

SIZE PRESS. Section of paper machine where surface treatments, such as starch solution, are applied to the sheet of paper to give it special qualities.

SKID. A platform, usually wooden, on which sheet paper is packed for shipment.

SLAKING. The mixing of green liquor and lime (CaO) prior to the causticizing reaction.

SLIME. An aggregation of heterogeneous material, sometimes having a slippery feel, found at various points within a pulp or papermaking system.

SLIME SPOTS. Semitransparent spots produced in paper caused by microorganisms which have grown in one of the many wet spots around the wet end of a paper machine.

SLITTER. A sharp disc which cuts paper into predetermined widths.

SLUSH. A mixture of pulp fibers and water.

SLUSH PULP. Wood fibers mixed with water so that the material can be pumped.

SMELT. The molten ash formed in the bed of the recovery furnace as a result of burning concentrated black liquor, consisting mainly of Na_2CO_3 and Na_2S, which is mixed with weak wash liquor in the dissolving tank to form green liquor.

SMOOTHNESS. The texture of the surface of the paper, also called its *finish*.

SNOWSTORM. Describes the appearance of paper with a wild, nonuniform formation.

SODA PROCESS. The process of making soda pulp. The principal chemical used in cooking liquor is sodium hydroxide, derived from sodium carbonate or soda ash.

SODA PULP. A chemical pulp produced by high-temperature digestion using sodium hydroxide.

SOFTWOODS. Cone-bearing trees. They usually keep their leaves or needles the year round and are popularly called evergreens.

SORTING. The operation of examining sheets of paper, generally on both sides, to remove those that are not salable as perfect paper.

SPECIES. Wood from a distinct type of tree or a group of trees having one or more characteristics in common that distinguish them from another tree or group of trees.

SPECIFIC SURFACE. Surface area of fibers per unit weight of dry fiber (cm^2/gm).

SPECIFIC VOLUME. Volume of fibers per unit weight of dry fiber (cc/gm).

SPENT LIQUORS. Chemical solutions which have been used in pulp manufacturing and, after proper treatment, are used again on fresh batches of wood chips.

SPLICE. The joining of the ends of paper to make a continuous roll.

SPLITTING. A refining action causing a physical separation of a fiber parallel to the axis.

STARCH. Material which is used as a sizing for paper.

STATIC ELECTRICITY. Charges of electricity which may be contained by paper which has been improperly dried or has sustained excessive pressure in calendering. It can also exist in properly dried paper which has become affected by local atmospheric conditions after shipment.

STOCK. Pulp which has been beaten and refined, treated with sizing, color, filler, etc., and which after dilution is ready to be formed into a sheet of paper.

STOCK PREPARATION. A term for the several operations which occur between pulping (or bleaching) and formation of the web on a paper machine.

STRETCH. The elongation of a strip of paper when subjected to a tensile pull. The term is colloquially but incorrectly used to indicate expansion caused by moisture absorption of a sheet of paper.

STUFF. Papermaking material or a low-density stock as it is ready for the paper machine.

SUBSTANCE. The weight in pounds of a ream (either 480 or 500 sheets) of paper cut to a given size. The standard size ream varies with different grades of paper according to trade customs.

SUCTION BOX. A device at the wet end of the machine located under the paper machine wire, which removes water from the paper web by means of a vacuum pump.

SULFATE. Alkaline process of cooking pulp. More often referred to as the kraft process. Also, pulp cooked by this process.

SULFATE PROCESS. The process of making sulfate pulp; the principal chemical used in the cooking liquor is sodium sulfate (salt cake).

SULFIDITY. The Na_2S content divided by the active alkali content in sulfate cooking liquor expressed in percent:

$$\left(\frac{Na_2S}{NaOH + Na_2S}\right)\%$$

SULFITE. Acid process of cooking pulp. Also the pulp cooked by this process.

SULFITE PROCESS. The process of making sulfite pulp; the principal chemical used in the cooking liquor is calcium bisulfite.

SULFITE PULPING. The acid or neutral base pulping process.

SUPERCALENDER. A calender stack used to alter the surface properties and appearance of a paper. It is constructed on the same general principle as a calender, except that alternate chilled cast iron and soft rolls are used in the supercalender. The soft rolls are constructed of highly compressed cotton or paper. It is not part of the paper machine, whereas the calender is.

SURFACE-SIZED. Paper that has been treated with starch or other sizing material at the size press of the paper machine. This term is used interchangeably with the term *tub-sized,* although tub-size more properly refers to surface sizing applied as a separate operation where the paper is immersed in a tub of sizing (starch or glue), after which it passes between squeeze rolls and is air dried. *See also* Tub-sized.

SWEETENER. Long fiber or virgin pulp added to white water to improve its filtering properties in save-alls.

SWELLING. Increase in volume of fiber due to sorption of water.

TACK. The resistance of an ink film to being split between two surfaces, as between rollers, between plate and blanket, and between blanket and paper; stickiness.

TALL OIL. The mixture of resins, fatty acids, and other substances recovered in the skimmings of sulfate black liquor.

TEAR. The force expressed in grams required to tear a single sheet of paper through a distance of 1.69 inches using specified conditions and equipment.

TEMPERATURE OUT OF HEAT EXCHANGER (TOS). Temperature of the liquor after it has been heated to cooking temperature in external steam heaters. Expressed in degrees F.

TENSILE STRENGTH. The force parallel to the plane of the specimen required to produce failure in the specimen, given width and length under specified conditions of loading.

TEXT PAPERS. Those grades of printing papers specifically made for use in the manufacture of books.

TISSUE PAPER. A class of papers that are of gauzy texture and fairly transparent.

TITANIUM DIOXIDE. A filler made from titanium ores which has great opacifying and brightening properties and is of minute particle size.

TOTAL ALKALI. Alkali present in sulfate cooking liquor. ($NaOH + Na_2S + Na_2CO_3 + \frac{1}{2} Na_2SO_3$) expressed as Na_2O.

TOTAL CHEMICAL. All sodium salts in sulfate and cooking liquor expressed as Na_2O.

TOTAL TITRATABLE ALKALI. $NaOH + Na_2CO_3 + Na_2S$, all expressed as Na_2O.

TRANSLUCENCY. Ability to transmit light without being transparent.

TRANSLUCENTS. A soft cardboard, pasted or nonpasted, made of soda pulp with both sides clay coated.

TRIM. Indicates the maximum width of finished paper which can be made on a particular machine.

TRIMMER. Machine which cuts sheets to final size by the diagonal motion of the cutting blade or knife.

TUB-SIZED. Paper or board which has been impregnated with a sizing material (either starch or glue) in a tub or vat. *See also* surface-sized.

TUNNEL DRIER. A well-insulated sheet metal tunnel or large box through which paper or board is passed for the purpose of drying.

TURPENTINE. A mixture of terpenes, principally pinene, obtained by steam distillation of pine gum or recovered from the condensation of exhaust vapors from the cooking of softwoods by the sulfate process.

TWISTING. A torsional refining action producing fibers with a spiral form.

TWO-SIDED. A term applied to paper having large differences in color or finish on the wire or felt sides of the same sheet.

UNREDUCED SALT CAKE. Na_2SO_4 in the green liquor, expressed as Na_2SO_4 in grams per liter.

UNTRIMMED. Paper cut by the slitters with the grain and by rotary cutters across the grain on a sheeting machine.

WATERMARK. A translucent marking made in paper while it is still wet for purposes of identification of the paper.

WAVINESS. A warping effect occurring along the edges of paper, particularly across the grain of the paper, exposed to an excess of atmospheric moisture.

WEAKWASH. Overflow wash water from the lime mud washer. Recycled into the recausticizing process.

WEB. The continuous sheet of paper forming or having been formed and finished on a paper machine.

WELL-CLOSED. A term describing a closely formed or uniform distribution of fibers in a sheet of paper.

WET END. The beginning of the paper machine, comprising the headbox, wire, and press sections.

WET STRENGTH. The tensile strength of paper if it is wetted after manufacture. Wet strength is increased by adding certain synthetic resins to the furnish.

WETTING. An action which increases water physically sorbed by fibers.

WHITE LIQUOR. The name applied to liquors which are made by causticizing green liquors and are ready for use in the digester.

WHITE WATER. Filtrate from the paper machine wet end containing short fibers (fines) and fillers.

WILD. Irregular sheet formation.

WIRE. The moving "screen" at the wet end of a paper machine where the sheet is formed.

WIRE MARK. The impression left in a web of paper by the wire of a fourdrinier machine.

WIRE SIDE. The surface of a sheet of paper which was next to the wire when formed.

YANKEE MACHINE. A type of paper machine employing a single dryer of large circumference with a highly polished surface.

YIELD. The ratio of dry output pulp to dry input wood, expressed in percent.

Appendix C
Units and Conversion Tables

These conversion factors are among those found most useful to pulp and paper industry personnel involved with instrumentation. English units of measurement are still in common use in pulp and paper mills in the United States, particularly for nonelectrical quantities. However, the movement toward use of the International System of Units, commonly referred to as SI units (for Système International d'Unités) or metric system, is becoming rapid. Hence, these tables have been arranged to facilitate conversions into SI units by placing these conversions first, followed by the other useful conversions.

ATMOSPHERES — atm (Standard at sea-level pressure)
- x 101.325 = Kilopascals (kPa) absolute
- x 14.696 = Pounds-force per square inch absolute (psia)
- x 76.00 = Centimetres of mercury (cmHg) at 0°C
- x 29.92 = Inches of mercury (inHg) at 0°C
- x 33.96 = Feet of water (ftH_2O) at 68°F
- x 1.01325 = Bars (bar) absolute
- x 1.0332 = Kilograms-force per square centimetre (kg/cm^2) absolute
- x 1.0581 = Tons-force per square foot ($tonf/ft^2$) absolute
- x 760 = Torr (torr) (= mmHg at 0°C)

BARRELS, LIQUID, U.S. — bbl
- x 0.11924 = Cubic metres (m^3)
- x 31.5 = U.S. gallons (U.S. gal) liquid

BARRELS, PETROLEUM — bbl
- x 0.15899 = Cubic metres (m^3)
- x 42 = U.S. gallons (U.S. gal) oil

BARS — bar
- x 100 = Kilopascals (kPa)
- x 14.504 = Pounds-force per square inch (psi)
- x 33.52 = Feet of water (ftH_2O) at 68°F
- x 29.53 = Inches of mercury (inHg) at 0°C
- x 1.0197 = Kilograms-force per square centimetre (kg/cm^2)
- x 0.98692 = Atmospheres (atm) sea-level standard
- x 1.0443 = Tons-force per square foot ($tonf/ft^2$)
- x 750.06 = Torr (torr) (= mmHg at 0°C)

BRITISH THERMAL UNITS — Btu (See note)
- x 1055 = Joules (J)
- x 778 = Foot-pounds-force (ft • lbf)
- x 0.252 = Kilocalories (kcal)
- x 107.6 = Kilogram-force-metres (kgf • m)
- x 2.93×10^{-4} = Kilowatt-hours (kW • h)
- x 3.93×10^{-4} = Horsepower-hours (hp • h)

BRITISH THERMAL UNITS PER MINUTE — Btu/min (See note)
- x 17.58 = Watts (W)
- x 12.97 = Foot-pounds-force per second (ft • lbf/s)
- x 0.02358 = Horsepower (hp)

CENTARES
- x 1 = Square metres (m^2)

CENTIMETRES — cm
- x 0.3937 = Inches (in)

CENTIMETRES OF MERCURY — cmHg, at 0°C
- x 1.3332 = Kilopascals (kPa)
- x 0.013332 = Bars (bar)
- x 0.4468 = Feet of water (ftH_2O) at 68°F
- x 5.362 = Inches of water (inH_2O) at 68°F
- x 0.013595 = Kilograms-force per square centimetre (kg/cm^2)
- x 27.85 = Pounds-force per square foot (lbf/ft^2)
- x 0.19337 = Pounds-force per square inch (psi)
- x 0.013158 = Atmospheres (atm) standard
- x 10 = Torr (torr) (= mmHg at 0°C)

CENTIMETRES PER SECOND — cm/s
- x 1.9685 = Feet per minute (ft/min)
- x 0.03281 = Feet per second (ft/s)
- x 0.03600 = Kilometres per hour (km/h)
- x 0.6000 = Metres per minute (m/min)
- x 0.02237 = Miles per hour (mph)

CUBIC CENTIMETRES — cm^3
- x 3.5315×10^{-5} = Cubic feet (ft^3)
- x 6.1024×10^{-2} = Cubic inches (in^3)
- x 1.308×10^{-6} = Cubic yards (yd^3)
- x 2.642×10^{-4} = U.S. gallons (U.S. gal)
- x 2.200×10^{-4} = Imperial gallons (imp gal)
- x 1.000×10^{-3} = Litres (l)

CUBIC FEET — ft^3
- x 0.02832 = Cubic metres (m^3)
- x 2.832×10^4 = Cubic centimetres (cm^3)
- x 1728 = Cubic inches (in^3)
- x 0.03704 = Cubic yards (yd^3)
- x 7.481 = U.S. gallons (U.S. gal)
- x 6.229 = Imperial gallons (imp gal)
- x 28.32 = Litres (l)

CUBIC FEET PER MINUTE — cfm
- x 472.0 = Cubic centimetres per second (cm^3/s)
- x 1.699 = Cubic metres per hour (m^3/h)
- x 0.4720 = Litres per second (l/s)
- x 0.1247 = U.S. gallons per second (U.S. gps)
- x 62.30 = Pounds of water per minute (lbH_2O/min) at 68°F

CUBIC FEET PER SECOND — cfs
- x 0.02832 = Cubic metres per second (m^3/s)
- x 1.699 = Cubic metres per minute (m^3/min)
- x 448.8 = U.S. gallons per minute (U.S. gpm)
- x 0.6463 = Million U.S. gallons per day (U.S. gpd)

CUBIC INCHES — in^3
- x 1.6387×10^{-5} = Cubic metres (m^3)
- x 16.387 = Cubic centimetres (cm^3)
- x 0.016387 = Litres (l)
- x 5.787×10^{-4} = Cubic feet (ft^3)
- x 2.143×10^{-5} = Cubic yards (yd^3)
- x 4.329×10^{-3} = U.S. gallons (U.S. gal)
- x 3.605×10^{-3} = Imperial gallons (imp gal)

CUBIC METRES — m^3
- x 1000 = Litres (l)
- x 35.315 = Cubic feet (ft^3)
- x 61.024×10^3 = Cubic inches (in^3)
- x 1.3080 = Cubic yards (yd^3)
- x 264.2 = U.S. gallons (U.S. gal)
- x 220.0 = Imperial gallons (imp gal)

CUBIC METRES PER HOUR — m^3/h
- x 0.2778 = Litres per second (l/s)
- x 2.778×10^{-4} = Cubic metres per second (m^3/s)
- x 4.403 = U.S. gallons per minute (U.S. gpm)

CUBIC METRES PER SECOND — m^3/s
- x 3600 = Cubic metres per hour (m^3/h)
- x 15.85×10^3 = U.S. gallons per minute (U.S. gpm)

CUBIC YARDS — yd^3
x 0.7646 = Cubic metres (m^3)
x 764.6 = Litres (l)
x 7.646×10^5 = Cubic centimetres (cm^3)
x 27 = Cubic feet (ft^3)
x 46,656 = Cubic inches (in^3)
x 201.97 = U.S. gallons (U.S. gal)
x 168.17 = Imperial gallons (imp gal)

DEGREES, ANGULAR (°)
x 0.017453 = Radians (rad)
x 60 = Minutes (′)
x 3600 = Seconds (″)
x 1.111 = Grade (gon)

DEGREES PER SECOND, ANGULAR (°/s)
x 0.017453 = Radians per second (rad/s)
x 0.16667 = Revolutions per minute (r/min)
x 2.7778×10^{-3} = Revolutions per second (r/s)

DRAMS (dr)
x 1.7718 = Grams (g)
x 27.344 = Grains (gr)
x 0.0625 = Ounces (oz)

FATHOMS
x 1.8288 = Metres (m)
x 6 = Feet (ft)

FEET — ft
x 0.3048 = Metres (m)
x 30.480 = Centimetres (cm)
x 12 = Inches (in)
x 0.3333 = Yards (yd)

FEET OF WATER — ftH_2O, at 68°F
x 2.984 = Kilopascals (kPa)
x 0.02984 = Bars (bar)
x 0.8811 = Inches of mercury (inHg) at 0°C
x 0.03042 = Kilograms-force per square centimetre (kg/cm^2)
x 62.32 = Pounds-force per square foot (lbf/ft^2)
x 0.4328 = Pounds-force per square inch (psi)
x 0.02945 = Standard atmospheres

FEET PER MINUTE — ft/min
x 0.5080 = Centimetres per second (cm/s)
x 0.01829 = Kilometres per hour (km/h)
x 0.3048 = Metres per minute (m/min)
x 0.016667 = Feet per second (ft/s)
x 0.01136 = Miles per hour (mph)

FEET PER SECOND PER SECOND — ft/s^2
x 0.3048 = Metres per second per second (m/s^2)
x 30.48 = Centimetres per second per second (cm/s^2)

FOOT-POUNDS-FORCE — ft • lbf
x 1.356 = Joules (J)
x 1.285×10^{-3} = British thermal units (Btu) (see note)
x 3.239×10^{-4} = Kilocalories (kcal)
x 0.13825 = Kilogram-force-metres (kgf • m)
x 5.050×10^{-7} = Horsepower-hours (hp • h)
x 3.766×10^{-7} = Kilowatt-hours (kW • h)

GALLONS, U.S. — U.S. gal
x 3785.4 = Cubic centimetres (cm^3)
x 3.7854 = Litres (l)
x 3.7854×10^{-3} = Cubic metres (m^3)
x 231 = Cubic inches (in^3)
x 0.13368 = Cubic feet (ft^3)
x 4.951×10^{-3} = Cubic yards (yd^3)
x 8 = Pints (pt) liquid
x 4 = Quarts (qt) liquid
x 0.8327 = Imperial gallons (imp gal)
x 8.328 = Pounds of water at 60°F in air
x 8.337 = Pounds of water at 60°F in vacuo

GALLONS, IMPERIAL — imp gal
x 4546 = Cubic centimetres (cm^3)
x 4.546 = Litres (l)
x 4.546×10^{-3} = Cubic metres (m^3)
x 0.16054 = Cubic feet (ft^3)
x 5.946×10^{-3} = Cubic yards (yd^3)
x 1.20094 = U.S. gallons (U.S. gal)
x 10.000 = Pounds of water at 62°F in air

GALLONS, PER MINUTE, U.S. — U.S. gpm
x 0.22715 = Cubic metres per hour (m^3/h)
x 0.06309 = Litres per second (l/s)
x 8.021 = Cubic feet per hour (cfh)
x 2.228×10^{-3} = Cubic feet per second (cfs)

GRAINS — gr av. or troy
x 0.0648 = Grams (g)

GRAINS PER U.S. GALLON — gr/U.S. gal at 60°F
x 17.12 = Grams per cubic metre (g/m^3)
x 17.15 = Parts per million by weight in water
x 142.9 = Pounds per million gallons

GRAINS PER IMPERIAL GALLON — gr/imp gal at 62°F
x 14.25 = Grams per cubic metre (g/m^3)
x 14.29 = Parts per million by weight in water

GRAMS — g
x 15.432 = Grains (gr)
x 0.035274 = Ounces (oz) av.
x 0.032151 = Ounces (oz) troy
x 2.2046×10^{-3} = Pounds (lb)

GRAMS-FORCE — gf
x 9.807×10^{-3} = Newtons (N)

GRAMS-FORCE PER CENTIMETRE — gf/cm
x 98.07 = Newtons per metre (N/m)
x 5.600×10^{-3} = Pounds-force per inch (lbf/in)

GRAMS PER CUBIC CENTIMETRE — g/cm^3
x 62.43 = Pounds per cubic foot (lb/ft^3)
x 0.03613 = Pounds per cubic inch (lb/in^3)

GRAMS PER LITRE — g/l
x 58.42 = Grains per U.S. gallon (gr/U.S. gal)
x 8.345 = Pounds per 1000 U.S. gallons
x 0.06243 = Pounds per cubic foot (lb/ft^3)
x 1002 = Parts per million by mass (weight) in water at 60°F

HECTARES — ha
x 1.000×10^4 = Square metres (m^2)
x 1.0764×10^5 = Square feet (ft^2)

HORSEPOWER — hp
x 745.7 = Watts (W)
x 0.7457 = Kilowatts (kW)
x 33,000 = Foot-pounds-force per minute (ft • lbf/min)
x 550 = Foot-pounds-force per second (ft • lbf/s)
x 42.43 = British thermal units per minute (Btu/min) (see note)
x 10.69 = Kilocalories per minute (kcal/min)
x 1.0139 = Horsepower (metric)

HORSEPOWER — hp boiler
x 33,480 = British thermal units per hour (Btu/h) (see note)
x 9.809 = Kilowatts (kW)

HORSEPOWER-HOURS — hp • h
x 0.7457 = Kilowatt-hours (kW • h)
x 1.976×10^6 = Foot-pounds-force (ft • lbf)
x 2545 = British thermal units (Btu) (see note)
x 641.5 = Kilocalories (kcal)
x 2.732×10^5 = Kilogram-force-metres (kgf • m)

INCHES — in
x 2.540 = Centimetres (cm)

INCHES OF MERCURY — inHg at 0°C
x 3.3864 = Kilopascals (kPa)
x 0.03386 = Bars (bar)
x 1.135 = Feet of water (ftH_2O) at 68°F
x 13.62 = Inches of water (inH_2O) at 68°F
x 0.03453 = Kilograms-force per square centimetre (kg/cm^2)
x 70.73 = Pounds-force per square foot (lbf/ft^2)
x 0.4912 = Pounds-force per square inch (psi)
x 0.03342 = Standard atmospheres

INCHES OF WATER — inH_2O at 68°F
x 0.2487 = Kilopascals (kPa)
x 2.487×10^{-3} = Bars (bar)
x 0.07342 = Inches of mercury (inHg) at 0°C
x 2.535×10^{-3} = Kilograms-force per square centimetre (kg/cm^2)
x 0.5770 = Ounces-force per square inch (ozf/in^2)
x 5.193 = Pounds-force per square foot (lbf/ft^2)
x 0.03606 = Pounds-force per square inch (psi)
x 2.454×10^{-3} = Standard atmospheres

JOULES — J
x 0.9484×10^{-3} = British thermal units (Btu) (see note)
x 0.2390 = Calories (cal) thermochemical
x 0.7376 = Foot-pounds-force (ft • lbf)
x 2.778×10^{-4} = Watt-hours (W • h)

KILOGRAMS — kg
- x 2.2046 = Pounds (lb)
- x 1.102×10^{-3} = Tons (ton) short

KILOGRAMS-FORCE — kgf
- x 9.807 = Newtons (N)
- x 2.205 = Pounds-force (lbf)

KILOGRAMS-FORCE PER METRE — kgf/m
- x 9.807 = Newtons per metre (N/m)
- x 0.6721 = Pounds-force per foot (lbf/ft)

KILOGRAMS-FORCE PER SQUARE CENTIMETRE — kg/cm^2
- x 98.07 = Kilopascals (kPa)
- x 0.9807 = Bars (bar)
- x 32.87 = Feet of water (ftH_2O) at 68°F
- x 28.96 = Inches of mercury (inHg) at 0°C
- x 2048 = Pounds-force per square foot (lbf/ft^2)
- x 14.223 = Pounds-force per square inch (psi)
- x 0.9678 = Standard atmospheres

KILOGRAMS-FORCE PER SQUARE MILLIMETRE — kgf/mm^2
- x 9.807 = Megapascals (MPa)
- x 1.000×10^6 = Kilograms-force per square metre (kgf/m^2)

KILOMETRES PER HOUR — km/h
- x 27.78 = Centimetres per second (cm/s)
- x 0.9113 = Feet per second (ft/s)
- x 54.68 = Feet per minute (ft/min)
- x 16.667 = Metres per minute (m/min)
- x 0.53996 = International knots (kn)
- x 0.6214 = Miles per hour (mph)

KILOMETRES PER HOUR PER SECOND — $km \cdot h^{-1} \cdot s^{-1}$
- x 0.2778 = Metres per second per second (m/s^2)
- x 27.78 = Centimetres per second per second (cm/s^2)
- x 0.9113 = Feet per second per second (ft/s^2)

KILOMETRES PER SECOND — km/s
- x 37.28 = Miles per minute (mi/min)

KILOPASCALS — kPa
- x 10^3 = Pascals (Pa) or newtons per square metre (N/m^2)
- x 0.1450 = Pounds-force per square inch (psi)
- x 0.010197 = Kilograms-force per square centimetre (kg/cm^2)
- x 0.2953 = Inches of mercury (inHg) at 32°F
- x 0.3351 = Feet of water (ftH_2O) at 68°F
- x 4.021 = Inches of water (inH_2O) at 68°F

KILOWATTS — kW
- x 4.425×10^4 = Foot-pounds-force per minute (ft • lbf/min)
- x 737.6 = Foot-pounds-force per second (ft • lbf/s)
- x 56.90 = British thermal units per minute (Btu/min) (see note)
- x 14.33 = Kilocalories per minute (kcal/min)
- x 1.3410 = Horsepower (hp)

KILOWATT-HOURS — kW • h
- x 3.6×10^6 = Joules (J)
- x 2.655×10^6 = Foot-pounds-force (ft • lbf)
- x 3413 = British thermal units (Btu) (see note)
- x 860 = Kilocalories (kcal)
- x 3.671×10^5 = Kilogram-force metres (kgf • m)
- x 1.3410 = Horsepower-hours (hp • h)

KNOTS — kn (International)
- x 0.5144 = Metres per second (m/s)
- x 1.151 = Miles per hour (mph)

LITRES — l
- x 1000 = Cubic centimetres (cm^3)
- x 0.035315 = Cubic feet (ft^3)
- x 61.024 = Cubic inches (in^3)
- x 1.308×10^{-3} = Cubic yards (yd^3)
- x 0.2642 = U.S. gallons (U.S. gal)
- x 0.2200 = Imperial gallons (imp gal)

LITRES PER MINUTE — l/min
- x 0.01667 = Litres per second (l/s)
- x 5.885×10^{-4} = Cubic feet per second (cfs)
- x 4.403×10^{-3} = U.S. gallons per second (U.S. gal/s)
- x 3.666×10^{-3} = Imperial gallons per second (imp gal/s)

LITRES PER SECOND — l/s
- x 10^{-3} = Cubic metres per second (m^3/s)
- x 3.600 = Cubic metres per hour (m^3/h)
- x 60 = Litres per minute (l/min)
- x 15.85 = U.S. gallons per minute (U.S. gpm)
- x 13.20 = Imperial gallons per minute (imp gpm)

MEGAPASCALS — MPa
- x 10^6 = Pascals (Pa) or newtons per square metre (N/m^2)
- x 10^3 = Kilopascals (kPa)
- x 145.0 = Pounds-force per square inch (psi)
- x 0.1020 = Kilograms-force per square millimetre (kgf/mm^2)

METRES — m
- x 3.281 = Feet (ft)
- x 39.37 = Inches (in)
- x 1.0936 = Yards (yd)

METRES PER MINUTE — m/min
- x 1.6667 = Centimetres per second (cm/s)
- x 0.0600 = Kilometres per hour (km/h)
- x 3.281 = Feet per minute (ft/min)
- x 0.05468 = Feet per second (ft/s)
- x 0.03728 = Miles per hour (mph)

METRES PER SECOND — m/s
- x 3.600 = Kilometres per hour (km/h)
- x 0.0600 = Kilometres per minute (km/min)
- x 196.8 = Feet per minute (ft/min)
- x 3.281 = Feet per second (ft/s)
- x 2.237 = Miles per hour (mph)
- x 0.03728 = Miles per minute (mi/min)

MICROMETRES — μm formerly micron
- x 10^{-6} = Metres (m)

MILES — mi
- x 1.6093×10^3 = Metres (m)
- x 1.6093 = Kilometres (km)
- x 5280 = Feet (ft)
- x 1760 = Yards (yd)

MILES PER HOUR — mph
- x 44.70 = Centimetres per second (cm/s)
- x 1.6093 = Kilometres per hour (km/h)
- x 26.82 = Metres per minute (m/min)
- x 88 = Feet per minute (ft/min)
- x 1.4667 = Feet per second (ft/s)
- x 0.8690 = International knots (kn)

MILES PER MINUTE — mi/min
- x 1.6093 = Kilometres per minute (km/min)
- x 2682 = Centimetres per second (cm/s)
- x 88 = Feet per second (ft/s)
- x 60 = Miles per hour (mph)

MINUTES, ANGULAR — (′)
- x 2.909×10^{-4} = Radians (rad)

NEWTONS — N
- x 0.10197 = Kilograms-force (kgf)
- x 0.2248 = Pounds-force (lbf)
- x 7.233 = Poundals
- x 10^5 = Dynes

OUNCES — oz av.
- x 28.35 = Grams (g)
- x 2.835×10^{-5} = Tonnes (t) metric ton
- x 16 = Drams (dr) av.
- x 437.5 = Grains (gr)
- x 0.06250 = Pounds (lb) av.
- x 0.9115 = Ounces (oz) troy
- x 2.790×10^{-5} = Tons (ton) long

OUNCES — oz troy
- x 31.103 = Grams (g)
- x 480 = Grains (gr)
- x 20 = Pennyweights (dwt) troy
- x 0.08333 = Pounds (lb) troy
- x 0.06857 = Pounds (lb) av.
- x 1.0971 = Ounces (oz) av.

OUNCES — oz U.S. fluid
- x 0.02957 = Litres (l)
- x 1.8046 = Cubic inches (in)

OUNCES-FORCE PER SQUARE INCH — ozf/in^2
- x 43.1 = Pascals (Pa)
- x 0.06250 = Pounds-force per square inch (psi)
- x 4.395 = Grams-force per square centimetre (gf/cm^2)

PARTS PER MILLION BY MASS — mass (weight) in water
- x 0.9991 = Grams per cubic metre (g/m^3) at 15°C
- x 0.0583 = Grains per U.S. gallon (gr/U.S. gal) at 60°F
- x 0.0700 = Grains per imperial gallon (gr/imp gal) at 62°F
- x 8.328 = Pounds per million U.S. gallons at 60°F

PASCALS — Pa
- x 1 = Newtons per square metre (N/m^2)
- x 1.450×10^{-4} = Pounds-force per square inch (psi)
- x 1.0197×10^{-5} = Kilograms-force per square centimetre (kg/cm^2)
- x 10^{-3} = Kilopascals (kPa)

PENNYWEIGHTS — dwt troy
- x 1.5552 = Grams (g)
- x 24 = Grains (gr)

POISES — P
- x 0.1000 = Newton-seconds per square metre ($N \cdot s/m^2$)
- x 100 = Centipoises (cP)
- x 2.0886×10^{-3} = Pound-force-seconds per square foot ($lbf \cdot s/ft^2$)
- x 0.06721 = Pounds per foot second ($lb/ft \cdot s$)

POUNDS-FORCE — lbf av.
- x 4.448 = Newtons(N)
- x 0.4536 = Kilograms-force (kgf)

POUNDS — lb av.
- x 453.6 = Grams (g)
- x 16 = Ounces (oz) av.
- x 256 = Drams (dr) av.
- x 7000 = Grains (gr)
- x 5×10^{-4} = Tons (ton) short
- x 1.2153 = Pounds (lb) troy

POUNDS — lb troy
- x 373.2 = Grams (g)
- x 12 = Ounces (oz) troy
- x 240 = Pennyweights (dwt) troy
- x 5760 = Grains (gr)
- x 0.8229 = Pounds (lb) av.
- x 13.166 = Ounces (oz) av.
- x 3.6735×10^{-4} = Tons (ton) long
- x 4.1143×10^{-4} = Tons (ton) short
- x 3.7324×10^{-4} = Tonnes (t) metric tons

POUNDS-MASS OF WATER AT 60°F
- x 453.98 = Cubic centimetres (cm^3)
- x 0.45398 = Litres (l)
- x 0.01603 = Cubic feet (ft^3)
- x 27.70 = Cubic inches (in^3)
- x 0.1199 = U.S. gallons (U.S. gal)

POUNDS OF WATER PER MINUTE AT 60°F
- x 7.576 = Cubic centimetres per second (cm^3/s)
- x 2.675×10^{-4} = Cubic feet per second (cfs)

POUNDS PER CUBIC FOOT — lb/ft^3
- x 16.018 = Kilograms per cubic metre (kg/m^3)
- x 0.016018 = Grams per cubic centimetre (g/cm^3)
- x 5.787×10^{-4} = Pounds per cubic inch (lb/in^3)

POUNDS PER CUBIC INCH — lb/in^3
- x 2.768×10^4 = Kilograms per cubic metre (kg/m^3)
- x 27.68 = Grams per cubic centimetre (g/cm^3)
- x 1728 = Pounds per cubic foot (lb/ft^3)

POUNDS-FORCE PER FOOT — lbf/ft
- x 14.59 = Newtons per metre (N/m)
- x 1.488 = Kilograms-force per metre (kgf/m)
- x 14.88 = Grams-force per centimetre (gf/cm)

POUNDS-FORCE PER SQUARE FOOT — lbf/ft^2
- x 47.88 = Pascals (Pa)
- x 0.01605 = Feet of water (ftH_2O) at 68°F
- x 4.882×10^{-4} = Kilograms-force per square centimetre (kg/cm^2)
- x 6.944×10^{-3} = Pounds-force per square inch (psi)

POUNDS-FORCE PER SQUARE INCH — psi
- x 6.895 = Kilopascals (kPa)
- x 0.06805 = Standard atmospheres
- x 2.311 = Feet of water (ftH_2O) at 68°F
- x 27.73 = Inches of water (inH_2O) at 68°F
- x 2.036 = Inches of mercury (inHg) at 0°C
- x 0.07031 = Kilograms-force per square centimetre (kg/cm^2)

QUARTS — qt dry
- x 1101 = Cubic centimetres (cm^3)
- x 67.20 = Cubic inches (in^3)

QUARTS — qt liquid
- x 946.4 = Cubic centimetres (cm^3)
- x 57.75 = Cubic inches (in^3)

QUINTALS — obsolete metric mass term
- x 100 = Kilograms (kg)
- x 220.46 = Pounds (lb) U.S. av.
- x 101.28 = Pounds (lb) Argentina
- x 129.54 = Pounds (lb) Brazil
- x 101.41 = Pounds (lb) Chile
- x 101.47 = Pounds (lb) Mexico
- x 101.43 = Pounds (lb) Peru

RADIANS — rad
- x 57.30 = Degrees (°) angular

RADIANS PER SECOND — rad/s
- x 57.30 = Degrees per second (°/s) angular

STANDARD CUBIC FEET PER MINUTE — scfm (at 14.696 psia and 60°F)
- x 0.4474 = Litres per second (l/s) at standard conditions (760 mmHg and 0°C)
- x 1.608 = Cubic metres per hour (m^3/h) at standard conditions (760 mmHg and 0°C)

STOKES — St
- x 10^{-4} = Square metres per second (m^2/s)
- x 1.076×10^{-3} = Square feet per second (ft^2/s)

TONS-MASS — tonm long
- x 1016 = Kilograms (kg)
- x 2240 = Pounds (lb) av.
- x 1.1200 = Tons (ton) short

TONNES — t metric ton, millier
- x 1000 = Kilograms (kg)
- x 2204.6 = Pounds (lb)

TONNES-FORCE — tf metric ton-force
- x 980.7 = Newtons (N)

TONS — ton short
- x 907.2 = Kilograms (kg)
- x 0.9072 = Tonnes (t)
- x 2000 = Pounds (lb) av.
- x 32000 = Ounces (oz) av.
- x 2430.6 = Pounds (lb) troy
- x 0.8929 = Tons (ton) long

TONS OF WATER PER 24 HOURS AT 60°F
- x 0.03789 = Cubic metres per hour (m^3/h)
- x 83.33 = Pounds of water per hour ($lb/h\ H_2O$) at 60°F
- x 0.1668 = U.S. gallons per minute (U.S. gpm)
- x 1.338 = Cubic feet per hour (cfh)

WATTS — W
- x 0.05690 = British thermal units per minute (Btu/min) (see note)
- x 44.25 = Foot-pounds-force per minute ($ft \cdot lbf/min$)
- x 0.7376 = Foot-pounds-force per second ($ft \cdot lbf/s$)
- x 1.341×10^{-3} = Horsepower (hp)
- x 0.01433 = Kilocalories per minute (kcal/min)

WATT-HOURS — $W \cdot h$
- x 3600 = Joules (J)
- x 3.413 = British thermal units (Btu) (see note)
- x 2655 = Foot-pounds-force ($ft \cdot lbf$)
- x 1.341×10^{-3} = Horsepower-hours ($hp \cdot h$)
- x 0.860 = Kilocalories (kcal)
- x 367.1 = Kilogram-force-metres ($kgf \cdot m$)

NOTE: SIGNIFICANT FIGURES The precision to which a given conversion factor is known, and its application, determine the number of significant figures which should be used. While many handbooks and standards give factors contained in this table to six or more significant figures, the fact that different sources disagree, in many cases, in the fifth or further figure indicates that four or five significant figures represent the precision for these factors fairly. At present the accuracy of process instrumentation, analog or digital, is in the tenth percent region at best, thus needing only three significant figures. Hence this table is confined to four or five significant figures. The advent of the pocket calculator (and the use of digital computers in process instrumentation) tends to lead to use of as many figures as the calculator will handle. However, when this exceeds the precision of the data, or the accuracy of the application, such a practice is misleading and timewasting.

NOTE: BRITISH THERMAL UNIT When making calculations involving Btu it must be remembered that there are several definitions of the Btu. The first three significant figures of the conversion factors given in this table are common to most definitions of the Btu. However, if four or more significant figures are needed in the calculation, the appropriate handbooks and standards should be consulted to be sure the proper definition and factor are being used.

TEMPERATURE CONVERSION TABLES

Fahrenheit and Celsius (Centigrade)

C	*	F	C	*	F	C	*	F	C	*	F	C	*	F
—273.15	—459.67		—17.2	1	33.8	10.6	51	123.8	43	110	230	266	510	950
—268	—450		—16.7	2	35.6	11.1	52	125.6	49	120	248	271	520	968
—262	—440		—16.1	3	37.4	11.7	53	127.4	54	130	266	277	530	986
—257	—430		—15.6	4	39.2	12.2	54	129.2	60	140	284	282	540	1004
—251	—420		—15.0	5	41.0	12.8	55	131.0	66	150	302	288	550	1022
—246	—410		—14.4	6	42.8	13.3	56	132.8	71	160	320	293	560	1040
—240	—400		—13.9	7	44.6	13.9	57	134.6	77	170	338	299	570	1058
—234	—390		—13.3	8	46.4	14.4	58	136.4	82	180	356	304	580	1076
—229	—380		—12.8	9	48.2	15.0	59	138.2	88	190	374	310	590	1094
—223	—370		—12.2	10	50.0	15.6	60	140.0	93	200	392	316	600	1112
—218	—360		—11.7	11	51.8	16.1	61	141.8	99	210	410	321	610	1130
—212	—350		—11.1	12	53.6	16.7	62	143.6				327	620	1148
—207	—340		—10.6	13	55.4	17.2	63	145.4				332	630	1166
—201	—330		—10.0	14	57.2	17.8	64	147.2				338	640	1184
—196	—320		—9.4	15	59.0	18.3	65	149.0				343	650	1202
—190	—310		—8.9	16	60.8	18.9	66	150.8	100	212	413	349	660	1220
—184	—300		—8.3	17	62.6	19.4	67	152.6				354	670	1238
—179	—290		—7.8	18	64.4	20.0	68	154.4				360	680	1256
—173	—280		—7.2	19	66.2	20.6	69	156.2				366	690	1274
—169	—273	—459.4	—6.7	20	68.0	21.1	70	158.0				371	700	1292
—168	—270	—454	—6.1	21	69.8	21.7	71	159.8	104	220	428	377	710	1310
—162	—260	—436	—5.6	22	71.6	22.2	72	161.6	110	230	446	382	720	1328
—157	—250	—418	—5.0	23	73.4	22.8	73	163.4	116	240	464	388	730	1346
—151	—240	—400	—4.4	24	75.2	23.3	74	165.2	121	250	482	393	740	1364
—146	—230	—382	—3.9	25	77.0	23.9	75	167.0				399	750	1382
—140	—220	—364	—3.3	26	78.8	24.4	76	168.8	127	260	500	404	760	1400
—134	—210	—346	—2.8	27	80.6	25.0	77	170.6	132	270	518	410	770	1418
—129	—200	—328	—2.2	28	82.4	25.6	78	172.4	138	280	536	416	780	1436
—123	—190	—310	—1.7	29	84.2	26.1	79	174.2	143	290	554	421	790	1454
—118	—180	—292	—1.1	30	86.0	26.7	80	176.0	149	300	572	427	800	1472
—112	—170	—274	—0.6	31	87.8	27.2	81	177.8	154	310	590	432	810	1490
—107	—160	—256	0	32	89.6	27.8	82	179.6	160	320	608	438	820	1508
—101	—150	—238	0.6	33	91.4	28.3	83	181.4	166	330	626	443	830	1526
—95.6	—140	—220	1.1	34	93.2	28.9	84	183.2	171	340	644	449	840	1544
—90.0	—130	—202	1.7	35	95.0	29.4	85	185.0	177	350	662	454	850	1562
—84.4	—120	—184	2.2	36	96.8	30.0	86	186.8	182	360	680	460	860	1580
—78.9	—110	—166	2.8	37	98.6	30.6	87	188.6	188	370	698	466	870	1598
—73.3	—100	—148	3.3	38	100.4	31.1	88	190.4	193	380	716	471	880	1616
—67.8	—90	—130	3.9	39	102.2	31.7	89	192.2	199	390	734	477	890	1634
—62.2	—80	—112	4.4	40	104.0	32.2	90	194.0	204	400	752	482	900	1652
—56.7	—70	—94	5.0	41	105.8	32.8	91	195.8	210	410	770	488	910	1670
—51.1	—60	—76	5.6	42	107.6	33.3	92	197.6	216	420	788	493	920	1688
—45.6	—50	—58	6.1	43	109.4	33.9	93	199.4	221	430	806	499	930	1706
—40.0	—40	—40	6.7	44	111.2	34.4	94	201.2	227	440	824	504	940	1724
—34.4	—30	—22	7.2	45	113.0	35.0	95	203.0	232	450	842	510	950	1742
—28.9	—20	—4	7.8	46	114.8	35.6	96	204.8	238	460	860	516	960	1760
—23.3	—10	14	8.3	47	116.6	36.1	97	206.6	243	470	878	521	970	1778
—17.8	0	32	8.9	48	118.4	36.7	98	208.4	249	480	896	527	980	1796
			9.4	49	120.2	37.2	99	210.2	254	490	914	532	990	1814
			10.0	50	122.0	37.8	100	212.0	260	500	932	538	1000	1832

*In the center column, find the temperature to be converted. The equivalent temperature is in the left column, if converting to Celsius, and in the right column, if converting to Fahrenheit.

INTERPOLATION VALUES

C	*	F	C	*	F
0.56	1	1.8	3.33	6	10.8
1.11	2	3.6	3.89	7	12.6
1.67	3	5.4	4.44	8	14.4
2.22	4	7.2	5.00	9	16.2
2.78	5	9.0	5.56	10	18.0

Fahrenheit and Celsius (Centigrade) *continued*

C	*	F	C	*	F	C	*	F	C	*	F
543	**1010**	1850	821	**1510**	2750	1099	**2010**	3650	1377	**2510**	4550
549	**1020**	1868	827	**1520**	2768	1104	**2020**	3668	1382	**2520**	4568
554	**1030**	1886	832	**1530**	2786	1110	**2030**	3686	1388	**2530**	4586
560	**1040**	1904	838	**1540**	2804	1116	**2040**	3704	1393	**2540**	4604
566	**1050**	1922	843	**1550**	2822	1121	**2050**	3722	1399	**2550**	4622
571	**1060**	1940	849	**1560**	2840	1127	**2060**	3740	1404	**2560**	4640
577	**1070**	1958	854	**1570**	2858	1132	**2070**	3758	1410	**2570**	4658
582	**1080**	1976	860	**1580**	2876	1138	**2080**	3776	1416	**2580**	4676
588	**1090**	1994	866	**1590**	2894	1143	**2090**	3794	1421	**2590**	4694
593	**1100**	2012	871	**1600**	2912	1149	**2100**	3812	1427	**2600**	4712
599	**1110**	2030	877	**1610**	2930	1154	**2110**	3830	1432	**2610**	4730
604	**1120**	2048	882	**1620**	2948	1160	**2120**	3848	1438	**2620**	4748
610	**1130**	2066	888	**1630**	2966	1166	**2130**	3866	1443	**2630**	4766
616	**1140**	2084	893	**1640**	2984	1171	**2140**	3884	1449	**2640**	4784
621	**1150**	2102	899	**1650**	3002	1177	**2150**	3902	1454	**2650**	4802
627	**1160**	2120	904	**1660**	3020	1182	**2160**	3920	1460	**2660**	4820
632	**1170**	2138	910	**1670**	3038	1188	**2170**	3938	1466	**2670**	4838
638	**1180**	2156	916	**1680**	3056	1193	**2180**	3956	1471	**2680**	4856
643	**1190**	2174	921	**1690**	3074	1199	**2190**	3974	1477	**2690**	4874
649	**1200**	2192	927	**1700**	3092	1204	**2200**	3992	1482	**2700**	4892
654	**1210**	2210	932	**1710**	3110	1210	**2210**	4010	1488	**2710**	4910
660	**1220**	2228	938	**1720**	3128	1216	**2220**	4028	1493	**2720**	4928
666	**1230**	2246	943	**1730**	3146	1221	**2230**	4046	1499	**2730**	4946
671	**1240**	2264	949	**1740**	3164	1227	**2240**	4064	1504	**2740**	4964
677	**1250**	2282	954	**1750**	3182	1232	**2250**	4082	1510	**2750**	4982
682	**1260**	2300	960	**1760**	3200	1238	**2260**	4100	1516	**2760**	5000
688	**1270**	2318	966	**1770**	3218	1243	**2270**	4118	1521	**2770**	5018
693	**1280**	2336	971	**1780**	3236	1249	**2280**	4136	1527	**2780**	5036
699	**1290**	2354	977	**1790**	3254	1254	**2290**	4154	1532	**2790**	5054
704	**1300**	2372	982	**1800**	3272	1260	**2300**	4172	1538	**2800**	5072
710	**1310**	2390	988	**1810**	3290	1266	**2310**	4190	1543	**2810**	5090
716	**1320**	2408	993	**1820**	3308	1271	**2320**	4208	1549	**2820**	5108
721	**1330**	2426	999	**1830**	3326	1277	**2330**	4226	1554	**2830**	5126
727	**1340**	2444	1004	**1840**	3344	1282	**2340**	4244	1560	**2840**	5144
732	**1350**	2462	1010	**1850**	3362	1288	**2350**	4262	1566	**2850**	5162
738	**1360**	2480	1016	**1860**	3380	1293	**2360**	4280	1571	**2860**	5180
743	**1370**	2498	1021	**1870**	3398	1299	**2370**	4298	1577	**2870**	5198
749	**1380**	2516	1027	**1880**	3416	1304	**2380**	4316	1582	**2880**	5216
754	**1390**	2534	1032	**1890**	3434	1310	**2390**	4334	1588	**2890**	5234
760	**1400**	2552	1038	**1900**	3452	1316	**2400**	4352	1593	**2900**	5252
766	**1410**	2570	1043	**1910**	3470	1321	**2410**	4370	1599	**2910**	5270
771	**1420**	2588	1049	**1920**	3488	1327	**2420**	4388	1604	**2920**	5288
777	**1430**	2606	1054	**1930**	3506	1332	**2430**	4406	1610	**2930**	5306
782	**1440**	2624	1060	**1940**	3524	1338	**2440**	4424	1616	**2940**	5324
788	**1450**	2642	1066	**1950**	3542	1343	**2450**	4442	1621	**2950**	5342
793	**1460**	2660	1071	**1960**	3560	1349	**2460**	4460	1627	**2960**	5360
799	**1470**	2678	1077	**1970**	3578	1354	**2470**	4478	1632	**2970**	5378
804	**1480**	2696	1082	**1980**	3596	1360	**2480**	4496	1638	**2980**	5396
810	**1490**	2714	1088	**1990**	3614	1366	**2490**	4514	1643	**2990**	5414
816	**1500**	2732	1093	**2000**	3632	1371	**2500**	4532	1649	**3000**	5432

Temperature Conversion Formulas

Degrees Celsius (formerly Centigrade) C

$C + 273.15 = K$ Kelvin
$(C \times {}^{9}/_{5}) + 32 = F$ Fahrenheit
$C \times {}^{4}/_{5} = R$ Réaumur

Degrees Fahrenheit — F

$F + 459.67 =$ Rankine
$(F - 32) \times {}^{5}/_{9} = C$ Celsius
$(F - 32) \times {}^{4}/_{9} = R$ Réaumur

Degrees Réaumur — R

$R \times {}^{5}/_{4} = C$ Celsius
$(R \times {}^{9}/_{4}) + 32 = F$ Fahrenheit

PROPERTIES OF SATURATED STEAM AND SATURATED WATER

Press.	Temp.	Volume, ft³/lbm			Enthalpy, Btu/lbm			Entropy, Btu/lbm x F			Energy, Btu/lbm	
psia	F	Water	Evap.	Steam	Water	Evap.	Steam	Water	Evap.	Steam	Water	Steam
		v_f	v_{fg}	v_g	h_f	h_{fg}	h_g	s_f	s_{fg}	s_g	u_f	u_g
3208.2	705.47	0.05078	0.00000	0.05078	906.0	0.0	906.0	1.0612	0.0000	1.0612	875.9	875.9
3094.3	700.0	0.03662	0.03857	0.07519	822.4	172.7	995.2	0.9901	0.1490	1.1390	801.5	952.2
3000.0	695.33	0.03428	0.05073	0.08500	801.8	218.4	1020.3	0.9728	0.1891	1.1619	782.8	973.1
2708.6	680.0	0.03037	0.08080	0.11117	758.5	310.1	1068.5	0.9365	0.2720	1.2086	743.2	1012.8
2500.0	668.11	0.02859	0.10209	0.13068	731.7	361.6	1093.3	0.9139	0.3206	1.2345	718.5	1032.9
2365.7	660.0	0.02768	0.11663	0.14431	714.9	392.1	1107.0	0.8995	0.3502	1.2498	702.8	1043.9
2059.9	640.0	0.02595	0.15427	0.18021	679.1	454.6	1133.7	0.8686	0.4134	1.2821	669.2	1065.0
2000.0	635.80	0.02565	0.16266	0.18831	672.1	466.2	1138.3	0.8625	0.4256	1.2881	662.6	1068.6
1786.9	620.0	0.02466	0.19615	0.22081	646.9	506.3	1153.2	0.8403	0.4689	1.3092	638.8	1080.2
1543.2	600.0	0.02364	0.24384	0.26747	617.1	550.6	1167.7	0.8134	0.5196	1.3330	610.4	1091.4
1500.0	596.20	0.02346	0.25372	0.27719	611.7	558.4	1170.1	0.8085	0.5288	1.3373	605.2	1093.1
1326.17	580.0	0.02279	0.29937	0.32216	589.1	589.9	1179.0	0.7876	0.5673	1.3550	583.5	1099.9
1200.0	567.19	0.02232	0.34013	0.36245	571.9	613.0	1184.8	0.7714	0.5969	1.3683	566.9	1104.3
1133.38	560.0	0.02207	0.36507	0.38714	562.4	625.3	1187.7	0.7625	0.6132	1.3757	557.8	1106.5
1000.0	544.58	0.02159	0.42436	0.44596	542.6	650.4	1192.9	0.7434	0.6476	1.3910	538.6	1110.4
962.79	540.0	0.02146	0.44367	0.46513	536.8	657.5	1194.3	0.7378	0.6577	1.3954	532.9	1111.4
812.53	520.0	0.02091	0.53864	0.55956	512.0	687.0	1199.0	0.7133	0.7013	1.4146	508.8	1115.0
800.0	518.21	0.02087	0.54809	0.56896	509.8	689.6	1199.4	0.7111	0.7051	1.4163	506.7	1115.2
680.86	500.0	0.02043	0.65448	0.67492	487.9	714.3	1202.2	0.6890	0.7443	1.4333	485.4	1117.2
600.0	486.20	0.02013	0.74962	0.76975	471.7	732.0	1203.7	0.6723	0.7738	1.4461	469.5	1118.2
566.15	480.0	0.02000	0.79716	0.81717	464.5	739.6	1204.1	0.6648	0.7871	1.4518	462.4	1118.5
500.0	467.01	0.01975	0.90787	0.92762	449.5	755.1	1204.7	0.6490	0.8148	1.4639	447.7	1118.8
466.87	460.0	0.01961	0.97463	0.99424	441.5	763.2	1204.8	0.6405	0.8299	1.4704	439.8	1118.9
400.0	444.60	0.01934	1.1416	1.610	424.2	780.4	1204.6	0.6217	0.8630	1.4847	422.7	1118.7
381.54	440.0	0.01926	1.1976	1.2169	419.0	785.4	1204.4	0.6161	0.8729	1.4890	417.6	1118.5
308.780	420.0	0.01894	1.4808	1.4997	396.9	806.2	1203.1	0.5915	0.9165	1.5080	395.8	1117.5
300.0	417.35	0.01889	1.5238	1.5427	394.0	808.9	1202.9	0.5882	0.9223	1.5105	392.9	1117.2
250.0	400.97	0.01865	1.8245	1.8432	376.1	825.0	1201.1	0.5679	0.9585	1.5264	375.3	1115.8
247.259	400.0	0.01864	1.8444	1.8630	375.1	825.9	1201.0	0.5667	0.9607	1.5274	374.3	1115.7
200.0	381.80	0.01839	2.2689	2.2873	355.5	842.8	1198.3	0.5438	1.0016	1.5454	354.8	1113.7
195.729	380.0	0.01836	2.3170	2.3353	353.6	844.5	1198.0	0.5416	1.0057	1.5473	352.9	1113.5
153.010	360.0	0.01811	2.9392	2.9573	332.3	862.1	1194.4	0.5161	1.0517	1.5678	331.8	1110.6
150.0	358.43	0.01809	2.9958	3.0139	330.6	863.4	1194.1	0.5141	1.0554	1.5695	330.1	1110.4
120.0	341.27	0.01789	3.7097	3.7275	312.6	877.8	1190.4	0.4919	1.0960	1.5879	312.2	1107.6
117.992	340.0	0.01787	3.7699	3.7878	311.3	878.8	1190.1	0.4902	1.0990	1.5892	310.9	1107.4
100.0	327.82	0.01774	4.4133	4.4310	298.5	888.6	1187.2	0.4743	1.1284	1.6027	298.2	1105.2
89.643	320.0	0.01766	4.8961	4.9138	290.4	894.8	1185.2	0.4640	1.1477	1.6116	290.1	1103.7
80.0	312.04	0.01757	5.4536	5.4711	282.1	900.9	1183.1	0.4534	1.1675	1.6208	281.9	1102.1
70.0	302.93	0.01748	6.1875	6.2050	272.7	907.8	1180.6	0.4411	1.1905	1.6316	272.5	1100.2
67.005	300.0	0.01745	6.4483	6.4658	269.7	910.0	1179.7	0.4372	1.1979	1.6351	269.5	1099.6
60.0	292.71	0.017383	7.1562	7.1736	262.2	915.4	1177.6	0.4273	1.2167	1.6440	262.0	1098.0
50.0	281.02	0.017274	8.4967	8.5140	250.2	923.9	1174.1	0.4112	1.2474	1.6586	250.1	1095.3
49.200	280.0	0.017264	8.6267	8.6439	249.2	924.6	1173.8	0.4098	1.2501	1.6599	249.1	1095.1
40.0	267.25	0.017151	10.479	10.496	236.1	933.6	1169.8	0.3921	1.2844	1.6765	236.0	1092.1
35.427	260.0	0.017089	11.745	11.762	228.8	938.6	1167.4	0.3819	1.3043	1.6862	228.6	1090.3
30.0	250.34	0.017009	13.727	13.744	218.9	945.2	1164.1	0.3682	1.3313	1.6995	218.8	1087.9
25.0	240.07	0.016927	16.284	16.301	208.52	952.1	1160.6	0.3535	1.3607	1.7141	208.4	1085.2
24.968	240.0	0.016926	16.304	16.321	208.45	952.1	1160.6	0.3533	1.3609	1.7142	208.3	1085.2
20.0	227.96	0.016834	20.070	20.087	196.27	960.1	1156.3	0.3358	1.3962	1.7320	196.21	1082.0
17.186	220.0	0.016775	23.131	23.148	188.23	965.2	1153.4	0.3241	1.4201	1.7442	188.18	1079.8
15.0	213.03	0.016726	26.274	26.290	181.21	969.7	1150.9	0.3137	1.4415	1.7552	181.16	1077.9
14.696	212.00	0.016719	26.782	26.799	180.17	970.3	1150.5	0.3121	1.4447	1.7568	180.12	1077.6
11.526	200.0	0.016637	33.622	33.639	168.09	977.9	1146.0	0.2940	1.4824	1.7764	168.05	1074.2
10.0	193.21	0.016592	38.404	38.420	161.26	982.1	1143.3	0.2836	1.5043	1.7879	161.23	1072.3
8.0	182.86	0.016527	47.328	47.345	150.87	988.5	1139.3	0.2676	1.5384	1.8060	150.84	1069.2
7.5110	180.0	0.016510	50.208	50.225	148.00	990.2	1138.2	0.2631	1.5480	1.8111	147.98	1068.4
6.0	170.05	0.016451	61.967	61.984	138.03	996.2	1134.2	0.2474	1.5820	1.8294	138.01	1065.4
5.0	162.24	0.016407	73.515	73.532	130.20	1000.9	1131.1	0.2349	1.6094	1.8443	130.18	1063.1
4.7414	160.0	0.016395	77.27	77.29	127.96	1002.2	1130.2	0.2313	1.6174	1.8487	127.94	1062.4
4.0	152.96	0.016358	90.63	90.64	120.92	1006.4	1127.3	0.2199	1.6428	1.8626	120.90	1060.2
3.0	141.47	0.016300	118.71	118.73	109.42	1013.2	1122.6	0.2009	1.6854	1.8864	109.41	1056.7
2.8892	140.0	0.016293	122.98	123.00	107.95	1014.0	1122.0	0.1985	1.6910	1.8895	107.94	1056.2
2.0	126.07	0.016230	173.74	173.76	94.03	1022.1	1116.2	0.1750	1.7450	1.9200	94.03	1051.8
1.6927	120.0	0.016204	203.25	203.26	87.97	1025.6	1113.6	0.1646	1.7693	1.9339	87.96	1049.9
1.0	101.74	0.016136	333.59	333.60	69.732	1036.1	1105.8	0.1326	1.8455	1.9781	69.73	1044.1
0.94924	100.0	0.016130	350.4	350.4	67.999	1037.1	1105.1	0.1295	1.8530	1.9825	68.00	1043.5
0.50683	80.0	0.016072	633.3	633.3	48.037	1048.4	1096.4	0.0932	1.9426	2.0359	48.036	1037.0
0.25611	60.0	0.016033	1207.6	1207.6	28.060	1059.7	1087.7	0.0555	2.0391	2.0946	28.060	1030.5
0.12163	40.0	0.016019	2445.8	2445.8	8.027	1071.0	1079.0	0.0162	2.1432	2.1594	8.027	1024.0
0.08865	32.018	0.016022	3302.4	3302.4	0.0003	1075.5	1075.5	0.0000	2.1872	2.1872	0.000	1021.3

LIQUID GRAVITY TABLES AND WEIGHT FACTORS

Liquid Lighter than Water

Sp Gr 60 F/60 F	°Be	°API	Lb per gal at 60 F in vacuo	Lb per cu ft at 60 F in vacuo
.500	150.00	151.50	4.169	31.18
.505	147.23	148.70	4.210	31.50
.510	144.51	145.95	4.252	31.81
.515	141.84	143.26	4.294	32.12
.520	139.23	140.62	4.335	32.43
.525	136.67	138.02	4.377	32.74
.530	134.15	135.48	4.419	33.05
.535	131.68	132.99	4.460	33.37
.540	129.26	130.54	4.502	33.68
.545	126.88	128.13	4.544	33.99
.550	124.55	125.77	4.585	34.30
.555	122.25	123.45	4.627	34.61
.560	120.00	121.18	4.669	34.93
.565	117.79	118.94	4.711	35.24
.570	115.61	116.75	4.752	35.55
.575	113.48	114.59	4.794	35.86
.580	111.38	112.47	4.836	36.17
.585	109.32	110.38	4.877	36.48
.590	107.29	108.33	4.919	36.80
.595	105.29	106.32	4.961	37.11
.600	103.33	104.33	5.002	37.42
.605	101.40	102.38	5.044	37.73
.610	99.51	100.47	5.086	38.04
.615	97.64	98.58	5.127	38.36
.620	95.81	96.73	5.169	38.67
.625	94.00	94.90	5.211	38.98
.630	92.22	93.10	5.252	39.29
.635	90.47	91.33	5.294	39.60
.640	88.75	89.59	5.336	39.91
.645	87.05	87.88	5.377	40.23
.650	85.38	86.19	5.419	40.54
.655	83.74	84.53	5.461	40.85
.660	82.12	82.89	5.503	41.16
.665	80.53	81.28	5.544	41.47
.670	78.96	79.69	5.586	41.79
.675	77.41	78.13	5.628	42.10
.680	75.88	76.59	5.669	42.41
.685	74.38	75.07	5.711	42.72
.690	72.90	73.57	5.753	43.03
.695	71.44	72.10	5.794	43.34
.700	70.00	70.64	5.836	43.66
.705	68.58	69.21	5.878	43.97
.710	67.18	67.80	5.919	44.28
.715	65.80	66.40	5.961	44.59
.720	64.44	65.03	6.003	44.90
.725	63.10	63.67	6.044	45.22
.730	61.78	62.34	6.086	45.53
.735	60.48	61.02	6.128	45.84
.740	59.19	59.72	6.170	46.15
.745	57.92	58.43	6.211	46.46
.750	56.67	57.17	6.253	46.77
.755	55.43	55.92	6.295	47.09
.760	54.21	54.68	6.336	47.40
.765	53.01	53.47	6.378	47.71
.770	51.82	52.27	6.420	48.02
.775	50.65	51.08	6.461	48.33
.780	49.49	49.91	6.503	48.65
.785	48.34	48.75	6.545	48.96
.790	47.22	47.61	6.586	49.27
.795	46.10	46.49	6.628	49.58
.800	45.00	45.38	6.670	49.89
.805	43.91	44.28	6.711	50.21
.810	42.84	43.19	6.753	50.52
.815	41.78	42.12	6.795	50.83
.820	40.73	41.06	6.836	51.14
.825	39.70	40.02	6.878	51.45
.830	38.67	38.98	6.920	51.76
.835	37.66	37.96	6.962	52.08
.840	36.67	36.95	7.003	52.39
.845	35.68	35.96	7.045	52.70
.850	34.71	34.97	7.087	53.01
.855	33.74	34.00	7.128	53.32
.860	32.79	33.03	7.170	53.64
.865	31.85	32.08	7.212	53.95
.870	30.92	31.14	7.253	54.26
.875	30.00	30.21	7.295	54.57
.880	29.09	29.30	7.337	54.88
.885	28.19	28.39	7.378	55.19
.890	27.30	27.49	7.420	55.51
.895	26.42	26.60	7.462	55.82
.900	25.56	25.72	7.503	56.13
.905	24.70	24.85	7.545	56.44
.910	23.85	23.99	7.587	56.75
.915	23.01	23.14	7.629	57.07
.920	22.17	22.30	7.670	57.38
.925	21.35	21.47	7.712	57.69
.930	20.54	20.65	7.754	58.00
.935	19.73	19.84	7.795	58.31
.940	18.94	19.03	7.837	58.62
.945	18.15	18.24	7.879	58.94
.950	17.37	17.45	7.920	59.25
.955	16.60	16.67	7.962	59.56
.960	15.83	15.90	8.004	59.87
.965	15.08	15.13	8.045	60.18
.970	14.33	14.38	8.087	60.50
.975	13.59	13.63	8.129	60.81
.980	12.86	12.89	8.170	61.12
.985	12.13	12.15	8.212	61.43
.990	11.41	11.43	8.254	61.74
.995	10.70	10.71	8.295	62.05
1.000	10.00	10.00	8.337	62.37

Liquid Heavier than Water

Sp Gr 60 F/60 F	°Be	°Tw	Lb per gal at 60 F in vacuo	Lb per cu ft at 60 F in vacuo	Sp Gr 60 F/60 F	°Be	°Tw	Lb per gal at 60 F in vacuo	Lb per cu ft at 60 F in vacuo
1.000	0.00	0	8.337	62.37	1.500	48.33	100	12.506	93.55
1.005	.72	1	8.379	62.68	1.505	48.65	101	12.547	93.86
1.010	1.44	2	8.421	62.99	1.510	48.97	102	12.589	94.17
1.015	2.14	3	8.462	63.30	1.515	49.29	103	12.631	94.49
1.020	2.84	4	8.504	63.61	1.520	49.61	104	12.673	94.80
1.025	3.54	5	8.546	63.93	1.525	49.92	105	12.714	95.11
1.030	4.22	6	8.587	64.24	1.530	50.23	106	12.756	95.42
1.035	4.90	7	8.629	64.55	1.535	50.54	107	12.798	95.73
1.040	5.58	8	8.671	64.86	1.540	50.84	108	12.839	96.04
1.045	6.24	9	8.712	65.17	1.545	51.15	109	12.881	96.36
1.050	6.90	10	8.754	65.48	1.550	51.45	110	12.923	96.67
1.055	7.56	11	8.796	65.80	1.555	51.75	111	12.964	96.98
1.060	8.21	12	8.837	66.11	1.560	52.05	112	13.006	97.29
1.065	8.85	13	8.879	66.42	1.565	52.35	113	13.048	97.60
1.070	9.49	14	8.921	66.73	1.570	52.64	114	13.089	97.92
1.075	10.12	15	8.962	67.04	1.575	52.94	115	13.131	98.23
1.080	10.74	16	9.004	67.36	1.580	53.23	116	13.173	98.54
1.085	11.36	17	9.046	67.67	1.585	53.52	117	13.214	98.85
1.090	11.97	18	9.088	67.98	1.590	53.81	118	13.256	99.16
1.095	12.58	19	9.129	68.29	1.595	54.09	119	13.298	99.47
1.100	13.18	20	9.171	68.60	1.600	54.37	120	13.339	99.79
1.105	13.78	21	9.213	68.91	1.605	54.66	121	13.381	100.10
1.110	14.37	22	9.254	69.23	1.610	54.94	122	13.423	100.41
1.115	14.96	23	9.296	69.54	1.615	55.22	123	13.465	100.72
1.120	15.54	24	9.338	69.85	1.620	55.49	124	13.506	101.03
1.125	16.11	25	9.379	70.16	1.625	55.77	125	13.548	101.35
1.130	16.68	26	9.421	70.47	1.630	56.04	126	13.590	101.66
1.135	17.25	27	9.463	70.79	1.635	56.31	127	13.631	101.97
1.140	17.81	28	9.504	71.10	1.640	56.59	128	13.673	102.28
1.145	18.36	29	9.546	71.41	1.645	56.85	129	13.715	102.59
1.150	18.91	30	9.588	71.72	1.650	57.12	130	13.756	102.90
1.155	19.46	31	9.629	72.03	1.655	57.39	131	13.798	103.22
1.160	20.00	32	9.671	72.35	1.660	57.65	132	13.840	103.53
1.165	20.54	33	9.713	72.66	1.665	57.91	133	13.881	103.84
1.170	21.07	34	9.755	72.97	1.670	58.17	134	13.923	104.15
1.175	21.60	35	9.796	73.28	1.675	58.43	135	13.965	104.46
1.180	22.12	36	9.838	73.59	1.680	58.69	136	14.006	104.78
1.185	22.64	37	9.880	73.90	1.685	58.95	137	14.048	105.09
1.190	23.15	38	9.921	74.22	1.690	59.20	138	14.090	105.40
1.195	23.66	39	9.963	74.53	1.695	59.45	139	14.132	105.71
1.200	24.17	40	10.005	74.84	1.700	59.71	140	14.173	106.02
1.205	24.67	41	10.046	75.15	1.705	59.96	141	14.215	106.33
1.210	25.17	42	10.088	75.46	1.710	60.20	142	14.257	106.65
1.215	25.66	43	10.130	75.78	1.715	60.45	143	14.298	106.96
1.220	26.15	44	10.171	76.09	1.720	60.70	144	14.340	107.27
1.225	26.63	45	10.213	76.40	1.725	60.94	145	14.382	107.58
1.230	27.11	46	10.255	76.71	1.730	61.18	146	14.423	107.89
1.235	27.59	47	10.296	77.02	1.735	61.43	147	14.465	108.21
1.240	28.06	48	10.338	77.33	1.740	61.67	148	14.507	108.52
1.245	28.53	49	10.380	77.65	1.745	61.91	149	14.548	108.83
1.250	29.00	50	10.421	77.96	1.750	62.14	150	14.590	109.14
1.255	29.46	51	10.463	78.27	1.755	62.38	151	14.632	109.45
1.260	29.92	52	10.505	78.58	1.760	62.61	152	14.673	109.76
1.265	30.38	53	10.547	78.89	1.765	62.85	153	14.715	110.08
1.270	30.83	54	10.588	79.21	1.770	63.08	154	14.757	110.39
1.275	31.27	55	10.630	79.52	1.775	63.31	155	14.799	110.70
1.280	31.72	56	10.672	79.83	1.780	63.54	156	14.840	111.01
1.285	32.16	57	10.713	80.14	1.785	63.77	157	14.882	111.32
1.290	32.60	58	10.755	80.45	1.790	63.99	158	14.924	111.64
1.295	33.03	59	10.797	80.76	1.795	64.22	159	14.965	111.95
1.300	33.46	60	10.838	81.08	1.800	64.44	160	15.007	112.26
1.305	33.89	61	10.880	81.39	1.805	64.67	161	15.049	112.57
1.310	34.31	62	10.922	81.70	1.810	64.89	162	15.090	112.88
1.315	34.73	63	10.963	82.01	1.815	65.11	163	15.132	113.20
1.320	35.15	64	11.005	82.32	1.820	65.33	164	15.174	113.51
1.325	35.57	65	11.047	82.64	1.825	65.55	165	15.215	113.82
1.330	35.98	66	11.088	82.95	1.830	65.77	166	15.257	114.13
1.335	36.39	67	11.130	83.26	1.835	65.98	167	15.299	114.44
1.340	36.79	68	11.172	83.57	1.840	66.20	168	15.340	114.75
1.345	37.19	69	11.214	83.88	1.845	66.41	169	15.382	115.07
1.350	37.59	70	11.255	84.19	1.850	66.62	170	15.424	115.38
1.355	37.99	71	11.297	84.51	1.855	66.83	171	15.465	115.69
1.360	38.38	72	11.339	84.82	1.860	67.04	172	15.507	116.00
1.365	38.77	73	11.380	85.13	1.865	67.25	173	15.549	116.31
1.370	39.16	74	11.422	85.44	1.870	67.46	174	15.591	116.63
1.375	39.55	75	11.464	85.75	1.875	67.67	175	15.632	116.94
1.380	39.93	76	11.505	86.07	1.880	67.87	176	15.674	117.25
1.385	40.31	77	11.547	86.38	1.885	68.08	177	15.716	117.56
1.390	40.68	78	11.589	86.69	1.890	68.28	178	15.757	117.87
1.395	41.06	79	11.630	87.00	1.895	68.48	179	15.799	118.18
1.400	41.43	80	11.672	87.31	1.900	68.68	180	15.841	118.50
1.405	41.80	81	11.714	87.62	1.905	68.88	181	15.882	118.81
1.410	42.16	82	11.755	87.94	1.910	69.08	182	15.924	119.12
1.415	42.53	83	11.797	88.25	1.915	69.28	183	15.966	119.43
1.420	42.89	84	11.839	88.56	1.920	69.48	184	16.007	119.74
1.425	43.25	85	11.880	88.87	1.925	69.68	185	16.049	120.06
1.430	43.60	86	11.922	89.18	1.930	69.87	186	16.091	120.37
1.435	43.95	87	11.964	89.50	1.935	70.06	187	16.132	120.68
1.440	44.31	88	12.006	89.81	1.940	70.26	188	16.174	120.99
1.445	44.65	89	12.047	90.12	1.945	70.45	189	16.216	121.30
1.450	45.00	90	12.089	90.43	1.950	70.64	190	16.258	121.61
1.455	45.34	91	12.131	90.74	1.955	70.83	191	16.299	121.93
1.460	45.68	92	12.172	91.06	1.960	71.02	192	16.341	122.24
1.465	46.02	93	12.214	91.37	1.965	71.21	193	16.383	122.55
1.470	46.36	94	12.256	91.68	1.970	71.40	194	16.424	122.86
1.475	46.69	95	12.297	91.99	1.975	71.58	195	16.466	123.17
1.480	47.03	96	12.339	92.30	1.980	71.77	196	16.508	123.49
1.485	47.36	97	12.381	92.61	1.985	71.95	197	16.549	123.80
1.490	47.68	98	12.422	92.93	1.990	72.14	198	16.591	124.11
1.495	48.01	99	12.464	93.24	1.995	72.32	199	16.633	124.42
1.500	48.33	100	12.506	93.55	2.000	72.50	200	16.674	124.73

NOTE: To compute weight of water in air at 60 F, subtract weight of air at 60 F, 0.010 pounds per gallon or 0.076 pounds per cubic foot.

When weighing water on an equal arm balance using brass weights having a specific gravity of 8.4, add 145 parts per million by weight to compensate for the volume of air displaced by the brass weights.

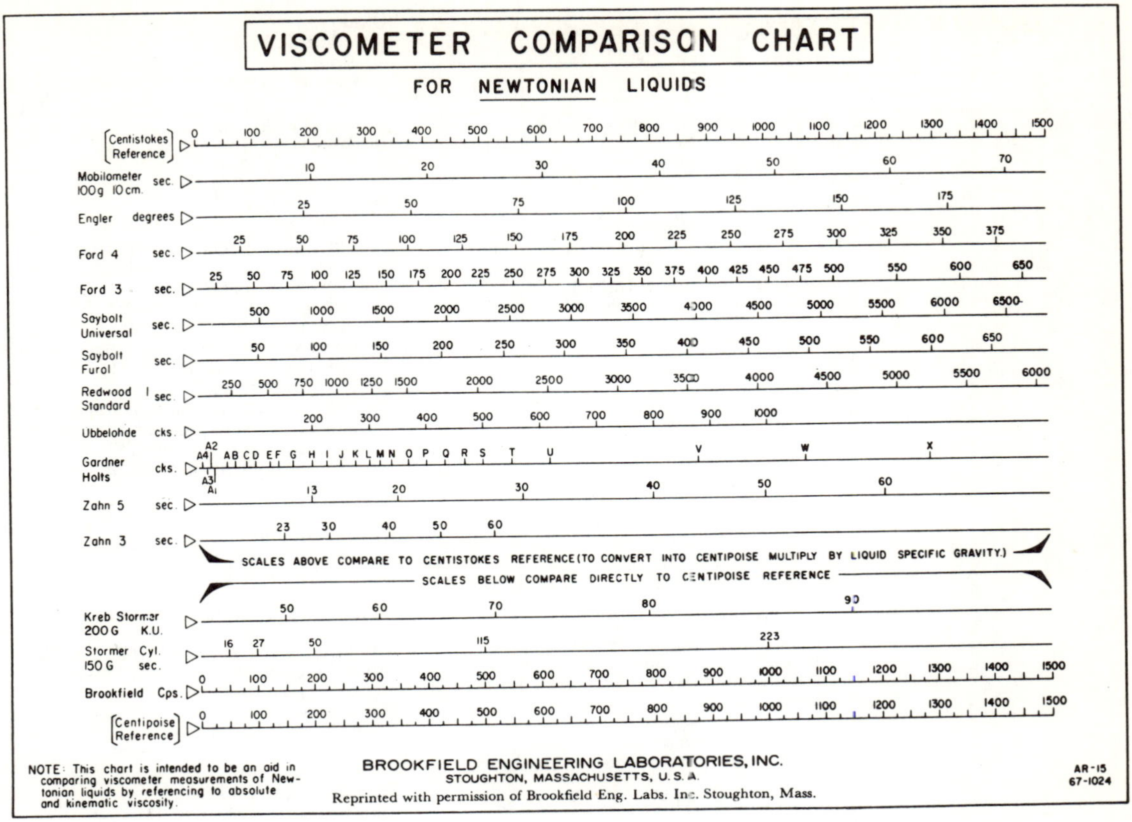
VISCOMETER COMPARISON CHART
FOR NEWTONIAN LIQUIDS
Centistokes Reference
0 100 200 300 400 500 600 700 800 900 1000 1100 1200 1300 1400 1500
Mobilometer 100g 10cm. sec.
10 20 30 40 50 60 70
Engler degrees
25 50 75 100 125 150 175
Ford 4 sec.
25 50 75 100 125 150 175 200 225 250 275 300 325 350 375
Ford 3 sec.
25 50 75 100 125 150 175 200 225 250 275 300 325 350 375 400 425 450 475 500 550 600 650
Saybolt Universal sec.
500 1000 1500 2000 2500 3000 3500 4000 4500 5000 5500 6000 6500
Saybolt Furol sec.
50 100 150 200 250 300 350 400 450 500 550 600 650
Redwood I Standard sec.
250 500 750 1000 1250 1500 2000 2500 3000 3500 4000 4500 5000 5500 6000
Ubbelohde cks.
200 300 400 500 600 700 800 900 1000
Gardner Holts cks.
A2 A4 A3 A1 A B C D E F G H I J K L M N O P Q R S T U V W X
Zahn 5 sec.
13 20 30 40 50 60
Zahn 3 sec.
23 30 40 50 60
SCALES ABOVE COMPARE TO CENTISTOKES REFERENCE (TO CONVERT INTO CENTIPOISE MULTIPLY BY LIQUID SPECIFIC GRAVITY.)
SCALES BELOW COMPARE DIRECTLY TO CENTIPOISE REFERENCE
Kreb Stormer 200 G K.U.
50 60 70 80 90
Stormer Cyl. 150 G sec.
16 27 50 115 223
Brookfield Cps.
0 100 200 300 400 500 600 700 800 900 1000 1100 1200 1300 1400 1500
Centipoise Reference
0 100 200 300 400 500 600 700 800 900 1000 1100 1200 1300 1400 1500
NOTE: This chart is intended to be an aid in comparing viscometer measurements of Newtonian liquids by referencing to absolute and kinematic viscosity.
BROOKFIELD ENGINEERING LABORATORIES, INC.
STOUGHTON, MASSACHUSETTS, U. S. A.
Reprinted with permission of Brookfield Eng. Labs. Inc. Stoughton, Mass.
AR-15
67-1024

Index